Data-Centric Systems and Applications

Elzbieta Malinowski · Esteban Zimányi

Advanced Data Warehouse Design

From Conventional to Spatial
and Temporal Applications

With 152 Figures and 10 Tables

 Springer

Elzbieta Malinowski
Universidad de Costa Rica
School of Computer &
Information Science
San Pedro, San José
Costa Rica
emalinow@cariari.ucr.ac.cr

Esteban Zimányi
Université Libre Bruxelles
Dept. of Computer &
Decision Engineering (CoDE)
Avenue F.D. Roosevelt 50
1050 Bruxelles
Belgium
ezimanyi@ulb.ac.be

ISBN: 978-3-642-09383-8   e-ISBN: 978-3-540-74405-4

ACM Computing Classification: H.2, H.3, H.4, J.1, J.2

© 2008 Springer-Verlag Berlin Heidelberg
Softcover reprint of the hardcover 1st edition 2008

Cover design: KünkelLopka, Heidelberg

Printed on acid-free paper

9 8 7 6 5 4 3 2 1

springer.com

To Yamil,
my one and only love and husband,
who throughout 25 years has been
a patient friend, an untiring supporter,
and a wise adviser
E.M.

To Fabienne and Elena,
with all my love and tenderness,
for all the joy and the beautiful
moments that I share with them
E.Z.

Foreword

When I was asked to write the foreword of this book, I was both honored and enthusiastic about contributing to the first book covering the topic of spatial data warehousing. In spite of over ten years of scientific literature regarding spatial data warehousing, spatial data mining, spatial online analytical processing, and spatial datacubes, Malinowski and Zimányi are the first to invest the energy necessary to provide a global view of spatial data warehousing in a coherent manner. This book introduces the reader to the basic concepts in an academic style that remains easy to read. Although one might think initially that there is nothing special about spatial data warehouses, this book shows the opposite by introducing fundamental extensions to the development of spatial data warehouses. On the one hand, there are extensions required for the design and implementation phases, and on the other hand there are extensions that improve the efficiency of these phases. This book offers a broad coverage of these extensions while at the same time guiding the reader to more detailed papers, thanks to a representative bibliography giving a sample of the main papers in the field. In addition, this book contains a glossary that is very useful for the reader since the datacube paradigm (also called analytical or multidimensional database paradigm) is very different from the traditional database paradigm (also called transactional or operational database paradigm).

Of particular interest in this book is the coverage of spatial datacubes, of their structure, and of implementation details in a data warehouse perspective. In particular, the different types of spatial hierarchies, spatial measures, and spatial data implementations are of interest. As a technology of choice to exploit data warehouses, OLAP technology is discussed with examples from Microsoft SQL Server Analysis Services 2005 and from Oracle 10g OLAP Option. Spatial OLAP (SOLAP) is also mentioned for spatial data warehouses.

In addition to defining spatial and temporal components of datacubes, the book proposes a method that is based on an extended entity-relationship approach, the spatial and temporal extensions of the MADS model, and a waterfall-based design approach. Readers already using the entity-relationship approach will immediately benefit from the proposed concepts. Nevertheless, all readers should benefit from reading this book, since the underlying fundamental concepts can be transposed to UML or other proprietary modeling languages. Considering past contributions of Malinowski and Zimányi in the field, this book provides an up-to-date solution that is coherent with their previous work. This fact alone is worth the interest of the reader since it contributes to the building and dissemination of knowledge in our field.

Typical of design method books where authors suggest their own recipe, this book proposes one approach covering both theoretical concepts and design phases in an extensive and coherent manner. This is a welcome contribution despite the fact that different approaches have always existed in system design and will always exist, explicitly or implicitly. This is necessary for method improvement and for different application contexts. Habitually, these methods converge towards the issues to tackle (the "what" and "why") while they typically diverge about the ways to solve these issues (the "how" and "when"). Consequently, one benefit of this book is to provide the readers with the necessary background to explore other elements of solutions beyond those cited. Typically, these solutions are not as well structured as this book, since knowledge of a research team is fragmented over different papers written at different times. Personally, I with my research team must confess to doing exactly such piecemeal dissemination regarding the Perceptory alternative in spite of years of practical usage in over 40 countries thanks to UML, geospatial standards, and adaptability to designers' own agile or waterfall methods. In this regard, the readers will be very grateful to Malinowski and Zimányi for having grouped under one cover, in a coherent and homogeneous manner, their views about conventional data warehouses, spatial data warehouses, and temporal data warehouses.

Finally, the arrival of the first book on a topic (other than a collaborative collection of chapters) can always be considered as a milestone which indicates that a field has reached a new level of maturity. As such, this book is a significant milestone on the SOLAP portal timeline (http://www.spatialbi.com) and recognizes Malinowski and Zimányi's contributions to the field.

Yvan Bédard
Creator of the acronym SOLAP and the first SOLAP commercial technology
Holder of the Canada NSERC Industrial Research Chair in Geospatial
Databases for Decision Support
Department of Geomatics Sciences & Centre for Research in Geomatics
Université Laval, Québec City, Canada

Preface

Data warehouses are databases of a specific kind that periodically collect information about the activities being performed by an organization. This information is then accumulated over a period of time for the purpose of analyzing how it evolves and, thus, for discovering strategic information such as trends, correlations, and the like. Data warehouses are increasingly being used by many organizations in many sectors to improve their operations and to better achieve their objectives. For example, a data warehouse application can be used in an organization to analyze customers' behavior. By understanding its customers, the organization is then able to better address their specific needs and expectations.

This book advocates a *conceptual* approach to designing data warehouse applications. For many decades, databases have been designed starting from a conceptual-design phase. This phase purposely leaves out any implementation considerations in order to capture as precisely as possible the requirements of the application from the users' perspective. When this has been done, a logical-design phase aims at translating these requirements according to a particular implementation paradigm, such as the relational or the object-relational paradigm. In the final stage, specific considerations concerning a particular implementation platform are taken into account in order to build an operational application. This separation of concerns offers many advantages and has contributed significantly to the widespread use of database applications in almost every aspect of our everyday life.

The situation is different for data warehouses, which are typically designed starting at a logical level, followed then by a physical-design phase. This state of affairs precludes an accurate description of users' analytical requirements, in particular because these requirements are driven by the technical limitations imposed by current data warehouse tools. There has recently been increased interest in the research community in devising conceptual models for data warehouse applications. Nevertheless, this area is still at an early stage and there is not yet a consensus about the specific characteristics that such a conceptual model should have.

This book presents the MultiDim model, a conceptual model for data warehouse design. We used as a starting point the classical entity-relationship model, thus taking into account the experience gained in more than four decades of applying conceptual modeling to traditional databases. The Multi-Dim model was designed in order to cope with the specific requirements of data warehouses and analytical applications. Therefore, it adopts a multidimensional view of information: *measures*, which are indicators that allow one to evaluate specific activities of an organization, are analyzed using various *dimensions* or perspectives. In addition, measures under analysis can be expressed at various levels of detail with the help of *hierarchies*, which define multiple abstraction levels for the dimensions.

The book covers the design of conventional data warehouses and also addresses two innovative domains that have recently been introduced to extend the capabilities of data warehouse systems, namely, the management of spatial and temporal (or time-varying) information.

Spatial information allows us to describe how objects are located in space: persons live in specific locations and go to other locations for their work and for their leisure activities. These locations belong to particular subdivisions of the Earth's surface, such as counties, states, countries, and continents. Spatial information has been successfully used for many years in a wide range of application domains, such as environmental applications, cadastral and land use management, utilities management, and transportation and logistics, to name but a few. However, current data warehouse systems do not provide support for managing spatial information and, thus, location information must be represented using traditional alphanumeric data types. The extension of data warehouses in order to support spatial information opens up a new spectrum of possibilities, since this allows the inherent semantics of spatial information to be taken into account. For example, this could allow one to monitor the evolution of urban green areas and their impact on biodiversity, and the evolution of the use of public transportation systems with respect to the use of private cars. In addition, displaying the results of analysis using cartographic representations enables the discovery of information that would be impossible to obtain otherwise. We define in this book a conceptual model for designing spatial data warehouses. This proposal has its roots in the work that has been performed in the fields of spatial databases and geographic information systems.

On the other hand, temporal information captures the evolution in time of real-world phenomena: persons change their homes or their jobs, which induces a modification of their behavior, new products and technologies appear while others are discontinued, etc. While current data warehouses allow one to keep track of the evolution of measures under analysis, they are not able to capture the evolution of the dimensions used for analyzing those measures. The need to support the evolution in time of all elements composing a data warehouse has been acknowledged for many years. For example, if reorganizations of sales districts are performed periodically to adapt to changing market conditions, it is necessary to keep track of these changes in order to analyze their impact on sales. However, the solutions provided so far for dealing with these issues are not satisfactory. We propose in this book a conceptual model for designing temporal data warehouses. Our proposal is based on work performed in the domain of temporal databases.

While this book proposes a conceptual approach to designing data warehouses, we also discuss how the conceptual specifications can be translated into logical and physical specifications for implementation with current data warehouse systems. We have chosen Microsoft Analysis Services 2005 and Oracle 10g as representative platforms for the implementation of our conceptual model. For conventional and temporal data warehouses, we have envisaged implementation on both platforms. However, for spatial data warehouses, we have used only Oracle 10g as the target platform, since a spatial extension of the database management system is required. By providing effective implementations of our conceptual model, we aim at promoting the use of the conceptual approach in data warehouse design.

As experience has shown, designing data warehouse applications is a complex and costly process. Therefore, providing methodological support for this task is of paramount importance. This book provides a method that incorporates conceptual design into the data warehouse design process. We cover two alternative, and complementary, approaches to dealing with this problem. The first one, called *analysis-driven design*, starts from the analysis requirements of decision-making users. The second approach, called *source-driven design*, starts by analyzing the information in the sources providing data to the warehouse. These two approaches can be combined into an iterative development process in order to ensure the correspondance of the analysis requirements with the available information. Obviously, the method must also take into account the specificities of the spatial or temporal information. However, since data warehouse design is a relatively recent domain, our proposal constitutes only a preliminary step in this respect; more research must be done as well as an evaluation of our method in a variety of data warehouse design projects.

This book is targeted at practitioners, graduate students, and researchers interested in the design issues of data warehouse applications. Practitioners and domain experts from industry with various backgrounds can benefit from the material covered in this book. It can be used by database experts who wish to approach the field of data warehouse design but have little knowledge about it. The book can also be used by experienced data warehouse designers who wish to enlarge the analysis possibilities of their applications by including spatial or temporal information. Furthermore, experts in spatial databases or in geographic information systems could profit from the data warehouse vision when they are building innovative spatial analytical applications. The book can also be used for teaching graduate or advanced undergraduate students, since it provides a clear and a concise presentation of the major concepts and results in the new fields of conceptual data warehouse design and of spatial and temporal data warehouses. Finally, researchers can find an introduction to the state of the art on the design of conventional, spatial, and temporal data warehouses; many references are provided to deepen their knowledge of these topics.

The book should allow readers to acquire both a broad perspective and an intimate knowledge of the field of data warehouse design. Visual notations and examples are intensively used to illustrate the use of the various constructs. The book is purposely written in an informal style that should make it readable without requiring a specific background in computer science, mathematics, or logic. Nevertheless, we have taken special care to make all definitions and arguments as precise as possible, and to provide the interested reader with formal definitions of all concepts.

We would like to thank the Cooperation Department of the Université Libre de Bruxelles, which funded the sojourn of Elzbieta Malinowski in Brussels; without its financial support, this book would never have been possible. Parts of the material included in this book have been previously presented in conferences or published in journals. At these conferences, we had the opportunity to have discussions with research colleagues from all around the world, and we exchanged various views about the subject with them. The anonymous reviewers of these conferences and journals provided us with insightful comments and suggestions that contributed significantly to improving the work presented in this book. We are also grateful to Serge Boucher, Boris Verhaegen, and Frédéric Servais, researchers in our department, who helped us to enhance considerably the figures of this book. Our special thanks to Yvan Bédard, professor at the Laval University, who kindly agreed to write the foreword for this book, even when it was only in draft form. Finally, we would like to warmly thank Ralf Gerstner from Springer for his continued interest in this book. The warm welcome given to our book proposal and his enthusiasm and encouragement throughout its writing helped significantly in giving us impetus to pursue our project to its end.

Elzbieta Malinowski
Esteban Zimányi
October 2007

Contents

Abstract

Decision support systems are interactive, computer-based information systems that provide data and analysis tools in order to assist managers at various levels of an organization in the process of decision making. Data warehouses have been developed and deployed as an integral part of decision support systems.

A data warehouse is a database that allows the storage of high volumes of historical data required for analytical purposes. This data is extracted from operational databases, transformed into a coherent whole, and loaded into a data warehouse during an extraction-transformation-loading (ETL) process.

Data in data warehouses can be dynamically manipulated using online analytical processing (OLAP) systems. Data warehouse and OLAP systems rely on a multidimensional model that includes measures, dimensions, and hierarchies. Measures are usually numeric values that are used for quantitative evaluation of aspects of an organization. Dimensions provide various analysis perspectives, while hierarchies allow measures to be analyzed at various levels of detail.

Currently, both designers and users find it difficult to specify the multidimensional elements required for analysis. One reason for this is the lack of conceptual models for data warehouse design, which would allow one to express data requirements on an abstract level without considering implementation details. Another problem is that many kinds of complex hierarchies that arise in real-world situations are not addressed by current data warehouse and OLAP systems.

In order to help designers to build conceptual models for decision support systems and to help users to better understand the data to be analyzed, we propose in this book the MultiDim model, a conceptual model that can be used to represent multidimensional data for data warehouse and OLAP applications. Our model is based mainly on the existing constructs of the entity-relationship (ER) model, such as entity types, relationship types, and attributes with their usual semantics, which allow us to represent the concepts of dimensions, hierarchies, and measures. The model also includes a conceptual classification of various kinds of hierarchies that exist in real-world situations and proposes graphical notations for them.

Users of data warehouse and OLAP systems increasingly require the inclusion of spatial data. The advantage of using spatial data in the analysis process is widely recognized, since it reveals patterns that are difficult to discover otherwise. However, although data warehouses typically include a spatial or location dimension, this dimension is usually represented in an alphanumeric format. Furthermore, there is still no systematic study that analyzes the inclusion and management of hierarchies and measures that are represented using spatial data.

With the aim of satisfying the growing requirements of decision-making users, we have extended the MultiDim model by allowing the inclusion of spatial data in the various elements composing the multidimensional model. The novelty of our contribution lies in the fact that a multidimensional model is seldom used for representing spatial data. To succeed in our aim, we applied the research achievements in the field of spatial databases to the specific features of a multidimensional model. The spatial extension of a multidimensional model raises several issues, to which we refer in this book, such as the influence of the various topological relationships between spatial levels in a hierarchy on the procedures required for measure aggregation, the aggregation of spatial measures, and the inclusion of spatial measures without the presence of spatial dimensions.

One of the essential characteristics of multidimensional models is the presence of a time dimension that allows one to keep track of changes in measures. However, the time dimension cannot be used for recording changes in other dimensions. This is a serious restriction of current multidimensional models, since in many cases users need to analyze how measures may be influenced by changes in dimension data. As a consequence, specific applications must be developed to cope with these changes. Further, there is still a lack of a comprehensive analysis to determine how the concepts developed for providing temporal support in conventional databases might be applied to data warehouses.

In order to allow users to keep track of temporal changes in all elements of a multidimensional model, we have introduced a temporal extension of the MultiDim model. This extension is based on research done in the area of temporal databases, which have been successfully used for modeling time-varying information for several decades. We propose the inclusion of various temporality types, such as valid time and transaction time, which are obtained from the source systems, in addition to the loading time generated in a data warehouse. We use this temporal support to provide a conceptual representation of time-varying dimensions, hierarchies, and measures. We also refer to specific constraints that should be imposed on time-varying hierarchies and to the problem when the time granularity in the source systems and the data warehouse differ.

The design of data warehouses is not an easy task. It requires one to consider all phases, from requirements specification to the final implementation, including the ETL process. Data warehouse design should also take into account the fact that the inclusion of data items in a data warehouse depends on both the users' needs and the availability of data in the source systems. However, currently, designers must rely on their experience, owing to the lack of a methodological framework that considers these aspects.

In order to assist developers during the data warehouse design process, we propose a method for the design of conventional, spatial, and temporal data warehouses. We refer to various phases, namely, requirements specification, conceptual design, logical design, and physical design. We include three different approaches to requirements specification depending on whether the users, the operational data sources, or both are the driving force in the process of requirements gathering. We show how each approach leads to the creation of a conceptual multidimensional model. We also present the logical- and physical-design phases that refer to data warehouse structures and the ETL process.

To ensure the correctness of the proposed conceptual model, we formally define the model providing its syntax and semantics. Throughout the book, we illustrate the concepts using real-world examples with the aim of demonstrating the usability of our conceptual model. Furthermore, we show how the conceptual specifications can be implemented in relational and object-relational databases. We do this using two representative data warehouse platforms, Microsoft Analysis Services 2005 and Oracle 10g with the OLAP and Spatial extensions.

1

Introduction

Organizations today are facing increasingly complex challenges in terms of management and problem solving in order to achieve their operational goals. This situation compels people in those organizations to utilize analysis tools that will better support their decisions. **Decision support systems** provide assistance to managers at various organizational levels for analyzing strategic information. These systems collect vast amount of data and reduce it to a form that can be used to analyze organizational behavior [53].

Since the early 1990s, **data warehouses** have been developed and deployed as an integral part of modern decision support systems [205, 269]. A data warehouse provides an infrastructure that enables users to obtain efficient and accurate responses to complex queries. Various systems and tools can be used for accessing and analyzing the data contained in data warehouses. For example, **online analytical processing (OLAP)** systems allow users to interactively query and automatically aggregate the data contained in a data warehouse. In this way, decision-making users can easily access the required information and analyze it at various levels of detail.

However, current data warehouse systems do not consider three aspects that are important for decision-making users.

The first aspect concerns the support for complex kinds of hierarchies. Current systems only allow simple hierarchies, typically those where the relationship between instances can be represented as a balanced tree. An example is a store–city–state hierarchy, where every store is related to a city, which in turn is related to a state. However, in many real-world situations the hierarchies correspond to unbalanced trees or to graphs. For example, depending on whether a client is a person or a company, different hierarchies may need to be considered, and these hierarchies may share some levels.

The second aspect relates to the inclusion of spatial data, which is not supported by current systems. In many situations, representing data spatially (e.g., on a map) could help to reveal patterns that are difficult to discover otherwise. For example, representing customers' locations on a map may help the user to discover that most customers do not necessarily buy products in nearby stores.

The third aspect relates to the inclusion of temporal (or time-varying) data. Current data warehouse systems allow one to keep track of the evolution in time of organizational aspects under analysis (i.e., measures), such as the amount of sales or the quantities of items in an inventory. Nevertheless, there is a need to store historical data for the other elements contained in a data warehouse (i.e., dimensions and hierarchies) to better establish cause-effect relationships. For example, storing changes to the ingredients of products may help one to analyze whether those changes have any influence on sales.

Typical data warehouse systems are very complex, and their development is a difficult and costly task. Data warehouse developer teams must establish what data and what kind of analysis the users require for supporting the decision-making process. For this it is necessary to provide a model that facilitates communication between users and data warehouse designers. In addition, a suitable method should be followed to ensure successful data warehouse development.

In this chapter, we refer to various aspects that motivated us to develop a conceptual multidimensional model with spatial and temporal extensions and to propose a method for data warehouse design. First, in Sect. 1.1 we briefly describe some general concepts related to data warehouses and to their spatial and temporal extensions. We also refer to conceptual modeling and the method used for designing conventional, spatial, and temporal data warehouses. Then, Sect. 1.2 presents the motivations behind this book. Finally, in Sect. 1.3 we describe the scope of this book and the contributions to research contained in it, and in Sect. 1.4 we refer to its organization.

1.1 Overview

1.1.1 Conventional Data Warehouses

As a consequence of today's increasingly competitive and rapidly changing world, organizations in all sectors need to perform sophisticated data analysis to support their decision-making processes. However, traditional databases, usually called **operational** or **transactional databases**, do not satisfy the requirements for data analysis. These databases support daily business operations, and their primary concern is to ensure concurrent access by multiple users and recovery techniques that guarantee data consistency. Typical operational databases contain detailed data, do not include historical data, and, since they are usually highly normalized, they perform poorly when executing complex queries that involve a join of many tables or aggregate large volumes of data. Furthermore, when users need to analyze the behavior of an organization as a whole, data from several different operational systems must be integrated. This can be a difficult task to accomplish because of the differences in data definition and content.

Data warehouses were proposed in order to better respond to the growing demands of decision-making users. A data warehouse is usually defined as a collection of subject-oriented, integrated, nonvolatile, and time-varying data to support management decisions [118]. This definition emphasizes some salient features of a data warehouse. **Subject-oriented** means that a data warehouse targets one or several subjects of analysis according to the analytical requirements of managers at various levels of the decision-making process. For example, a data warehouse in a retail company may contain data for the analysis of the purchase, inventory, and sales of products. **Integrated** expresses the fact that the contents of a data warehouse result from the integration of data from various operational and external systems. **Nonvolatile** indicates that a data warehouse accumulates data from operational systems for a long period of time. Thus, data modification and removal are not allowed in data warehouses: the only operation allowed is the purging of obsolete data that is no longer needed. Finally, **time-varying** underlines that a data warehouse keeps track of how its data has evolved over time; for instance, it may allow one to know the evolution of sales or inventory over the last several months or years.

Operational databases are typically designed using a **conceptual model**, such as the entity-relationship (ER) model, and **normalization** for optimizing the corresponding relational schema. Several authors (e.g., [146, 295]) have pointed out that these paradigms are not well suited for designing data warehouse applications. Data warehouses must be modeled in a way that ensures a better understanding of the data for analysis purposes and gives better performance for the complex queries needed for typical analysis tasks. In order to meet these expectations, a **multidimensional model** has been proposed.

The multidimensional model views data as consisting of facts linked to several dimensions. A **fact** represents a focus of analysis (for example analysis of sales in stores) and typically includes attributes called measures. **Measures** are usually numeric values that allow quantitative evaluation of various aspects of an organization to be performed. For example, measures such as the amount or quantity of sales might help to analyze sales activities in various stores. **Dimensions** are used to see the measures from different perspectives. For example, a time dimension can be used for analyzing changes in sales over various periods of time, whereas a location dimension can be used to analyze sales according to the geographic distribution of stores. Users may combine several different analysis perspectives (i.e., dimensions) according to their needs. For example, a user may require information about sales of computer accessories (the product dimension) in July 2006 (the time dimension) in all store locations (the store dimension). Dimensions typically include attributes that form **hierarchies**, which allow decision-making users to explore measures at various levels of detail. Examples of hierarchies are month–quarter–year in the time dimension and city–state–country in the location dimension.

Aggregation of measures takes place when a hierarchy is traversed. For example, moving in a hierarchy from a month level to a year level will yield aggregated values of sales for the various years.

The multidimensional model is usually represented by relational tables organized in specialized structures called **star schemas** and **snowflake schemas**. These relational schemas relate a fact table to several dimension tables. Star schemas use a unique table for each dimension, even in the presence of hierarchies, which yields denormalized dimension tables. On the other hand, snowflake schemas use normalized tables for dimensions and their hierarchies.

OLAP systems allow end users to perform dynamic manipulation and automatic aggregation of the data contained in data warehouses. They facilitate the formulation of complex queries that may involve very large amounts of data. This data is examined and aggregated in order to find patterns or trends of importance to the organization. OLAP systems have typically been implemented using two technologies: **relational OLAP (ROLAP)** and **multidimensional OLAP (MOLAP)**. The former stores data in a relational database management system, while the latter uses vendor-specific array data structures. **Hybrid OLAP (HOLAP)** systems combine both technologies, for instance using ROLAP for detailed fact data and MOLAP for aggregated data. Although current OLAP systems are based on a multidimensional model, i.e., they allow one to represent facts, measures, dimensions, and hierarchies, they are quite restrictive in the types of hierarchies that they can manage. This in an important drawback, since the specification of hierarchies in OLAP systems is important if one is to be able to perform automatic aggregation of measures while traversing hierarchies.

1.1.2 Spatial Databases and Spatial Data Warehouses

Over the years, **spatial data** has increasingly become part of operational and analytical systems in various areas, such as public administration, transportation networks, environmental systems, and public health, among others. Spatial data can represent either **objects** located on the Earth's surface, such as mountains, cities, and rivers, or geographic **phenomena**, such as temperature, precipitation, and altitude. Spatial data can also represent non-geographic data, i.e., data located in other spatial frames such as a human body, a house, or an engine. The amount of spatial data available is growing considerably owing to technological advances in areas such as remote sensing and global navigation satellite systems (GNSS), namely the Global Positioning System (GPS) and the forthcoming Galileo system.

The management of spatial data is carried out by **spatial databases** or **geographic information systems (GISs)**. Since the latter are used for storing and manipulating **geographic** objects and phenomena, we shall use the more general term "spatial databases" in the following. Spatial databases allow one to store spatial data whose location and shape are described in a two- or three-dimensional space. These systems provide a set of functions and operators that allow users to query and manipulate spatial data. Queries may refer to spatial characteristics of individual objects, such as their area or perimeter, or may require complex operations on two or more spatial objects. **Topological relationships** between spatial objects, such as "intersection", "inside", and "meet", are essential in spatial applications. For example, two roads may intersect, a lake may be inside a state, two countries may meet because they have a common border, etc. An important characteristic of topological relationships is that they do not change when the underlying space is distorted through rotation, scaling, and similar operations.

Although spatial databases offer sophisticated capabilities for the management of spatial data, they are typically targeted toward daily operations, similarly to conventional operational databases. Spatial databases are not well suited to supporting the decision-making process. As a consequence, a new field, called **spatial data warehouses**, has emerged as a combination of the spatial-database and data-warehouse technologies. As we have seen, data warehouses provide OLAP capabilities for analyzing data using several different perspectives, as well as efficient access methods for the management of high volumes of data. On the other hand, spatial databases provide sophisticated management of spatial data, including spatial index structures, storage management, and dynamic query formulation. Spatial data warehouses allow users to exploit the capabilities of both types of systems for improving data analysis, visualization, and manipulation.

1.1.3 Temporal Databases and Temporal Data Warehouses

Many applications require the storage, manipulation, and retrieval of **temporal data**, i.e., data that varies over time. These applications need to represent not only the changes to the data but also the time when these changes occurred. For example, land management systems need to represent the evolution of land distribution over time by keeping track of how and when land parcels have been split or merged, and when their owner changed. Similarly, healthcare systems may include historical information about patients' clinical records, including medical tests and the time when they were performed.

Temporal databases allow one to represent and manage time-varying information. These databases typically provide two orthogonal types of temporal support. The first one, called **valid time**, indicates the time when data was (or is or will be) valid in the modeled reality, for example the time when a student took some specific course. The second one, called **transaction time**, specifies the time when a data item was current in a database, for example when the information that the student took the course was stored in the database. An important characteristic of temporal data is its **granularity**, or level of detail. Considering valid time, employees' salaries may be captured at a granularity of a month, whereas stock exchange data may be captured at a granularity of a day or an hour. On the other hand, the granularity of transaction time is defined by the system, typically at a level of a millisecond.

Two approaches have been proposed for providing temporal support in relational databases, depending on whether modifications to the relational model are required or not. In one approach [273], temporality is represented in tables by means of hidden attributes, i.e., attributes that cannot be referenced by simple names in the usual way. Using this approach, temporal data is handled differently from traditional nontemporal data. Therefore, the various components of a database management system (DBMS) must be extended to support temporal semantics; this includes adequate storage mechanisms, specialized indexing methods, temporal query languages, etc. Another approach [54] does not involve any change to the classical relational model and considers temporal data just like data of any other kind, i.e., temporal support is implemented by means of explicit attributes. This approach introduces new interval data types and provides a set of new operators and extensions of existing ones for managing time-varying data.

Nevertheless, neither of the above approaches has been widely accepted, and current DBMSs do not provide yet facilities for manipulating time-varying data. Therefore, complex structures and significant programming effort are required for the correct management of data that varies over time.

Although temporal databases allow historical data to be managed, they do not offer facilities for supporting the decision-making process when aggregations of high volumes of historical data are required. Therefore, bringing together the research achievements in the fields of temporal databases and data warehouses has led to a new field, called **temporal data warehouses**. Temporal semantics forms an integral part of temporal data warehouses in a similar way to what is the case for temporal databases.

1.1.4 Conceptual Modeling for Databases and Data Warehouses

The conventional database design process includes the creation of database schemas at three different levels: conceptual, logical, and physical [66].

A **conceptual schema** is a concise description of the users' data requirements without taking into account implementation details. Conceptual schemas are typically expressed using the ER model [45] or the Unified Modeling Language (UML) [29].

A **logical schema** targets a chosen implementation paradigm, for example the relational, the object-oriented, or the object-relational paradigm. A logical schema is typically produced by applying a set of mapping rules that transform a conceptual schema. For example, a conceptual schema based on the ER model can be transformed into the relational model by applying well-known mapping rules (see, e.g., [66]).

Finally, a **physical schema** includes the definition of internal data structures, file organization, and indexes, among other things. The physical schema is highly technical and considers specific features of a particular DBMS in order to improve storage and performance.

In the database community, it has been acknowledged for several decades that conceptual models allow better communication between designers and users for the purpose of understanding application requirements. A conceptual schema is more stable than an implementation-oriented (logical) schema, which must be changed whenever the target platform changes [225]. Conceptual models also provide better support for visual user interfaces; for example, ER models have been very successful with users.

However, currently there is no well-established conceptual model for multidimensional data. There have been several proposals based on UML (e.g., [5, 169]), on the ER model (e.g., [263, 298]), and using specific notations (e.g., [82, 109, 251, 301]). Although these models include some features required for data warehouse applications, such as dimensions, hierarchies, and measures, they have several drawbacks, to which we shall refer in Sect. 1.2 and, in more detail, in Sect. 3.8.

1.1.5 A Method for Data Warehouse Design

The method used for designing operational databases includes well-defined phases. The first one, referred to as **requirements specification**, consists in gathering users' demands into a coherent and concisely written specification. The next phases, **conceptual design**, **logical design**, and **physical design**, consists in translating the requirements gathered in the first phase into database schemas that increasingly target the final implementation platform.

Since data warehouses are specialized databases aimed at supporting the decision-making process, the phases used in conventional database design have also been adopted for developing data warehouses. During the requirements-gathering process, users at various levels of management are interviewed to find out their analysis needs. The specification obtained serves as basis for creating a database schema capable of responding to users' queries. In many situations, owing to the lack of a well-accepted conceptual model for data warehouse applications, designers skip the conceptual-design phase and use instead a logical representation based on star and/or snowflake schemas. Afterwards, the physical design considers the facilities provided by current DBMSs for storing, indexing, and manipulating the data contained in the warehouse.

However, experience has shown that the development of data warehouse systems differs significantly from the development of conventional database systems. Therefore, modifications to the above-described method are necessary. For example, unlike conventional databases, the data in a warehouse is extracted from several source systems. As a consequence, some approaches for data warehouse design consider not only users' requirements but also the availability of data. Other approaches mainly take the underlying operational databases into account, instead of relying on users' requirements. Additionally, since in many cases the data extracted from the source systems must be transformed before being loaded into the data warehouse, it is necessary to consider during data warehouse design the **extraction-transformation-loading (ETL)** process.

On the other hand, there is not yet a well-established method for designing spatial or temporal databases. The usual practice is to design a conventional database ignoring the spatial and temporal aspects and, later on, extend the conventional schema produced so far with spatial or temporal features. A similar situation arises for spatial and temporal data warehouses, although this is compounded by the fact that, being very recent research areas, much less experience is available to assist in developing them.

1.2 Motivation for the Book

In this section, we explain the reasons that motivated us to explore the field of conceptual modeling for data warehouse and OLAP applications. We also refer to the importance of including spatial and temporal data in a conceptual multidimensional model. Finally, we show the significance of defining a method for designing conventional, spatial, and temporal data warehouses.

The domain of conceptual design for data warehouse applications is still at a research stage. The analysis presented by Rizzi [259] shows the small interest of the research community in conceptual multidimensional modeling. Even though conceptual models are closer than logical models to the way users perceive an application domain, in the current state of affairs data warehouses are designed using mostly logical models. Therefore, when building data warehouses, users have difficulties in expressing their requirements, since specialized knowledge related to technical issues is required. Further, logical models limit users to defining only those elements that the underlying implementation systems can manage. As a typical example, users are constrained to use only the simple hierarchies that are implemented in current data warehouse tools.

Even though there are some conceptual models for data warehouse design, they either provide a graphical representation based on the ER model or UML with little or no formal definition, or provide only a formal definition without any graphical support. In addition, they do not allow users to express analysis requirements in an unambiguous way; in particular, they do not distinguish the various kinds of hierarchies that exist in real-world applications. This situation is considered as a shortcoming of existing models for data warehouses [103].

Furthermore, although location information, such as address, city, or state, is included in many data warehouses, this information is usually represented in an alphanumeric manner. Representing location information as spatial data leads to the emerging field of spatial data warehouses. However, this field raises several research issues.

First, there is no consensus in the literature about the meaning of a spatial data warehouse. The term is used in several different situations, for example when there are high volumes of spatial data, when it is required to integrate or aggregate spatial data, or when the decision-making process uses spatial data. Even though it is important to consider all the above aspects in spatial data warehouses, what is still missing is a proposal for spatial data warehouses with the clearly distinguished spatial elements required for multidimensional modeling. Such a proposal should take account of the fact that although both spatial databases and spatial data warehouses manage spatial data, their purposes are different. Spatial databases are used for answering queries that involve spatial locations. Examples of such queries are "where is the closest store to my house?", "which highways connect Brussels and Warsaw?", and "how do I get to a specific place from my current position given by a GPS device?" In contrast, spatial data warehouses use spatial data to support the decision-making process. Examples of analytical queries are "what is the best location for a new store?", "what roads should be constructed to relieve traffic congestion?", and "how can we divide a specific area into sales regions to increase sales?".

Second, applications that include spatial data are usually complex and need to be modeled taking user requirements into account. However, the lack of a conceptual approach for data warehouse and OLAP systems, joined with the absence of a commonly accepted conceptual model for spatial applications, makes the modeling task difficult. Further, although some existing conceptual models for spatial databases allow relationships among spatial objects to be represented, they are not adequate for multidimensional modeling, since they do not include the concepts of dimensions, hierarchies, and measures. To the best of our knowledge, there is as yet no proposal for a conceptual model for spatial-data-warehouse applications.

Finally, any proposal for a conceptual multidimensional model including spatial data should take account of various aspects not present in conventional multidimensional models, such as topological relationships and aggregation of spatial measures, among others. Some of these aspects are briefly mentioned in the existing literature, for example spatial aggregation, while others are neglected, for example the influence on aggregation procedures of topological relationships between spatial objects forming hierarchies.

Another important characteristic of data warehouses is that data is stored for long periods of time. In accordance with the "time-varying" and "non-volatility" features of data warehouses (see Sect. 1.1.1), changes to data cannot overwrite existing values. However, although the usual multidimensional models allow changes in measures to be represented, they are not able to represent changes in dimension data. Consequently, the features of data warehouse mentioned above apply only to measures, leaving the representation of changes in dimensions to the applications.

Several solutions have been proposed for representing changes to dimension data in the context of relational databases, called **slowly changing dimensions** [145]. The first solution, called type 1 or the overwrite model, corresponds to the usual situation where, upon an update, old dimension data is replaced by new data, thus losing track of data changes. The second solution, called type 2 or the conserving-history model, inserts a new record each time the value of a dimension attribute changes. However, this approach is complex, since it requires the generation of different keys for records representing the same instance. It also introduces unnecessary data redundancy, since the values of attributes that have not changed are repeated. Finally, this solution also demands significant programming effort for querying time-varying dimension data. The last solution, called type 3 or the limited-history model, introduces an additional column for every attribute for which changes in value must be kept. These columns store the old and new values of the attribute. In this case the history of changes is limited to the number of additional columns. As can be seen, these solutions are not satisfactory, since they either do not preserve the entire history of the data or are complex to implement. Further, they do not take account of research related to managing time-varying information in temporal databases. Therefore, temporal data warehouses were proposed as a solution to the above problems.

Currently, the research related to temporal data warehouses raises many issues, mostly dealing with implementation aspects, such as special aggregation procedures, special storage, and indexing methods, among others. However, very little attention has been given by the research community to conceptual modeling for temporal data warehouses and to the analysis of what temporal support should be included in these systems. Such analysis should take account of the fact that temporal databases and conventional data warehouses have some similarities, in particular both of them manage historical data, but should also take into account their differences. For example, data in data warehouses is integrated from existing source systems, whereas data in temporal databases is introduced by users in the context of operational applications; data warehouses support the decision-making process, whereas temporal databases reflect data changes in reality and in the database content; and data in data warehouses is neither modified nor deleted[1] but in contrast, users of temporal databases change data directly. Therefore, it is necessary to propose some form of temporal support adequate for data warehouse applications, and a conceptual model with time-varying elements that takes into account the specific semantics of data warehouses.

The development of conventional, spatial, and temporal data warehouses is a very complex task that requires several phases for its realization. It is therefore necessary to specify a method in order to assist developers in the execution of these phases. Existing methods offer a variety of solutions, especially for the requirements specification phase. However, such a diversity of approaches can be confusing for designers. For example, one approach first considers users' requirements and then checks them against the availability of data in the source systems, while another approach first develops the data

---

[1] We ignore modifications due to errors during data loading, and deletions for purging data in data warehouses.

warehouse schema on the basis of the underlying source systems and then adapts it to users' requirements. It is not clear in which situations it is better to use either one of these approaches. As a consequence, many data warehouse projects do not give much importance to the requirements specification phase, and focus instead on more technical aspects [220]. This may lead to a situation where the data warehouse may not meet users' expectations for supporting the decision-making process. In addition, some of the existing proposals include additional phases corresponding to particular characteristics of the conceptual multidimensional model in use. In this way, these methods impose on designers a conceptual model that may be unknown to them and may also be inadequate for a particular data warehouse project.

Moreover, even though many organizations may require a data warehouse for the purpose of enhancing their decision-making processes, they might not be able to afford external teams or consultants to build it. Nevertheless, these organization will typically have in them professionals highly skilled in the development of traditional operational databases, although inexperienced in the development of data warehouses. Therefore, providing a methodological framework for data warehouse development that is based on the one used for traditional databases may facilitate this transition process.

Finally, since spatial and temporal data warehouses are new research fields, there is not yet a methodological approach for developing them. However, it is well known that the inclusion of spatial data increases the power of the decision-making process by expanding the scope of the analysis and by providing enhanced visualization facilities. Similarly, temporal support expands the scope of the analysis by allowing all elements of a multidimensional model to be time-varying. Therefore, any proposed method should specify how and when spatial and temporal support may be included, considering all the phases required for the development of spatial and temporal data warehouses.

1.3 Objective of the Book and its Contributions to Research

The general objective of this book was to define a multidimensional conceptual model and an associated design method for data warehouse applications manipulating conventional, spatial, and temporal data. Our conceptual model, called MultiDim, includes both graphical and textual notation, as well as a formal definition.

This book summarizes original research conducted by the authors, which has been partly published previously in international journals, conference proceedings, and book chapters [173, 174, 175, 176, 177, 178, 179, 180, 181, 182, 183, 184]. Since the book covers several research areas, we describe next the contributions of this research to each of them.

1.3.1 Conventional Data Warehouses

The main contributions of our proposal in the field of conventional data warehouses include the following aspects.

The MultiDim model is based on well-known ER constructs, i.e., entity types, relationship types, and attributes, with their usual semantics. Therefore, it allows designers to apply the same modeling constructs as those used for operational database design, and it provides a conceptual representation independent of technical details. The model considers a multidimensional view of data in which the various elements of the model, such as facts, measures, dimensions, and hierarchies, can be clearly distinguished in a visual manner, thus facilitating the representation of the data required for data warehouse and OLAP applications.

Our proposal extends the ER model by including various types of hierarchies that exist in real-world applications. We classify these hierarchy types, taking into account their differences at the schema and instance levels. Since current data warehouse and OLAP systems support only a limited number of hierarchy types, our proposal has important implications for users, designers, and implementers. It will give users a better understanding of the data to be analyzed and help designers to build conceptual schemas for data warehouse applications. Additionally, the proposed classification will provide OLAP tool implementers with the requirements needed by business users for extending the functionality of current OLAP tools. The mappings to the relational and object-relational models show that the various hierarchy types can be implemented in current DBMSs, regardless of the fact that some of them are considered as advanced features of a multidimensional model [295].

Our proposal contains additional contributions related to the methodological framework proposed for conventional-data-warehouse design.

Despite the apparent similarity of the various phases proposed for designing databases and data warehouses, to account for the specificities of data warehouse design we both modify the content of these phases and include additional ones. In particular, on the basis of existing proposals, we consider three approaches to the requirements specification phase. We discuss the general advantages and disadvantages of these approaches, and include guidelines for choosing one of them depending on the characteristics of the users, the developer team, and the source systems. Further, we include an additional phase for the data extraction, transformation, and loading processes.

The proposed method for data warehouse design will provide the developer team with precise guidance on the required phases and their content. In addition, a comprehensive method combining various approaches will allow the most appropriate approach to be chosen, taking into consideration the experience of the developer team and the particularities of the data warehouse project. Since the proposed method is independent of any conceptual model, logical database model, or target implementation platform, it may be useful to both database developers who want to acquire basic knowledge for developing a data warehouse and to experienced data warehouse developers.

1.3.2 Spatial Data Warehouses

We extend the MultiDim model by the inclusion of spatial support. The novelty of our approach consists in the following aspects.

The multidimensional model is seldom used for representing spatial data even though this model provides a concise and organized representation for spatial data warehouses [18] and facilitates the delivery of data for spatial-OLAP systems. Our proposal provides spatial support for the various elements of the multidimensional model, i.e., facts, measures, dimensions, and hierarchies. In particular, the model includes a new feature, called the spatial fact relationship, which allows the topological relationships between spatial dimensions to be represented.

We also extend the previous classification of hierarchies and show its applicability to spatial hierarchies. We study how the topological relationships between spatial objects in a hierarchy influence the aggregation of measures. This gives insights to spatial-OLAP implementers for determining when measure aggregation can be done safely and when special aggregation procedures must be developed.

We provide a mapping for translating our spatially extended conceptual multidimensional model into the object-relational model. In this way, we show that schemas created using our model can be implemented in general-purpose DBMSs that include spatial extensions.

Finally, our proposal extends the method for conventional-data-warehouse design by taking into account the inclusion of spatial support in the various elements composing the multidimensional model. This method differs from the method proposed for spatial-database design, since it considers whether the users have knowledge of the analysis and manipulation of spatial data and whether the source systems include spatial data.

1.3.3 Temporal Data Warehouses

We also consider the inclusion of temporal support in the MultiDim model. The proposed temporal extension is based on previous research on temporal databases but takes account of the semantic differences between temporal data warehouses and temporal databases. Our proposal contributes in several ways to the field of temporal data warehouses.

We include in the MultiDim model several different temporality types that may be required for data warehouse applications but are currently ignored in research work. These temporality types allow us to extend the multidimensional model by including temporal support for dimensions, hierarchies, and measures. Therefore, we incorporate temporal semantics as an integral part of a conceptual multidimensional model. This should help users and designers to express precisely which elements of a data warehouse they want to be time-invariant and for which elements they want to keep track of changes over time. The proposed temporal extension allows changes to dimension data to be expressed in the same way as changes to measures, which cannot be achieved by current multidimensional models. An important consequence of this is that the temporal dimension is no longer required to indicate the validity of measures. We also show the usefulness of having several different temporality types for measures. This aspect is currently ignored in research work, which considers only valid time.

As for conventional and spatial data warehouses, we propose a mapping of temporal data warehouses to the object-relational model. Since current DBMSs are not temporally enhanced (even though some proposals exist, e.g., [54, 273]), this mapping will help implementers who use the MultiDim model for designing temporal data warehouses.

Finally, we propose a method for temporal-data-warehouse design that extends the method described for conventional-data-warehouse design. We refer to various cases, considering users' requirements for temporal support and the availability of historical data in the source systems.

1.4 Organization of the Book

Figure 1.1 illustrates the overall structure of the book and the interdependencies between the chapters. Readers may refer to this figure to tailor their use of this book to their own particular interests. Various paths may be followed, depending on whether the reader is interested in conventional, spatial, or temporal data warehouses. Readers interested only in conventional data warehouses may read the chapters following the path without labels, i.e., Chaps. 1, 2, 3, 6, and 8. Chapter 2 could be skipped by readers proficient in the database and data warehouse domains. On the other hand, readers interested in spatial or temporal data warehouses should include in addition Chaps. 4 and 7 or Chaps. 5 and 7, respectively.

We briefly describe next the remaining chapters of the book.

Chapter 2 provides an introduction to the fields of databases and data warehouses, covering the conceptual, logical, and physical design. This chapter thus provides the necessary background for the rest of the book. We provide also an introduction to two representative data warehouse platforms, Microsoft Analysis Services 2005 and Oracle 10g with the OLAP option.

Chapter 3 defines the MultiDim model, including the various kinds of hierarchies and their classification. For each of them, graphical notations and mappings to relational and object-relational databases are given.

Chapter 4 refers to the spatial extension of the MultiDim model, considering both the conceptual and the logical level. By means of examples, we discuss the inclusion of spatial characteristics in the various elements forming a multidimensional model, such as dimensions, hierarchies, and measures. Further,

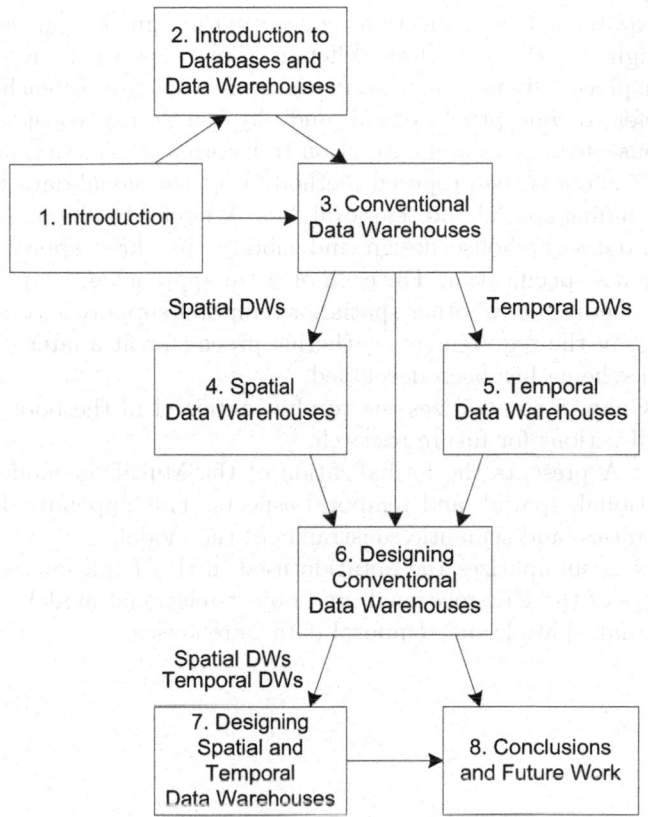

Fig. 1.1. Relationships between the chapters of this book

we study and classify the topological relationships that may exist between spatial levels forming a hierarchy, indicating when aggregation of measures can be done safely and when it requires special handling. We also present how the semantics can be preserved when a conceptual model is transformed to a logical model.

Chapter 5 presents the temporal extension of the MultiDim model. We propose the inclusion of various types of temporal support that are required for analysis purposes. Then, we discuss the inclusion of temporal support for dimensions and measures, taking into account the various roles that they play in a multidimensional model. We also consider the case where the time granularities in the source systems and the data warehouse are different. A logical representation of our conceptual model is also presented using an object-relational approach.

Chapter 6 specifies a method for conventional-data-warehouse design. Our proposal is in line with traditional database design methods that include the phases of requirements specification, conceptual design, logical design, and physical design. We describe three different approaches to the requirements specification phase, giving recommendations on when to use each of them. Then, we refer to conceptual, logical, and physical design, considering both data warehouse structures and extraction-transformation-loading processes.

Chapter 7 extends the proposed method for conventional-data-warehouse design by including spatial and temporal data. We refer to all phases used for conventional-data-warehouse design and modify the three approaches used for requirements specification. For each of these approaches, we propose two solutions, depending on whether spatial or temporal support is considered at the beginning of the requirements-gathering process or at a later stage, once a conceptual schema has been developed.

Finally, Chap. 8 summarizes the results contained in the book and indicates some directions for future research.

Appendix A presents the formalization of the MultiDim model, dealing with conventional, spatial, and temporal aspects. This appendix defines the syntax, semantics, and semantic constraints of the model.

Appendix B summarizes the notation used in this book for representing the constructs of the ER, relational, and object-relational models, as well as for conventional, spatial, and temporal data warehouses.

2

Introduction to Databases and Data Warehouses

This chapter introduces the basic concepts of databases and data warehouses. It compares the two fields and stresses the differences and complementarities between them. The aim of this chapter is to define the terminology and the framework used in the rest of the book, not to provide an extensive coverage of these fields. The outline of this chapter is as follows.

The first part of the chapter is devoted to databases. A database is a collection of logically related data that supports the activities of an organization. Many of the activities of our everyday life involve some interaction with a database. Section 2.1 begins by describing the basic concepts underlying database systems and describes the typical four-step process used for designing them, starting with requirements specification, and followed by conceptual, logical, and physical design. These steps allow a separation of concerns, where requirements specification gathers the requirements about the application and its environment, conceptual design targets the modeling of these requirements from the perspective of the users, logical design develops an implementation of the application according to a particular database technology, and physical design optimizes the application with respect to a particular implementation platform. We review in Sect. 2.2 the entity-relationship model, a popular conceptual model for designing databases. Sections 2.3.1 and 2.3.2 are devoted to two logical models of databases, the relational and the object-relational model. Finally, physical design considerations for databases are covered in Sect. 2.4.

The second part of the chapter is devoted to data warehouses. A data warehouse is a particular database targeted toward decision support. It takes data from various operational databases and other data sources and transforms it into new structures that fit better for the task of performing business analysis. We review in Sect. 2.5 the basic characteristics of data warehouses and compare them with operational databases. As explained in Sect. 2.6, data warehouses are based on a multidimensional model, where data is represented as hypercubes with dimensions corresponding to the various business perspectives and cube cells containing the measures to be analyzed. The architecture of data warehouse systems is described in detail in Sect. 2.9; as we shall see, in addition to the data warehouse itself, data warehouse systems are composed of back-end tools, which extract data from the various sources to populate the warehouse, and front-end tools, which are used to extract the information from the warehouse and present it to users. Sections 2.10 and 2.11 briefly present two representative data warehouse tools, SQL Server Analysis Services 2005 and Oracle OLAP 10g. Finally, Sect. 2.12 concludes this chapter.

2.1 Database Concepts

Databases constitute the core component of today's information systems. A **database** is a shared collection of logically related data, and a description of that data, designed to meet the information needs and support the activities of an organization. A database is deployed on a **database management system**(DBMS), which is a software system that allows users to define, create, manipulate, and administer a database.

Designing a database system is a complex undertaking that is typically divided into four phases:

- **Requirements specification**, which collects information about the users' needs with respect to the database system. A large number of approaches for requirements specification have been developed by both academia and practitioners. In general, these techniques help to elicit necessary and desirable system properties from prospective users and/or project managers, to homogenize requirements, and to assign priorities to them, i.e., separate necessary from "nice to have" system properties [314]. During this phase, active participation of users will increase customer satisfaction with the delivered system and avoid errors, which can be very expensive to correct if the subsequent phases have already been developed.

- **Conceptual design**, which aims at building a user-oriented representation of the database that does not contain any implementation considerations. This is done by using a **conceptual model** in order to identify the relevant entities, relationships, and attributes of the application domain. The entity-relationship model is one of the most often used conceptual models for designing database applications. Alternatively, object-oriented modeling techniques can also be applied, based on UML notation [29].

    Conceptual design can be performed using two different approaches, according to the complexity of the system and the developers' experience:

    - **Top-down design:** the requirements of the various users are merged before the design process begins, and a unique schema is built. Afterwards, a separation of the views corresponding to individual users' requirements can be performed. This approach can be difficult and expensive for large databases and inexperienced developers.

- **Bottom-up design:** a separate schema is built for each group of users with different requirements, and later, during the view integration phase, these schemas are merged to form a global conceptual schema for the entire database. This is the approach typically used for large databases.

- **Logical design**, which aims at translating the conceptual representation of the database obtained in the previous phase into a particular implementation model (or **logical model**) common to several DBMSs. Currently, the most common logical models are the relational model and the object-relational model. Other logical models include the object-oriented model and the semistructured (or XML-based) model. To ensure an adequate logical representation, suitable mapping rules must be specified. These ensure that the constructs included in the conceptual model can be transformed to the appropriate structures of the logical model.

- **Physical design**, which aims at customizing the logical representation of the database obtained in the previous phase to an implementation in a particular DBMS platform. Common DBMSs include SQL Server, Oracle, DB2, and MySQL, among others.

A major objective of this four-level process is to provide **data independence**, i.e., to ensure that schemas in upper levels are unaffected by changes to schemas in lower levels. There are two kinds of data independence. **Logical data independence** refers to immunity of the conceptual schema to changes in the logical schema. For example, rearranging the structure of relational tables should not affect the conceptual schema, provided that the requirements of the application remain the same. **Physical data independence** refers to immunity of the logical schema to changes in the physical schema. For example, physically sorting the records of a file on a computer disk does not affect the conceptual or logical schema, although this modification may be perceived by the user through a change in response time.

In the following sections we briefly describe one conceptual model, the entity-relationship model, and two logical models, the relational model and the object-relational model. We then briefly address physical design considerations.

2.2 The Entity-Relationship Model

The entity-relationship (ER) model [45] is one of the most often used conceptual models for designing database applications. Although there is general agreement about the meaning of the various concepts of the ER model, a number of different visual notations have been proposed for representing these concepts. In this book, we borrow some of the notations of MADS [227], a spatiotemporal conceptual model. Figure 2.1 shows a small excerpt from an ER schema for a hypothetical university application.

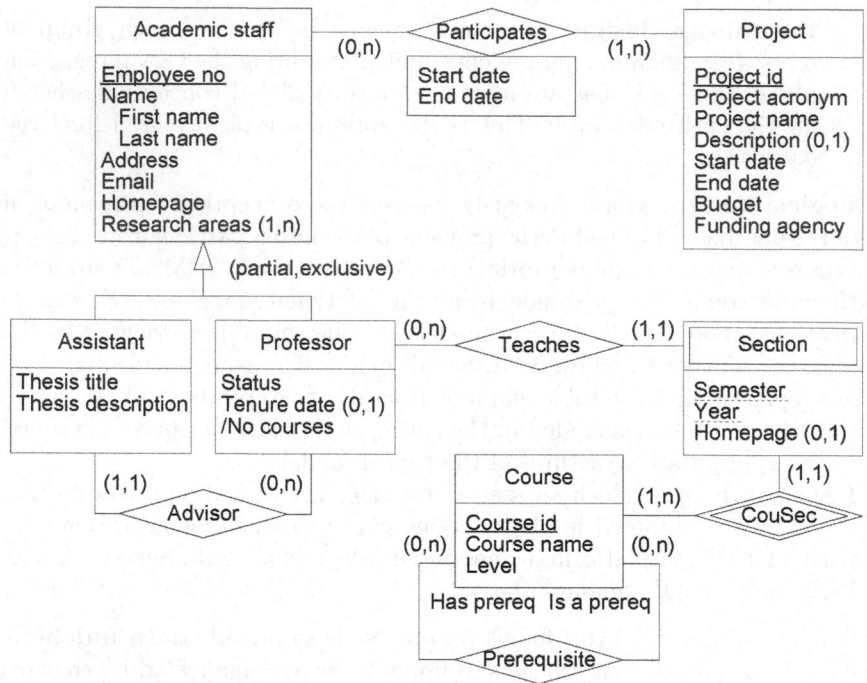

Fig. 2.1. An excerpt from an ER schema for a university application

Entity types are used to represent a set of real-world objects of interest to an application. In Fig. 2.1, Academic staff, Project, and Professor are examples of entity types. An object belonging to an entity type is called an **entity** or an **instance**. The set of instances of an entity type is called its **population**. From the application point of view, all entities of an entity type have the same characteristics.

In the real world, objects do not live in isolation; they are related to other objects. **Relationship types** are used to represent these associations between objects. In our example, Participates, Teaches, and Advisor are examples of relationship types. An association between objects of a relationship type is called a **relationship** or an **instance**. The set of associations of a relationship type is called its **population**.

The participation of an entity type in a relationship type is called a **role** and is diagrammatically represented by a line linking the two types. Each role of a relationship type has associated with it a pair of **cardinalities** describing the minimum and maximum number of times that an entity may participate in that relationship type. For example, the role between Participates and Project has cardinalities (1,n), meaning that each project participates between 1 and n (i.e., an undetermined number of times greater than 1) in the relationship. On the other hand, the cardinality (1,1) between Teaches and Course means that each course participates exactly once in the relationship type. A role is said to be **optional** or **mandatory** depending on whether its minimum cardinality is 0 or 1, respectively. Further, a role is said to be **monovalued** or **multivalued** depending on whether its maximum cardinality is 1 or n, respectively.

A relationship type may relate two or more object types: it is called **binary** if it relates two object types, and **n-ary** if it relates more than two object types. In Fig. 2.1, all relationship types are binary. Depending on the maximum cardinality of each role, binary relationship types can be categorized into **one-to-one**, **one-to-many**, and **many-to-many** relationship types. In Fig. 2.1, the relationship Teaches is one-to-many, since a course is related to at most one professor, while a professor may be related to several courses. On the other hand, Participates is a many-to-many relationship.

It may be the case that the same entity type is related more than once in a relationship type, as is the case for the Prerequisites relationship type. In this case the relationship type is called **recursive**, and **role names** are necessary to distinguish between different roles of the entity type. In Fig. 2.1, Has prereq and Is a prereq are role names.

In the real world, objects and the relationships between them have a series of structural characteristics that describe them. **Attributes** are used for recording these characteristics of entity or relationship types. For example, in Fig. 2.1 Address and Email are attributes of Academic staff, while Semester and Year are attributes of Teaches.

Like roles, attributes have associated **cardinalities**, defining the number of values that an attribute may take in each instance. Since most of the time the cardinality of an attribute is (1,1), we do not show this cardinality in our schema diagrams. Thus, each member of the academic staff will have exactly one Email, while they may have several (at least one) Research areas. As for roles, attributes are called **optional** or **mandatory** depending on whether its minimum cardinality is 0 or 1, respectively. Similarly, attributes are called **monovalued** or **multivalued** depending on whether they may take at most one or several values, respectively.

Further, attributes may be composed of other attributes, as shown by the attribute Name of Academic staff in our example, which is composed of First name and Last name. Such attributes are called **complex attributes**, while those that do not have components are called **simple attributes**. Finally, some attributes may be **derived**, as shown for the attribute No courses of Professor in our example. This means that the value of the attribute for each entity may be derived by use of a formula involving other elements of the schema. In our case, the derived attribute records the number of times that a particular professor participates in the relationship Teaches.

A common situation in the real world it that one or several attributes uniquely identify a particular object; such attributes are called **identifiers** or **keys**. In Fig. 2.1, identifiers are underlined; for example, Project id is the identifier of the entity type Project, meaning that every project has a unique value for this attribute, different from those for all other projects. In the figure, all identifiers are simple, i.e., they are composed of only one attribute, although it is common to have identifiers composed of two or more attributes.

Entity types that do not have an identifier of their own are called **weak entity types**. In contrast, regular entity types that do have an identifier are called **strong entity types**. In Fig. 2.1, Section is a weak entity type, as shown by the double line of its name box; all other entity types in the figure are strong entity types. A weak entity type is dependent on the existence of another entity type, called the **identifying** or **owner entity type**. The relationship type that relates a weak entity type to its owner is called the **identifying relationship type** of the weak entity type. In Fig. 2.1, Course and CouSec are, respectively, the identifying entity type and the identifying relationship type of Section. As shown in the figure, an identifying relationship is distinguished by the double line of its diamond. A relationship type that is not an identifying relationship type is called a **regular relationship type**. Note that identifying relationships types have a cardinality (1,1) in the role of the weak entity type and may have a (0,n) or (1,n) cardinality in the role of the owner. A weak entity type typically has a **partial key**, which is the set of attributes that can uniquely identify weak entities that are related to the same owner entity. In our example, we assume that a course has at most one associated section in a semester and a year. In Fig. 2.1, the partial key attributes are underlined with a dashed line.

Finally, owing to the complexity of conceptualizing the real world, human beings usually refer to the same concept using several different perspectives with different abstraction levels. The **generalization** (or **is-a**) **relationship** captures such a mental process. It relates two entity types, called the **supertype** and the **subtype**, meaning that both types represent the same concept at different levels of detail. In our example, there are two generalizations, between the supertype Academic staff and the subtypes Assistant and Professor. As shown in the figure, several generalization relationships sharing the same supertype and accounting for the same classification criterion (in this case job type) are usually grouped together.

Generalization has three essential characteristics. The first one is **population inclusion**, meaning that every instance of the subtype is also an instance of the supertype. In our example, this means that every professor is also a member of the academic staff. The second characteristic is **inheritance**, meaning that all characteristics of the supertype (for instance attributes and roles) are inherited by the subtype. Thus, in our example professors also have, for instance, a name and an email address. Finally, the third characteristic is **substitutability**, meaning that each time an instance of the supertype is required (for instance in an operation or in a query), an instance of the subtype can used instead.

Several generalization relationships can be grouped together when they account for the same criterion. For example, the two generalizations between the supertype Academic staff and the subtypes Assistant and Professor are connected together because they categorize staff members according to their position. Consequently, generalizations may be characterized according to two criteria. On the one hand, a generalization is either **total** or **partial**, depending on whether every instance of the supertype is also an instance of one of the subtypes. In our example, the generalization is partial, since there are members of the academic staff who are neither assistants nor professors. On the other hand, a generalization is also either **disjoint** or **overlapping**, depending on whether an instance may belong to one or several subtypes. In our example, the generalization is disjoint, since a member of the academic staff cannot be at the same time an assistant and a professor.

In many real-world situations, a subtype may have several supertypes. As an example, in Fig. 2.1 the entity type Assistant could also be a subtype of a Student entity type in addition to being a subtype of Academic staff. This is called **multiple inheritance**, since the subtype inherits properties from all its supertypes. Multiple inheritance may induce conflicts when a property with the same name is inherited from two supertypes. This would be the case if Assistant were to inherit the attribute Address from both Academic staff and Student. Various strategies have been devised for disambiguating such conflicts in the case of multiple inheritance. A typical strategy is to qualify the problematic property with the name of the type to which it belongs. In our example, it must be specified whether Academic staff.Address or Student.Address is meant when one is accessing the address of an assistant.

2.3 Logical Database Design

In this section, we shall briefly describe two logical data models for databases: the relational model and the object-relational model.

2.3.1 The Relational Model

Relational databases have been successfully used for several decades for storing information in many application domains. Nowadays, the relational model is the most often used approach for storing persistent information and is likely to remain so in the foreseeable future.

The relational model proposes a simple data structure, a **relation** (or **table**) composed of one or several **attributes** (or **columns**). Figure 2.2 shows a relational schema that corresponds to the schema of Fig. 2.1. As we will see later in this section, this relational schema is obtained by applying a set of translation rules to the corresponding ER schema. The relational schema is composed of a set of relations, such as Academic staff, Professor, and Assistant. Each of these relations is composed of several attributes. For example, Employee no, First name, and Last name are some attributes of the relation Academic staff.

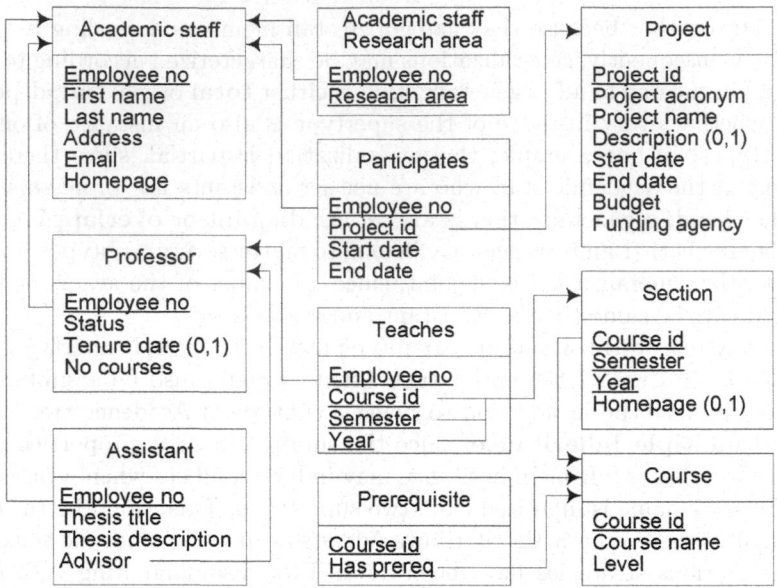

Fig. 2.2. A relational schema that corresponds to the schema of Fig. 2.1

In the relational model, each attribute is defined over a **domain**, or **data type**, i.e., a set of values with an associated set of operators. Typical relational implementations provide only a few basic domains, such as integer, float, date, and string. One important restriction in the relational model is that attributes must be atomic and monovalued; this is the case for all attributes in Fig. 2.2. Thus, the complex attribute Name of the entity type Academic staff in Fig. 2.1 has been split into the two atomic attributes First name and Last name in the table of the same name in Fig. 2.2. Therefore, a relation R is defined by a schema $R(A_1 : D_1, A_2 : D_2, \ldots, A_n : D_n)$, where the attribute A_i is defined over the domain D_i. The value of such a relation R is a set of **tuples** (or **rows**), i.e., a subset of the Cartesian product $D_1 \times D_2 \times \cdots \times D_n$.

The relational model allows several **integrity constraints** to be defined declaratively.

- An attribute may be defined as being **not null**, meaning that null values (or blanks) are not allowed in that attribute. In Fig. 2.2, only the attributes marked with a cardinality (0,1) allow null values.
- One or several attributes may be defined as a **key**, i.e., it is not allowed that two different tuples of the relation have identical values in such columns. In Fig. 2.2, keys are underlined. A key composed of several attributes is called a **composite key**, otherwise it is a **simple key**. In Fig. 2.2, the table Academic staff has a simple key, Employee no, while the table Prerequisite has a composite key, composed of Course id and Has prereq. In

the relational model each relation must have a **primary key** and may have other **alternative keys**. Further, the attributes composing the primary key do not accept null values.

- **Referential integrity** allows links between two tables (or twice the same table) to be specified, where a set of attributes in one table, called the **foreign key**, references the primary key of the other table. This means that the values accepted in the foreign key must also exist in the primary key. In Fig. 2.2, referential integrity constraints are represented by arrows from the foreign key to the primary key of the table that is referenced. For example, the attribute Employee no in table Professor references the primary key of the table Academic staff. This ensures, in particular, that every professor also appears in table Academic staff. Note also that referential integrity may involve foreign keys and primary keys composed of several attributes, as in the case of tables Teaches and Section.
- Finally, a **check constraint** defines a predicate that must be valid when adding or updating a tuple in a relation. For example, a check constraint could be used to verify that the only allowed values in the attribute status of table Professor are 'Assistant', 'Associate', and 'Full'. Most database management systems restrict check constraints to a single tuple: references to data stored in other tables or in other tuples of the same table are not allowed. Therefore, check constraints can be used only to verify simple constraints.

As can be seen, the above declarative integrity constraints are not sufficient to express the many constraints that exists in any application domain. Such constraints must then be implemented using triggers. A **trigger** is a named event-condition-action rule that is automatically activated when a relation is updated [321]. Throughout this book, we shall see several examples of integrity constraints implemented using triggers.

The translation of the ER model (or of the structural part of other ER-based conceptual models, including UML) to the relational model is well known and is implemented in many CASE tools. We define next the basic rules, based on [66].

Rule 1: A strong entity type is associated with a table containing the simple monovalued attributes and the simple components of the monovalued complex attributes of the entity type. This table also defines not null constraints for the mandatory attributes. The identifier of the entity type defines the primary key of the associated table.

Rule 2: A weak entity type is transformed as a strong entity type, with the exception that the associated table includes in addition the identifier of the owner entity type. A referential integrity constraint relates this identifier to the table associated with the owner entity type. The primary key of the table associated with the weak entity type is composed of the partial identifier of the weak entity type and the identifier of the owner entity type.

Rule 3: A binary one-to-one relationship type is represented by including the identifier of one of the participating entity types in the table associated with the other entity type. A referential integrity constraint relates this identifier to the table associated with the corresponding entity type. In addition, the simple monovalued attributes and the simple components of the monovalued complex attributes of the relationship type are also included in the table that represents the relationship type. This table also defines not null constraints for the mandatory attributes.

Rule 4: A regular binary one-to-many relationship type is represented by including the identifier of the entity type that participates in the relationship with a maximum cardinality of 1 in the table associated with the other entity type. A referential integrity constraint relates this identifier to the table associated with the corresponding entity type. In addition, the simple monovalued attributes and the simple components of the monovalued complex attributes of the relationship type are included in the table that represents the relationship type. This table also defines not null constraints for the mandatory attributes.

Rule 5: A binary many-to-many relationship type or an n-ary relationship type is associated with a table containing the identifiers of the entity types that participate in all its roles. A referential integrity constraint relates each one of these identifiers to the table associated with the corresponding entity type. The table also contains the simple monovalued attributes and the simple components of the monovalued complex attributes of the relationship type. This table also defines not null constraints for the mandatory attributes. The relationship identifier, if any, may define the key of the table, although a combination of all the role identifiers can also be used. Note that in order to preserve the semantics of associations, i.e., a relationship instance does not exist without its related entities, not null constraints are defined for all role identifiers.

Rule 6: A multivalued attribute of an entity or relationship type is associated with one additional table, which also includes the identifier of the entity or relationship type. A referential integrity constraint relates this identifier to the table associated with the entity or relationship type. The primary key of the table associated with the multivalued attribute is composed of all its attributes.

Rule 7: Generalization relationships can be dealt with in three different ways:

> **Rule 7a:** Both the supertype and the subtype are associated with tables, in which case the identifier of the supertype is propagated to the table associated with the subtype. A referential integrity constraint relates this identifier to the table associated with the supertype.
>
> **Rule 7b:** Only the supertype is associated with a table, in which case all attributes of the subtype become optional attributes of the supertype.

Rule 7c: Only the subtype is associated with a table, in which case all attributes of the supertype are inherited in the subtype.

Note that the generalization type (total vs. partial and disjoint vs. overlapping) may preclude one of the above three approaches. For example, the third possibility is not applicable for partial generalizations.

Applying these rules to the schema given in Fig. 2.1 yields the relational schema shown in Fig. 2.2. Note that the above rules apply in the general case; however, other mappings are possible. For example, binary one-to-one and one-to-many relationships may be represented by a table of its own, using Rule 5 above. The choice between alternative representation depends on the characteristics of the particular application at hand.

SQL (structured query language) is the most common language for creating, manipulating, and retrieving data from relational database management systems. SQL is composed of several sublanguages. The **data definition language** (DDL) allows the schema of a database to be defined. The **data manipulation language** (DML) is used to query a database and to modify its content (i.e., to add, update, and delete data in a database). Throughout this book, we consider the latest version of the SQL standard, SQL:2003 [193].

The set of SQL DDL commands defining the relational schema of Fig. 2.2 is as follows.

```
create table AcademicStaff as (
     EmployeeNo integer primary key,
     FirstName character varying (30) not null,
     LastName character varying (30) not null,
     Address character varying (50) not null,
     Email character varying (30) not null,
     Homepage character varying (64) not null );
create table AcadStaffResArea as (
     EmployeeNo integer not null,
     ResearchArea character varying (30) not null,
     primary key (EmployeeNo,ResearchArea),
     foreign key EmployeeNo references AcademicStaff(EmployeeNo) );
create table Professor as (
     EmployeeNo integer primary key,
     Status character varying (10) not null,
     TenureDate date,
     NoCourses integer not null,
     constraint Professor_Status
          check ( Status in ( 'Assistant', 'Associate', 'Full' ) ),
     foreign key EmployeeNo references AcademicStaff(EmployeeNo) );
create table Assistant as (
     EmployeeNo integer primary key,
     ThesisTitle character varying (64) not null,
     ThesisDescription text not null,
     Advisor integer not null,
     foreign key EmployeeNo references AcademicStaff(EmployeeNo),
```

```
            foreign key Advisor references Professor(EmployeeNo) );
    create table Participates as (
        EmployeeNo integer not null,
        ProjectId integer not null,
        StartDate date not null,
        EndDate date not null,
        primary key (EmployeeNo,ProjectId),
        foreign key EmployeeNo references AcademicStaff(EmployeeNo),
        foreign key ProjectId references Project(ProjectId) );
    create table Project as (
        ProjectId integer primary key,
        ProjectAcronym character (15) not null,
        ProjectName character varying (30) not null,
        Description character varying (30),
        StartDate date not null,
        EndDate date not null,
        Budget character varying (30) not null,
        FundingAgency character varying (30) not null );
    create table Teaches as (
        EmployeeNo integer not null,
        CourseId integer not null,
        Semester integer not null,
        Year integer not null,
        primary key (EmployeeNo,CourseId,Semester,Year),
        foreign key EmployeeNo references Professor(EmployeeNo),
        foreign key (CourseId,Semester,Year)
            references Section(CourseId,Semester,Year) );
    create table Section as (
        CourseId integer not null,
        Semester integer not null,
        Year integer not null,
        Homepage character varying (64),
        primary key (CourseId,Semester,Year),
        foreign key CourseId references Course(CourseId) );
    create table Course as (
        CourseId integer primary key,
        CourseName character varying (30) not null,
        Level character varying (30) not null );
    create table Prerequisite as (
        CourseId integer not null,
        HasPrereq integer not null,
        primary key (CourseId,HasPrereq),
        foreign key CourseId references Course(CourseId),
        foreign key HasPrereq references Course(CourseId) );
```

As shown in this simple schema excerpt, there is a significant difference in expressive power between the ER model and the relational model. This difference may be explained by the fact that the ER model is a *conceptual* model aimed at expressing concepts as closely as possible to the users' perspective, whereas the relational model is a *logical* model targeted toward particular implementation platforms. Several ER concepts do not have a correspondence in the relational model and must be expressed using the only concepts allowed in the model, i.e., relations, attributes, and the related constraints. This translation implies a semantic loss in the sense that data that is invalid in an ER schema is allowed in the corresponding relational schema, unless the relational schema is supplemented by additional constraints. In addition, many such constraints must be hand-coded by the user using mechanisms such as triggers or stored procedures. Furthermore, from the users' perspective, the relational schema is much less readable than the corresponding ER schema. This is crucial when one is considering schemas with hundreds of entity or relationship types and thousands of attributes. This is not a surprise, since this was exactly the reason for devising conceptual models back in the 1970s, i.e., the aim was to better understand the semantics of large relational schemas.

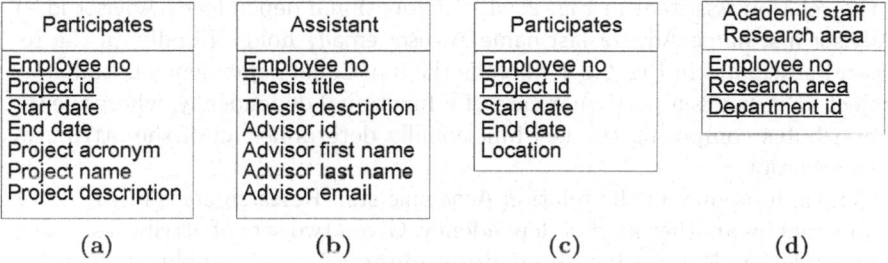

Fig. 2.3. Examples of relations that are not normalized

When one is considering a relational schema, it must be determined whether the relations in the schema have redundancies, and thus may induce anomalies in the presence of insertions, updates, and deletions. Consider for example the relation Participates in Fig. 2.3.1, which is a variation of the relation with the same name in Fig. 2.2. We can easily verify that the information about a project such as its name, acronym, and description is repeated for each staff member who works on that project. Therefore, when for example the description of a project is to be updated, it must be ensured that all tuples in the relation Participates concerning that particular project are given the modified description, otherwise there will be inconsistencies. Similarly, the relation Assistant in Fig. 2.3.1 is also redundant, since the first name, last name, and email address of every professor are repeated for all assistants who have the same advisor. Consider now relation Participates in Fig. 2.3.1, in which the additional attribute Location stores the location of the project. Suppose now that each location is associated with at most one project. In this case, the location information will be repeated for each staff member that works on the project of that location. Finally, consider the relation Academic staff Research area in Fig. 2.3.1, where an additional attribute Department id has been added with respect to the relation with the same name in Fig. 2.2. Suppose that members of the academic staff works in several different departments. Since the research areas of staff members are independent of the departments in which they work, there is a redundancy in the above table. Indeed, the information about the research areas of a staff member will be repeated as many times as the number of departments in which he/she works.

Dependencies and normal forms are used to precisely describe the redundancies above. A **functional dependency** is a constraint between two sets of attributes in a relation. Given a relation R and two sets of attributes X and Y in R, a functional dependency $X \to Y$ holds if and only if, in all the tuples of the relation, each value of X is associated with at most one value of Y. In this case it is said that X *determines* Y. The redundancies in Figs. 2.3.1,b,c can be expressed by means of functional dependencies. For example, in the relation Participates in Fig. 2.3.1, we have the functional dependency Project id $\to$ {Project acronym, Project name, Project description}. Also, in the relation Assistant in Fig. 2.3.1, the functional dependency Advisor id $\to$ {Advisor first name, Advisor last name, Advisor email} holds. Finally, in the relation Participates in Fig. 2.3.1, there is the functional dependency Location $\to$ Project id. A key is a particular case of a functional dependency, where the set of attributes composing the key functionally determines all of the attributes in the relation.

The redundancy in the relation Academic staff Research areas in Fig. 2.3.1 is captured by another kind of dependency. Given two sets of attributes X and Y in a relation R, a **multivalued dependency** $X \to\to Y$ holds if the value of X determines a set of values for Y, independently of any other attributes. In this case it is said that X *multidetermines* Y. In the relation in Fig. 2.3.1, we have the multivalued dependencies Employee no $\to\to$ Research area, and consequently Employee no $\to\to$ Department id. It is well known that functional dependencies are special cases of multivalued dependencies, i.e., every functional dependency is also a multivalued dependency. A multivalued dependency $X \to\to Y$ is said to be trivial if either $Y \subseteq X$ or $X \cup Y = R$, otherwise it is nontrivial.

A **normal form** is an integrity constraint certifying that a relational schema satisfies particular properties. Since the beginning of the relational model in the 1970s, many types of normal forms have been defined. In addition, normal forms have also been defined for other models, such as the entity-relationship model and the object-relational model. In the following, we consider only four normal forms that are widely used in relational databases.

As already said, the relational model allows only attributes that are atomic and monovalued. This restriction is called the **first normal form**. As we shall see in Sect. 2.3.2, the object-relational model removes this restriction and allows composite and multivalued attributes.

The **second normal form** avoids redundancies such as those in the table Participates in Fig. 2.3.1. In order to define the second normal form, we must define the following concepts:

- A **prime attribute** is an attribute that is part of a key.
- A **full functional dependency** is a dependency $X \to Y$ in which the removal of an attribute from X invalidates the dependency.

Now we can give the definition of the second normal form: A relation schema is in the second normal form if every nonprime attribute is fully functionally dependent on every key. As we can see, the table Participates above is not in the second normal form, since Project acronym, Project name, and Project description are nonprime attributes (they do not belong to a key) and are dependent on Project id, i.e., on part of the key of the relation. To make the relation comply with the second normal form, the nonprime attributes dependent on Project id must be removed from the table and an additional table Project(Project id, Project acronym, Project name, Project description) must be added to store the information about projects.

The **third normal form** avoids redundancies such as those in the table Assistant in Fig. 2.3.1. In order to define the third normal form, we must define one additional concept:

- A dependency $X \to Z$ is **transitive** if there is a set of attributes Y such that the dependencies $X \to Y$ and $Y \to Z$ hold.

Now we can give the definition of the third normal form: A relation is in the third normal form if it is in the second normal form and there are no transitive dependencies between a key and a nonprime attribute. The table Assistant above is not in the third normal form, since there is a transitive dependency from Employee no to Advisor id, and from Advisor id to Advisor first name, Advisor last name, and Advisor email. To make the relation comply with the third normal form, the attributes dependent on Advisor id must be removed from the table and an additional table Advisor(Advisor id, Advisor acronym, Advisor first name, Advisor last name) must be added to store the information about advisors.

The **Boyce-Codd normal form** avoids redundancies such as those in the table Participates in Fig. 2.3.1. Recall that in this case it is supposed that there is a functional dependency Location $\to$ Project id. A relation is in the Boyce-Codd normal form with respect to a set of functional dependencies F if, for every nontrivial dependency $X \to Y$ that can be derived from F, X is a key or contains a key of R. The table Participates above is not in the Boyce-Codd normal form, since the above functional dependency holds and Location is not a key of the relation. To make the relation comply with the Boyce-Codd form, the attribute Location must be removed from the table, and an additional table LocationProject(Location, Project id) must be added to store the information about the project associated with each location. Note that all relations in Fig. 2.2 are in the Boyce-Codd normal form.

The **fourth normal form** avoids redundancies such as those in the table Academic staff Research area in Fig. 2.3.1. A relation is in the fourth normal form with respect to a set of functional and multivalued dependencies F if, for every nontrivial dependency $X \rightarrow\rightarrow Y$ that can be derived from F, X is a key or contains a key of R. The table above is not in the fourth normal form, since there are multivalued dependencies from Employee no to Research area, and from Employee no to Department id, and Employee no is not a key of the relation. To make the relation comply with the fourth normal form, the attribute Department id must be removed from the table, and an additional table AcademicStaffDepart(Employee, Department id) must be added to store the information about the departments in which a member of the academic staff works.

2.3.2 The Object-Relational Model

As shown in the previous section, the relational model suffers from several weaknesses that become evident when we deal with complex applications.

- The relational model provides a very simple data structure (i.e., a relation), which disallows multivalued and complex attributes. Therefore, in a relational database, complex objects must be split into several tables. This induces performance problems, since assembly and disassembly operations using joins are needed for retrieving and storing complex objects in a relational database.
- The set of types provided by relational DBMSs is very restrictive. It includes only some basic types such as integer, float, string, and date, and uninterpreted binary streams that must be manipulated explicitly by the user. Such a restricted set of types does not fit complex application domains.
- There is no integration of operations with data structures, i.e., there is no encapsulation, and no methods associated with a table.
- Since there is no possibility to directly reference an object by use of a surrogate or a pointer, every link between tables is based on comparison of values. Therefore, joins represent a bottleneck with respect to performance.

During the 1980s, a considerable amount of research addressed the issue of relaxing the assumption that relations must satisfy the first normal form. Many results for the relational model were generalized to such an extended model, called **non-first-normal-form model**, or NFNF or NF2 model (e.g., [11, 262]). Such research has been introduced into the database standard SQL:2003 [192, 193] under the name of the object-relational model. In addition, current database management systems such as Oracle, Informix, DB2, and PostgreSQL have also introduced object-relational extensions, although these do not necessarily comply with the SQL:2003 standard.

The object-relational model preserves the foundations of the relational model, while extending its modeling power by organizing data using an object model [154]. The object-relational model allows attributes to have complex types, i.e., it inherently groups related facts into a single row [48]. This can be done by introducing abstract data types [256, 268]. These kinds of models are more expressive and more versatile [69]. In what follows we review several object-relational extensions to the traditional relational model, as follows:

- complex and/or multivalued attributes,
- user-defined types with associated methods,
- system-generated identifiers, and
- inheritance among types.

In addition to a set of predefined types (those of SQL-92 and a few new ones), SQL:2003 allows three **composite types**: row, array, and multiset. The **row type** allows structured values (i.e., composite attributes) to be stored in a column of a table. The **array type** allows variable-sized vectors of values of the same type to be stored in a column, and the **multiset type** allows an unordered collection of values, with duplicates permitted (i.e., bags). The array and multiset types can thus be used for storing multivalued attributes. Multisets have no declared maximum cardinality, unlike arrays, which have a user-specified or implementation-defined maximum cardinality. The composite types can be combined, allowing nested collections, although this is considered as an "advanced feature" in the standard.

SQL:2003 supports two sorts of **user-defined types. Distinct types** are types that can be represented internally by the same SQL predefined type (for instance the char type) but cannot be mixed in expressions (e.g., one cannot test whether the name and the social security number of an employee are equal, even if they are both represented by the same type). On the other hand, **structured user-defined types** are analogous to class declarations in object-oriented languages. Structured types may have attributes, which can be of any SQL type (including other structured types, at any nesting). Structured types may also have methods, which can be instance or static (class) methods. The attributes of structured types are encapsulated through observer and mutator functions. Both distinct and structured types can be used as a domain in a column of a table, a domain of an attribute of another type, or a domain of a table. Comparison and ordering of values of user-defined types are done only through user-defined methods.

SQL:2003 supports single inheritance, by which a subtype inherits attributes and methods from its supertype. The subtype can define additional attributes and methods, and can overload and override inherited methods. An **overloaded method** provides another definition of the inherited method with a different signature. An **overridden method** provides another implementation of the inherited method that has the same signature. Types and methods can be final or not. A **final type** cannot have subtypes, while a **final method** cannot be redefined. Types may be declared as instantiable or not. An **instantiable type** includes constructor methods that are used for creating instances.

SQL:2003 provides two types of tables. **Relational tables** are tables of the usual kind, although the domains for attributes are all predefined or user-defined types. **Typed tables** are tables that use structured user-defined types for their definition. However, these tables must declare keys and other **integrity constraints** in the table declaration. In addition, typed tables have a **self-referencing column** that contains a value that uniquely identifies each row. Such a column may be the primary key of the table, or it could be a column whose values are generated by the DBMS, i.e., surrogates. Row identifiers can be used for establishing links between tables. For this purpose, SQL:2003 provides a special **ref type** whose values are those unique identifiers. A ref type is always associated with a specified structured type.

There are several possible ways to translate a conceptual model such as the ER model (or the class diagram of UML) to the object-relational model (e.g. [58, 60, 84, 186]). We now describe one of them, which consists in using typed tables. For this, a two-step process is needed: define the structured types, and then define the corresponding table. We define below the basic rules of this transformation.

Rule 1: A strong entity type is associated with a structured type containing all its attributes. Multivalued attributes are defined using either the array or the multiset type. Composite attributes are defined using the row type or the structured type. Further, a table of the above structured type must be declared, in which the not null constraints for the mandatory attributes are also defined. The identifier of the entity type defines the key of the associated table.

Rule 2: A weak entity type is transformed as a strong entity type, with the exception that the associated structured type contains in addition the identifier of the owner entity type. Further, the combination of the partial identifier of the weak entity type and the identifier of its owner entity type defines the key of the associated table.

Rule 3: A regular relationship type is associated with a structured type containing the identifiers of all the entity types that participate in its roles, and containing all the attributes of the relationship. Further, a table of the above structured type must be declared in which the key of the table and not null constraints for the mandatory attributes must be declared.

In addition, inverse references may be defined for each role, i.e., a ref type pointing to the structured type of the relationship may be added to the structured type of the entity playing that role. Depending on whether the role is monovalued or multivalued, the corresponding ref type may be simple or a multiset.

Rule 4: As the object-relational model supports only single inheritance between types, it is necessary to remove multiple inheritance by transforming some generalization relationships into binary one-to-one relationships, to which Rule 3 above is applied. Then, all the remaining generalization relationships are implemented by making the subtype inherit from the supertype.

The above rules apply in the general case, but additional optimizations can be performed, for example to reduce the number of tables in the object-relational schema. A well-known example is the embedding of relationships without attributes into the types corresponding to the linked entity types.

Applying the above rules to the conceptual schema in Fig. 2.1 yields the types for the object-relational schema given in Fig. 2.4. With each of these types is associated a corresponding table. In the schema, arrows represent ref types pointing to a referenced type. Note that the relationship type Teaches in Fig. 2.1 is represented in Fig. 2.4 in the types corresponding to professors and sections. All other entity and relationship types of the conceptual schema are represented as structured types in the object-relational schema. In addition, pairs of reference attributes are used for linking the type representing the relationship type to the types representing the participating entity types. It is the users' responsibility to ensure the consistency of these cross-referenced attributes.

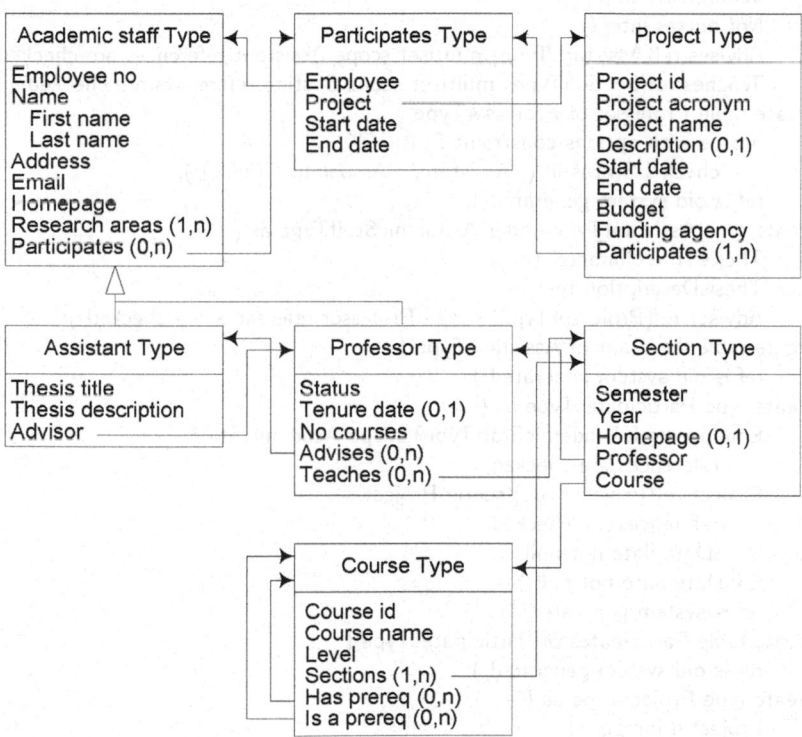

Fig. 2.4. Types for the SQL:2003 schema corresponding to the ER schema of Fig. 2.1

The SQL:2003 commands defining the schema are as follows.

```
create type AcademicStaffType as (
    EmployeeNo integer,
    Name row (
        FirstName character varying (30),
        LastName character varying (30) ),
    Address character varying (50),
    Email character varying (30),
    Homepage character varying (64),
    ResearchAreas character varying (30) multiset,
    Participates ref(ParticipatesType) multiset scope Participates
        references are checked )
    ref is system generated;
create table AcademicStaff of AcademicStaffType (
    constraint AcademicStaffPK primary key (EmployeeNo),
    Email with options constraint AcademicStaffEmailNN Email not null,
    ref is oid system generated );
create type ProfessorType under AcademicStaffType as (
    Status character (10),
    TenureDate date,
    NoCourses integer,
    Advises ref(AssistantType) multiset scope Assistant references are checked,
    Teaches ref(SectionType) multiset scope Section references are checked );
create table Professor of ProfessorType (
    Status with options constraint StatusCK
        check ( Status in ( 'Assistant', 'Associate', 'Full' ) ),
    ref is oid system generated );
create type AssistantType under AcademicStaffType as (
    ThesisTitle character (64),
    ThesisDescription text,
    Advisor ref(ProfessorType) scope Professor references are checked );
create table Assistant of AssistantType (
    ref is oid system generated );
create type ParticipatesType as (
    Employee ref(AcademicStaffType) scope AcademicStaff
        references are checked,
    Project ref(ProjectType) scope Project
        references are checked,
    StartDate date not null,
    EndDate date not null )
    ref is system generated;
create table Participates of ParticipatesType (
    ref is oid system generated );
create type ProjectType as (
    ProjectId integer,
    ProjectAcronym character (15),
    ProjectName character varying (30),
    Description character varying (30),
```

```
        StartDate date,
        EndDate date,
        Budget character varying (30),
        FundingAgency character varying (30),
        Participates ref(ParticipatesType) multiset scope Participates
            references are checked )
        ref is system generated;
create table Project of ProjectType (
        constraint ProjectPK primary key (ProjectId),
        ref is oid system generated );
create type SectionType as (
        Professor ref(ProfessorType) scope Professor references are checked,
        Course ref(CourseType) scope Course references are checked,
        Semester integer,
        Year integer,
        Homepage character varying (64) )
        ref is system generated;
create table Section of SectionType (
        ref is oid system generated );
create type CourseType as (
        CourseId integer,
        CourseName character varying (30),
        Level character varying (30),
        Sections ref(SectionType) multiset scope Section references are checked,
        HasPrereq ref(CourseType) multiset scope Course references are checked,
        IsAPrereq ref(CourseType) multiset scope Course references are checked )
        ref is system generated;
create table CourseType of CourseTypeType (
        constraint CourseTypePK primary key (CourseId),
        ref is oid system generated );
```

Note that we have implemented the above schema using typed tables, instead of using classical relational tables. In particular, this allows us to define methods for the types associated with the tables. Note also that constraints associated with the types (for instance primary key constraints) are specified in the corresponding table declarations. For example, a check constraint in the Professor table restricts the values that can take the attribute Status. For brevity, in the above SQL commands we have not specified many constraints, in particular not null constraints for mandatory attributes. An example of such a constraint is given for the attribute Email in the AcademicStaff table.

As shown in this example, the object-relational schema is closer to the users' perspective than is the relational schema. However, the corresponding ER schema is more expressive: for instance, associations in the ER model are represented by pairs of reference attributes. In addition, implementation considerations appear in the object-relational schema, such as the reference attributes.

2.4 Physical Database Design

The objective of **physical database design** is to specify how database records are stored, accessed, and related in order to ensure adequate performance of a database application. Physical database design thus requires one to know the specificities of the given application, in particular the properties of the data and the usage patterns of the database. The latter involves analyzing the transactions or queries that are run frequently and will have a significant impact on performance, the transactions that are critical to the operations of the organization, and the periods of time during which there will be a high demand on the database (called the **peak load**). This information is used to identify the parts of the database that may cause performance problems.

There are a number of factors that can be used to measure the performance of database applications. **Transaction throughput** is the number of transactions that can be processed in a given time interval. In some systems, such as electronic payment systems, a high transaction throughput is critical. **Response time** is the elapsed time for the completion of a single transaction. Minimizing response time is essential from the user's point of view. Finally, **disk storage** is the amount of disk space required to store the database files. However, a compromise usually has to be made among these factors. From a general perspective, this compromise implies the following factors:

1. **Space-time trade-off:** It is often possible to reduce the time taken to perform an operation by using more space, and vice versa. For example, using a compression algorithm allows one to reduce the space occupied by a large file but implies extra time for the decompression process.
2. **Query-update trade-off:** Access to data can be made more efficient by imposing some structure upon it. However, the more elaborate the structure, the more time is taken to build it and to maintain it when its contents change. For example, sorting the records of a file according to a key field allows them to be located more easily but there is a greater overhead upon insertions.

Further, once an initial physical design has been implemented, it is necessary to monitor the system and to tune it as a result of the observed performance and any changes in requirements. Many DBMSs provide utilities to monitor and tune the operations of the system.

As the functionality provided by current DBMSs varies widely, physical design requires one to know the various techniques for storing and finding data that are implemented in the particular DBMS that will be used.

A database is organized on **secondary storage** (for instance magnetic or optical disks) into one or more **files**, where each file consists of one or several **records** and each record consists of one or several **fields**. Typically, each tuple in a relation corresponds to a record in a file. When a user requests a particular tuple, the DBMS maps this logical record into a physical disk address and retrieves the record into main memory using the operating-system file access routines.

Data is stored on a computer disk in **disk blocks** (or **pages**) that are set by the operating system during disk formatting. Transfer of data between main memory and the disk and vice versa takes place in units of disk blocks. One important aspect of physical database design is the need to provide a good match between disk blocks and logical units such as tables and records. DBMSs store data on **database blocks** (or **pages**) and most DBMSs provide the ability to specify a database block size. The selection of a database block size depends on several issues. For example, in some DBMSs the finest locking granularity is at the block level, not at the record level. Therefore, as the number of records that can be contained in a block increases, the chance that two transactions will request access to entries in the same block also increases. On the other hand, for optimal disk efficiency, the database block size must be equal to or be a multiple of the disk block size.

File organization is the physical arrangement of data in a file into records and blocks on secondary storage. There are three main types of file organization. **Heap files** (or **unordered files**) have the simplest type of file organization, where records are placed in the file in the same order as they are inserted. This makes insertion very efficient. However, retrieval is relatively slow, since the various blocks of the file must be read in sequence until the required record is found. **Sequential files** (or **ordered files**) have their records sorted on the values of one or more fields (called **ordering fields**). Ordered files allow fast retrieving of records, provided that the search condition is based on the ordering fields. However, inserting and deleting records in a sequential file are problematic, since the ordering must be maintained. Finally, **hash files** use a **hash function** that calculates the address of the block (or **bucket**) in which a record is to be stored on the basis of one or more fields of the record. Within a bucket, records are placed in order of arrival. A **collision** arises when a bucket is filled to its capacity and a new record must be inserted into that bucket. Hashing provides the fastest possible access for retrieving an arbitrary record given the value of its hash field. However, collision management degrades the overall performance.

Independently of the particular file organization, additional access structures called **indexes** are typically used to speed up the retrieval of records in response to search conditions. Indexes provide alternative ways of accessing the records and enable efficient access to them based on the **indexing fields** that are used to construct the index. Any field(s) of the file can be used to create an index, and multiple indexes on different fields can be constructed in the same file.

There are many different types of indexes. We describe below some categories of indexes according to various criteria.

- One categorization of indexes distinguishes between **clustering** and **non-clustering indexes**, also called **primary** and **secondary indexes**. In a clustering index, the records in the data file are physically ordered according to the field(s) on which the index is defined. This is not the case for a nonclustering index. A file can have at most one clustering index, and in addition can have several nonclustering indexes. Nonclustering indexes are always dense (see below).

- Indexes can be **single-column** or **multiple-column**, depending on the number of indexing fields on which they are based. When a multiple-column index is created, the order of columns in the index has an impact on data retrieval. Generally, the most restrictive value should be placed first for optimum performance.
- Another categorization of indexes is according to whether they are **unique** or **nonunique**: unique indexes do not allow duplicate values, while this is not the case for nonunique indexes.
- In addition, an index can be **sparse** or **dense**: a sparse index has index records for only some of the search values in the file, while a dense index has an index record for every search value in the file.
- Finally, indexes can be **single-level** or **multilevel**. When an index file becomes large and extends over many blocks, the search time required for the index increases. A multilevel index attempts to overcome this problem by splitting the index into a number of smaller indexes, and maintaining an index to the indexes. Although a multilevel index reduces the number of blocks accessed when one is searching for a record, it also has problems in dealing with insertions and deletions in the index because all index levels are physically ordered files. A **dynamic multilevel index** solves this problem by leaving some space in each of its blocks for inserting new entries. This type of index is often implemented by using data structures called **B-trees** and **B⁺-trees**, which are supported by most DBMSs.

Most DBMSs give the designer the option to set up indexes on any fields, thus achieving faster access at the expense of extra storage space for indexes, and overheads when updating. Because the indexed values are held in a sorted order, they can be efficiently exploited to handle partial matching and range searches, and in a relational system they can speed up join operations on indexed fields.

Some DBMSs, such as Oracle, support **clustering**. Clusters are groups of one or more tables physically stored together because they share common columns and are often used together. The related columns of the tables in a cluster are called the **cluster key**. Clustering improves data retrieval, since related records are physically stored together. Further, since the cluster key is stored only once, tables are stored more efficiently than if the tables were stored individually. Oracle supports two types of clusters, indexed and hash clusters, the difference being that in the latter a hash function is applied to the key value of a record.

2.5 Data Warehouses

The importance of data analysis has increased significantly in recent years as organizations in all sectors are being required to improve their decision-making processes in order to maintain their competitive advantage. Traditional database systems do not satisfy the requirements of data analysis. They support the daily operations of an organization and their primary concern is to ensure fast access to data in the presence of multiple users, which necessitates transaction processing, concurrency control, and recovery techniques that guarantee data consistency. These systems are known as **operational databases** or **online transaction processing** (**OLTP**) systems. Typical operational databases contain detailed data, do not include historical data, and since they are usually highly normalized, they perform poorly when executing complex queries that need to join many relational tables together or to aggregate large volumes of data. Further, when users wish to analyze the behavior of an organization as a whole, data from various operational systems must be integrated. This can be a difficult task to accomplish because of the differences in data definition and content.

Data warehouses were proposed in order to better respond to the growing demands of decision-making users. A **data warehouse** is a collection of subject-oriented, integrated, nonvolatile, and time-varying data to support management decisions [118]. We describe in detail below these distinctive aspects of data warehouses.

- **Subject-oriented** means that data warehouses focus on the analytical requirements of managers at various levels of the decision-making process, i.e., they are oriented toward particular subjects of analysis. These subjects vary depending on the kind of activities performed by the organization, for example analysis of sales in the case of a retail company, analysis of clients' behavior when using banking services, and analysis of the utilization of a railroad system in the case of a transportation company. This is to be contrasted with operational databases, where the focus is on specific functions that applications must perform, for example registering sales of products or depositing and withdrawing money. Within an organization, several subjects of analysis may be included in the same data warehouse. For example, a data warehouse in a retail company may contain data related to the analysis of the purchase, inventory, and sales of products.
- **Integrated** represents the complex effort of joining together data from several operational and external systems, which implies solving problems due to differences in data definition and content, such as differences in data format and data codification, synonyms (fields with different names but the same data), homonyms (fields with the same name but different meanings), multiplicity of occurrences of data, and many others. In operational databases these problems are typically solved in the database design phase.

- **Nonvolatile** means that durability of data is ensured by disallowing data modification and removal, thus expanding the scope of the data to a longer period of time than operational systems usually offer. A data warehouse gathers data encompassing several years, typically 5 to 10 years or beyond, while data in operational databases is often kept for only a short period of time, for example from 2 to 6 months, as required for daily operations, and it may be overwritten when necessary.
- **Time-varying** indicates the possibility of retaining different values for the same information, and the time when changes to these values occurred. For example, a data warehouse in a bank might store information about the average monthly balance of clients' accounts for a period covering several years. In contrast, an operational database may not have explicit temporal support, since either it is not necessary for day-to-day operations or it is difficult to implement. For example, it might be unnecessary to store information about the previous salaries of employees.

Table 2.1 shows several aspects that differentiate operational database systems from data warehouse systems.

Table 2.1. Comparison between operational databases and data warehouses

Aspect	Operational databases	Data warehouses
User type	Operators, office employees	Managers, high-ranking executives
Usage	Predictable, repetitive	Ad hoc, nonstructured
Data content	Current, detailed data	Historical, summarized data
Data organization	According to operational needs	According to the analysis problem
Data structures	Optimized for small transactions	Optimized for complex queries
Access frequency	High	From medium to low
Access type	Read, update, delete, insert	Read, append only
Number of records per access	Few	Many
Response time	Short	Can be long
Concurrency level	High	Low
Lock utilization	Necessary	Not necessary
Update frequency	High	None
Data redundancy	Low (normalized tables)	High (unnormalized tables)
Data modeling	ER model	Multidimensional model
Modeling and implementation	Entire system	Incremental

Data warehouse systems allow decision-making users to perform interactive analysis of data using **online analytical processing (OLAP)** systems [50]. In addition, data warehouses are also used for reporting, data mining, and statistical analysis.

2.6 The Multidimensional Model

Data warehouses and OLAP systems are based on a **multidimensional model**. This model allows a better understanding of data for analysis purposes and provides better performance for complex analytical queries. The multidimensional model views data in an n-dimensional space, usually called a **data cube** or a **hypercube** [85]. An example of a data cube is given in Fig. 2.5.

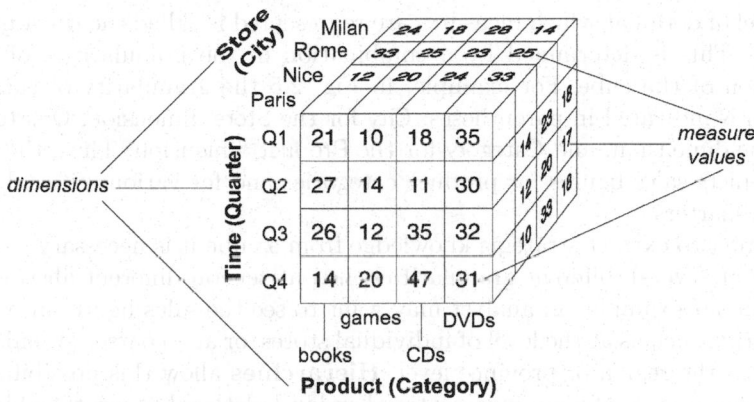

Fig. 2.5. A three-dimensional cube for sales data having dimensions Store, Time, and Product, and a measure amount

A data cube is defined by dimensions and facts. **Dimensions** are various perspectives that are used to analyze the data. For example, the data cube in Fig. 2.5 is used to analyze sales figures and has three dimensions: Store, Time, and Product. Instances of a dimension are called **members**. For example Paris, Nice, Rome, and Milan are members of the Store dimension. Dimensions have associated **attributes** describing the dimension. For example, the Product dimension could contain attributes such as Product Number, Product Name, Description, and Size, which are not shown in the figure.

On the other hand, the **cells** of a data cube, or **facts**, have associated with them numeric values, called **measures**. These measures allow quantitative evaluation of various aspects of the analysis problem at hand to be performed. For example, the numbers shown in the data cube in Fig. 2.5 represent a measure amount, indicating the total sales amount specified in, for instance, thousands of euros. A data cube typically contains several measures. For example, another measure, not shown in the cube in Fig. 2.5, could be quantity, representing the number of units sold.

A data cube may be **sparse** or **dense** depending on whether it has measures associated with each combination of dimension values. In the case of Fig. 2.5 this depends on whether all products are sold in all stores throughout the period of time considered. The figure shows two empty cells for the sales in Paris of books and CDs during the second and the third quarter, respectively. In real-world applications it is common to have sparse cubes. Adequately managing sparsity allows the required storage space to be reduced and improves query performance.

2.6.1 Hierarchies

The level of detail at which measures are represented is called the **data granularity**. This is determined by a combination of the granularities of each dimension of the cube. For example, in Fig. 2.5 the granularity of each dimension is indicated in parentheses: City for the Store dimension, Quarter for the Time dimension, and Category for the Product dimension. Thus, the data cube depicts sales figures for product categories and for various cities during various quarters.

In order to extract strategic knowledge from a cube it is necessary to view its data at several different granularities, i.e., at several different abstraction levels. In our example, an analyst may want to see the sales figures at a finer granularity, such as at the level of individual stores, or at a coarser granularity, such as at the region or province level. **Hierarchies** allow this possibility by defining a sequence of mappings relating low-level, detailed concepts to higher-level, more general concepts. Given two related levels in a hierarchy, the lower level is called the **child** and the higher level is called the **parent**. Figure 2.6 shows an example of a hierarchy for the Store dimension.[1] Each of the stores at the lowest level of the hierarchy can be mapped to a corresponding city, which in turn can be mapped to either a region or a province (this depends on the administrative subdivision of the individual country), and finally to a country. In the figure all countries, i.e., all members of the highest level of the hierarchy, are grouped under a member called All, which represents all of the hierarchy. As we shall see in the next section, the member All is used for obtaining the aggregation of measures for the whole hierarchy, i.e., for obtaining the total sales for all countries.

In real-world applications, there exist many kinds of hierarchies. For example, the hierarchy depicted in Fig. 2.6 is **balanced**, since there is the same number of levels from each individual store to the root of the hierarchy. Thus,

---

[1] Note that, as indicated by the ellipses, not all nodes of the hierarchy are shown.

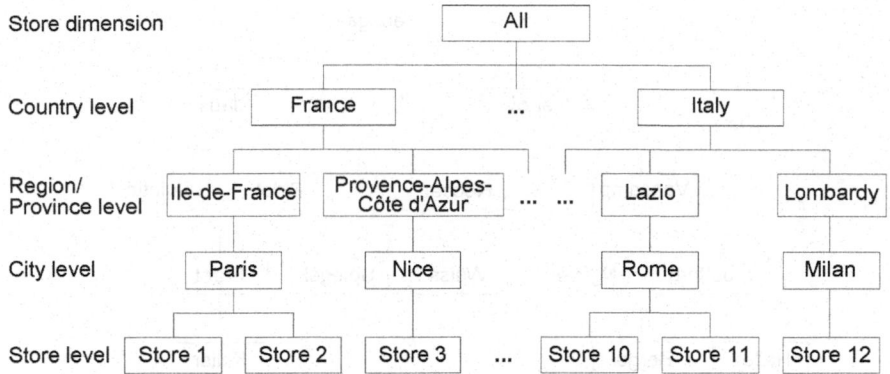

Fig. 2.6. Members for a hierarchy for the Store dimension

each member is related to a parent belonging to the level immediately above that of the member, and all branches of the hierarchy descend to the same level. However, this may not be the case in reality. Small European countries such as Luxembourg and Monaco do not have an administrative subdivision at the region or province level. Thus, for instance, the parent of the city Monaco is the country Monaco. These kinds of hierarchies are usually called **noncovering** or **ragged**, since the parent of at least one member does not belong to the level immediately above that of the member. On the other hand, some hierarchies may be **unbalanced** because they do not necessarily have instances at the lower levels. This may be the case, if in our sales cube example, some cities do not have associated stores but contain wholesalers who sell to distributors at the region/province level. As another example of an unbalanced hierarchy, consider the hierarchy in Fig. 2.7. This is a **recursive hierarchy**, or **parent-child hierarchy**, that captures the hierarchical relationships between employees and their supervisors. Such kinds of hierarchies are very common in real applications.

In Chap. 3, we shall study these and other kinds of hierarchies in detail, covering both their conceptual representation and their implementation in current data warehouse and OLAP systems.

2.6.2 Measure Aggregation

Aggregation of measures takes place when one changes the abstraction level at which data in a cube is visualized. This is performed by traversing the hierarchies of the dimensions. For example, if the hierarchy of Fig. 2.6 is used for changing the abstraction level of the data cube in Fig. 2.5 from City to Region/Province, then the sales figures for all cities in the same region will be aggregated using the sum operator. Similarly, visualizing the cube at the All level of the hierarchy will summarize the measures for all the countries.

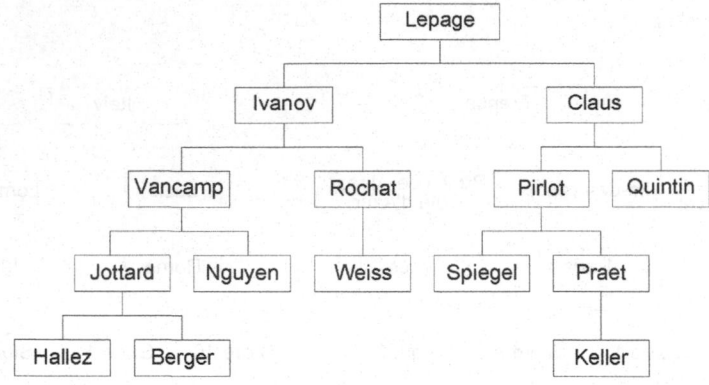

Fig. 2.7. Members of a parent-child hierarchy for the Employee dimension

In order to ensure correct aggregation of measures, summarizability conditions must hold [163]. **Summarizability** refers to the correct aggregation of measures in a higher level of the hierarchy (for example the Region/Province level in Fig. 2.6) taking into account existing aggregations in a lower level of the hierarchy (for example the City level in Fig. 2.6). Summarizability conditions include the following ones:

- **Disjointness of instances:** the grouping of instances in a level with respect to their parent in the next level must result in disjoint subsets. For example, in the hierarchy of Fig. 2.6 a city cannot belong to two regions or provinces.
- **Completeness:** all instances are included in the hierarchy and each instance is related to one parent in the next level. For example, the hierarchy in Fig. 2.6 contains all stores and each store is assigned to a city.
- **Correct use of aggregation functions:** measures can be of various types, and this determines the kind of aggregation function that can be applied.

A classification of measures according to additivity [146] is as follows.

- **Additive measures** (also called **flow** or **rate** measures) can be meaningfully summarized using addition along all dimensions. These are the most common type of measures. For example, the measure amount associated with the cube in Fig. 2.5 is an additive measure: it can be summarized when the hierarchies in the Store, Time, and Product dimensions are traversed.
- **Semiadditive measures** (also called **stock** or **level** measures) can be meaningfully summarized using addition along some, but not all, dimensions. A typical example is that of inventory quantities. These can be aggregated in the Store dimension in Fig. 2.6, thus allowing one to know, for instance, how many items in a particular category we have in a region

or a city, but cannot be aggregated in the Time dimension, for instance by adding the inventory quantities for two different quarters.

- **Nonadditive measures** (also called **value-per-unit** measures) cannot be meaningfully summarized using addition across any dimension. Typical examples are item price, cost per unit, and exchange rate.

Thus, when one is defining a measure it is necessary to determine the aggregation functions that will be used in the various dimensions. This is particularly important in the case of semiadditive and nonadditive measures. For example, inventory quantities, a semiadditive measure, can be aggregated using averaging along the Time dimension, and using addition for the other dimensions. Averaging can also be used for aggregating nonadditive measures such as item price or exchange rate. However, depending on the semantics of the application, other functions such as the minimum, maximum, or count could be used instead for aggregating these measures.

OLAP tools must implement fast aggregation mechanisms in order to allow users to interactively explore the cube data at different granularities. The efficiency of this aggregation depends on the kind of aggregation function used. This leads to another classification of measures as follows [98].

- **Distributive measures** are defined by an aggregation function that can be computed in a distributed way. Suppose that the data is partitioned into n sets, and that the aggregate function is applied to each set, giving n aggregated values. The function is distributive if the result of applying the function to the whole data set is the same as the result of applying a function (not necessarily the same) to the n aggregated values. The usual aggregation functions such as the count, sum, minimum, and maximum are distributive. Distributive measures can be computed efficiently.
- **Algebraic measures** are defined by an aggregation function that has a bounded number of arguments, each of which is obtained by applying a distributive function. A typical example is the average, which can be computed by dividing the sum by the count, the latter two functions being distributive.
- **Holistic measures** are measures that cannot be computed from other subaggregates. Typical examples include the median, the mode, and the rank. Holistic measures are expensive to compute, especially when data is modified, since they must be computed from scratch.

2.6.3 OLAP Operations

As already said, a fundamental characteristic of the multidimensional model is that it allows one to view data from multiple perspectives and at several different levels of detail. There are a number of OLAP operations that allow these perspectives and levels of detail to be materialized by exploiting the dimensions and their hierarchies, thus providing an interactive data analysis environment. A basic set of OLAP operations is described in the following.

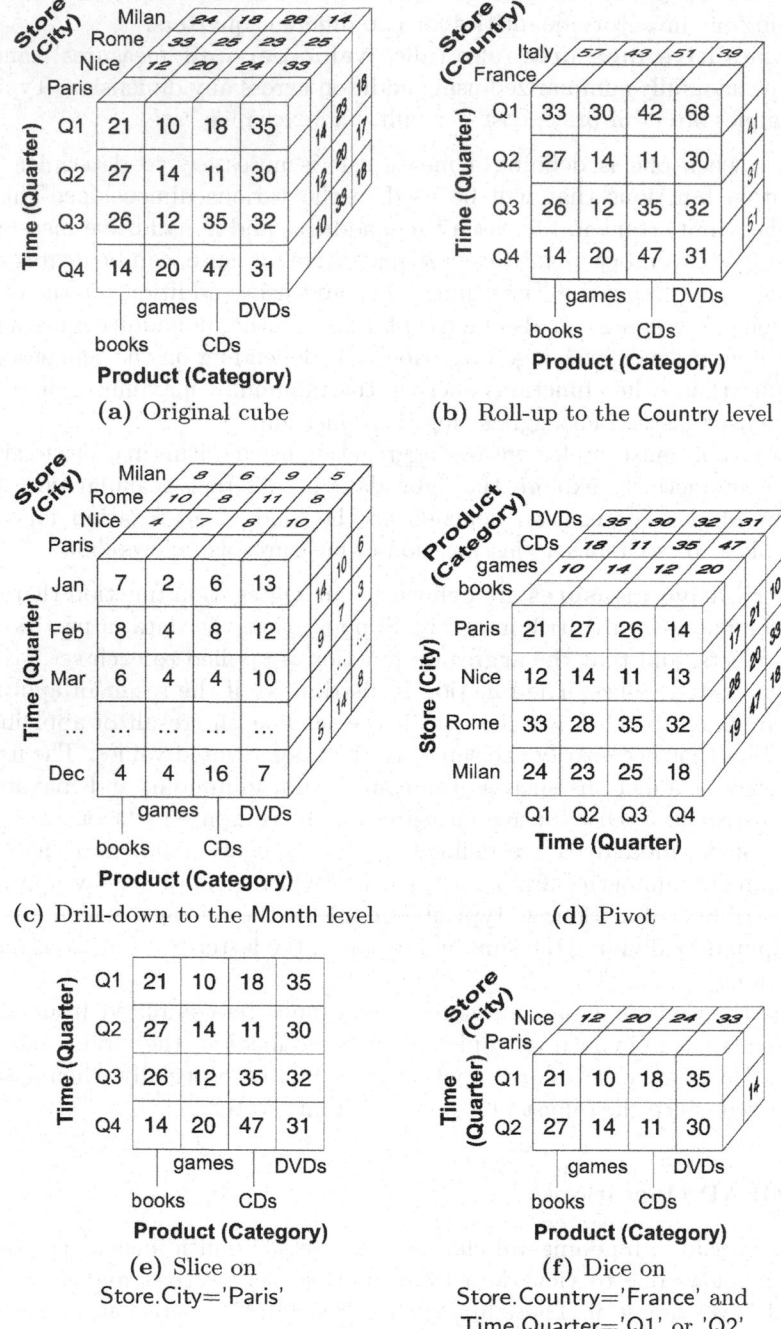

Fig. 2.8. OLAP operations

- The **roll-up** operation transforms detailed measures into summarized ones. This is done when on moves up in a hierarchy or reduces a dimension, i.e., one aggregates to the All level. The example given in Fig. 2.8b shows the result of rolling up the cube in Fig. 2.8a from the City to the Country level of the Store dimension. Thus, the values for cities located in the same country are aggregated in the result.
- The **drill-down** operation performs the operation opposite to the roll-up operation, i.e., it moves from a more general level to a detailed level in a hierarchy, thus presenting more detailed measures. An example of a drill-down from the Quarter to the Month level of the Time dimension is given in Fig. 2.8c.
- The **pivot** (or **rotate**) operation rotates the axes of a cube to provide an alternative presentation of the data. This is illustrated in Fig. 2.8d.
- The **slice** operation performs a selection on one dimension of a cube, resulting in a subcube. An example is given in Fig. 2.8e where only the data about Paris stores is shown.
- The **dice** operation defines a selection on two or more dimensions, thus again defining a subcube. An example is given in Fig. 2.8f where the cube shows sales data for Paris stores in the first six months.
- The **drill-across** operation executes queries involving more than one cube. This would be the case, for example, if we wanted to compare sales data in the cube in Fig. 2.8a with purchasing data located in another cube. This operation requires that the two cubes have at least one common dimension.
- The **drill-through** operation allows one to move from data at the bottom level in a cube to data in the operational systems from which the cube was derived. This could be used, for example, if one were trying to determine the reason for outlier values in a data cube.

In addition to the basic operations described above, OLAP tools provide a great variety of mathematical, statistical, and financial operators for computing ratios, variances, moving averages, ranks, interest, depreciation, currency conversions, etc.

2.7 Logical Data Warehouse Design

There are several different approaches to implementing a multidimensional model, depending on the way in which a data cube is stored.

- **Relational OLAP (ROLAP)** servers store data in relational databases and support extensions to SQL and special access methods to efficiently implement the multidimensional data model and the related operations.
- **Multidimensional OLAP (MOLAP)** servers directly store multidimensional data in special data structures (for instance, arrays) and implement the OLAP operations over those data structures. While MOLAP

systems offer less storage capacity than ROLAP systems, MOLAP systems provide better performance when multidimensional data is queried or aggregated.

• **Hybrid OLAP (HOLAP)** servers combine both technologies, benefiting from the storage capacity of ROLAP and the processing capabilities of MOLAP. For example, a HOLAP server may store large volumes of detailed data in a relational database, while aggregations are kept in a separate MOLAP store.

In ROLAP systems, multidimensional data is implemented as relational tables organized in specialized structures called star schemas, snowflake schemas, starflake schemas, and constellation schemas. We describe next these kinds of schemas.

In a **star schema**, there is only one central **fact table**, and a set of **dimension tables**, one for each dimension. An example is given in Fig. 2.9, where the fact table is depicted in gray and the dimension tables are depicted in white. As shown in the figure, **referential integrity** constraints are specified between the fact table and each of the dimension tables. In a star schema, the dimension tables may contain redundancy, especially in the presence of hierarchies: the tables are not necessarily normalized. This is the case for the dimensions Product and Store in Fig. 2.9. Indeed, all products belonging to the same category will have redundant information for the attributes describing the category and the department. The situation is similar for the dimension Store with respect to the attributes describing the city and the state.

A **snowflake schema** avoids the redundancy of star schemas by normalizing the dimension tables. Therefore, a dimension is represented by several tables related by **referential integrity** constraints. In addition, as in the case of star schemas, referential integrity constraints also relate the fact table and the dimension tables at the finest level of detail. An example of a snowflake schema is given in Fig. 2.10, where the dimensions Product and Store are represented by normalized tables. Normalized tables are easy to maintain and allow storage space to be optimized. However, a disadvantage is that performance is affected, since more joins need to be performed when executing queries that require hierarchies to be traversed.

A **starflake schema** is a combination of the star and the snowflake schemas where some dimensions are normalized while others are not. An example of a starflake schema arises when the dimension table Product in Fig. 2.9 is replaced by the tables Product, Category, and Department in Fig. 2.10.

Finally, a **constellation schema** has multiple fact tables that share dimension tables. The example given in Fig. 2.11 has two fact tables Sales and Purchases sharing the Time and Product dimension. Constellation schemas may include both normalized and unnormalized dimension tables.

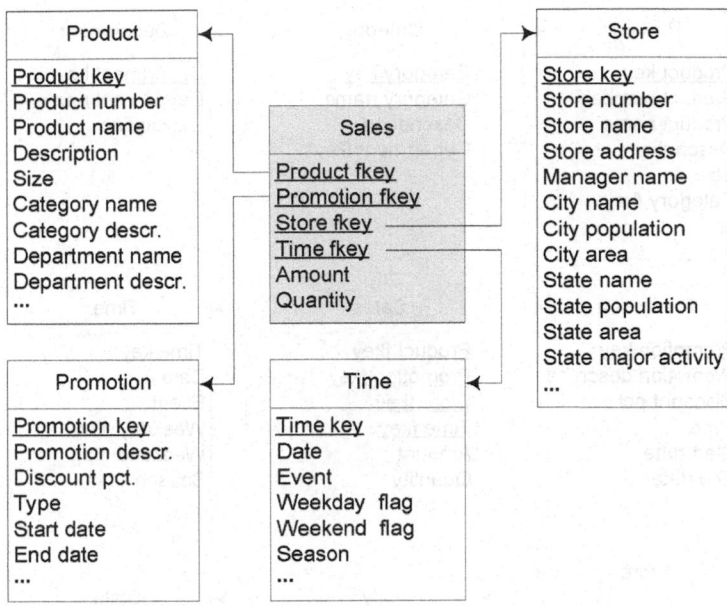

Fig. 2.9. Example of a star schema

2.8 Physical Data Warehouse Design

Data warehouses are typically several orders of magnitude larger than operational databases. As a consequence, the physical design of data warehouses is extremely crucial for ensuring adequate response time for complex ad hoc queries. In the following we discuss three common techniques for improving system performance: materialized views, indexing, and partitioning.

In the relational model, a **view** consists in a table that is derived from a query. This query can involve **base tables** (i.e., tables physically stored in the database) or other views. A **materialized view** is a view which is physically stored in a database. Materialized views allow query performance to be enhanced by precalculating costly operations such as joins and aggregations and storing the results in the database. Since some queries need only to access materialized views, they are executed faster. Obviously, the increased query performance is achieved at the expense of storage space.

One problem with materialized views is the **updating** of them: all modifications to the underlying base tables must be propagated into the view. In order to ensure efficiency, updates of materialized views must be performed in an incremental way. This implies capturing the modifications to the underlying tables and determining how they influence the content of the view. Much research work has been done in the area of view maintenance (see, for example, [90]). Three strategies have been proposed for view maintenance.

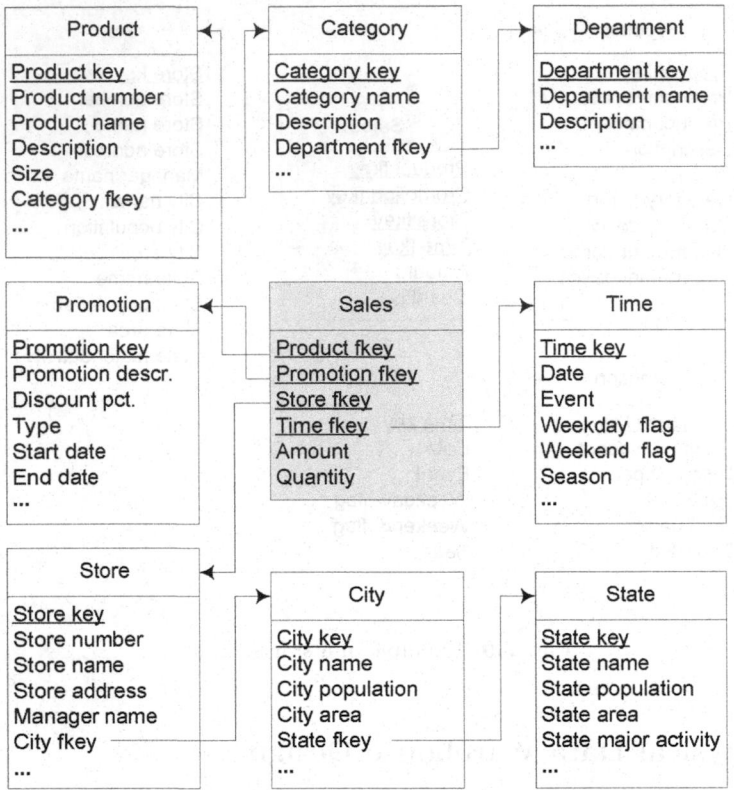

Fig. 2.10. Example of a snowflake schema

In the first one, the views are updated periodically: in this case they can be considered as snapshots taken at a particular instant in time. In the second strategy, the views are updated immediately at the end of each transaction. Finally, in the third strategy, the updates are propagated in a deferred way: in this case, views are updated only when they are needed in a query.

In a data warehouse, not all possible aggregations can be precalculated and materialized, since the number of aggregates grows exponentially with the number of dimensions and hierarchies [270]. Although optimized data structures, such as Dwarfs [271], have been proposed to deal with this problem, an important decision in designing a data warehouse is the **selection of materialized views** to be defined. This amounts to identifying a set of queries to be prioritized, which determines the corresponding views to be materialized. The goal is to select an appropriate set of views that minimizes the total query response time and the cost of maintaining the selected views, given a limited amount of resources, such as storage space or materialization time [91]. Many algorithms have been designed for selection of materialized

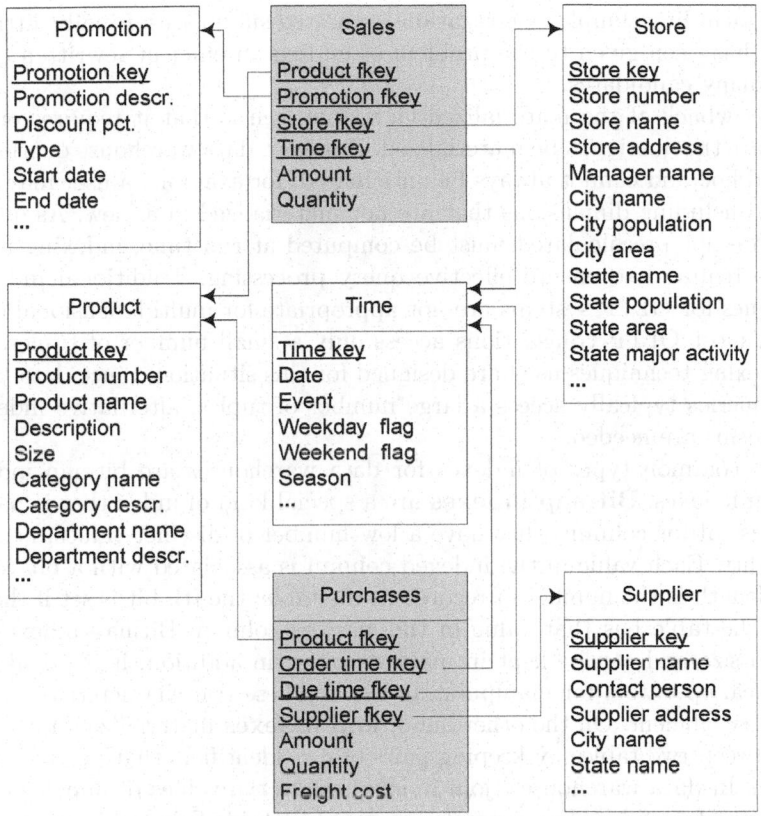

Fig. 2.11. Example of a constellation schema

views and currently some commercial DBMSs are providing tools that tune the selection of materialized views on the basis of previous queries to the data warehouse.

The queries addressed to a data warehouse must be rewritten in order to best exploit the existing materialized views to improve query response time. This process, known as **query rewriting**, tries to use the materialized views as much as possible, even if they only partially fulfill the query conditions. However, selecting the best rewriting for a query is a complex process, in particular for queries involving aggregations. Many algorithms have been proposed for query rewriting in the presence of materialized views (see, e.g., [228]). However, these algorithms impose various restrictions on the given query and the potential materialized views so that the rewriting can be done. A common restriction is that all tables included in the definition of the materialized view must also appear in the definition of the query. The existing algorithms also have limitations on the class of queries that can be handled, for instance queries including complex aggregations or nested subqueries. Finally, little attention has been given to the problem of finding an efficient rewritten query out of many candidates.

A drawback of the materialized-view approach is that it requires one to anticipate the queries to be materialized. However, data warehouse queries are often ad hoc and cannot always be anticipated: for example, a user may pose a query including dimensions that are not materialized in a view. As queries which are not precalculated must be computed at run time, indexing methods are required to ensure effective query processing. Traditional indexing techniques for OLTP systems are not appropriate for multidimensional data. Indeed, most OLTP transactions access only a small number of tuples, and the indexing techniques used are designed for this situation. Since data warehouse queries typically access a large number of tuples, alternative indexing mechanisms are needed.

Two common types of indexes for data warehouses are bitmap indexes and join indexes. **Bitmap indexes** are a special kind of index that is particularly useful for columns that have a low number of distinct values, i.e., low cardinality. Each value in the indexed column is associated with a bit vector whose length is the number of records in the table: the ith bit is set if the ith row of the table has that value in the indexed column. Bitmap indexes are small in size and can be kept in main memory; in addition, it is possible to use logical operators for manipulating them. These two characteristics make them very efficient. On the other hand, **join indexes** materialize a relational join between two tables by keeping pairs of row identifiers that participate in the join. In data warehouses, join indexes relate the values of dimensions to rows in the fact table. For example, given a fact table Sales and a dimension Client, a join index maintains for each distinct client a list of row identifiers of the tuples recording the sales to this client.

With the selection of materialized views, **index selection** is an important decision when one is designing a data warehouse. Given a constrained resource such as storage space or maintenance time, the process consists in selecting a set of indexes so as to minimize the execution cost of a set of queries and updates. This problem is challenging for several reasons [44]. First, since database schemas of real applications can be large and indexes can be defined on a sequence of one or more columns, the space of the indexes that are relevant to the set of queries and updates can be very large. In addition, it is important to take into account the interactions between indexes. Various algorithms for index selection have been proposed, and nowadays major DBMS vendors are providing tools to assist database administrators in this task.

It is worth noting that the problems of materialized-view selection and index selection are strongly dependent on each other. Some publications have considered the two problems conjointly [7, 261, 322]. Automatically selecting an appropriate set of materialized views and indexes is a nontrivial task. Indeed, searching the space of all relevant indexes and materialized views for a set of queries and updates is infeasible in practice, particularly when this set is large or complex. Therefore, it is crucial to reduce the solution space by focusing the search on a smaller subset. In addition, as shown in [7], searching over the combined space of indexes and materialized views provides better solutions than does selecting the views and the indexes independently.

Fragmentation or **partitioning** is a mechanism typically used in relational databases to reduce the execution time of queries. It consists in dividing the contents of a relation into several files that can be more efficiently processed in this way. There are two ways of fragmenting a relation: vertically and horizontally. **Vertical fragmentation** splits the attributes of a table into groups that can be independently stored. For example, a table can be partitioned such that the most often used attributes are stored in one partition, while other, less often used attributes are kept in another partition. Since vertical fragmentation implies smaller records, more records can be brought into main memory, which reduces their processing time. On the other hand, **horizontal fragmentation** divides the records of a table into groups according to a particular criterion. There are several methods for performing horizontal fragmentation. In range partitioning, the rows of a table are partitioned according to a range of values. For example, one common partitioning scheme in data warehouses is based on time, where each partition contains data about a particular time period, for instance a particular year or month. This has immediate advantages for the process of refreshing the data warehouse, since only the last partition must be accessed. In hashing partitioning, the rows of a table are partitioned according to a hash function applied to an attribute of the table. Finally, composite partitioning combines these two methods, i.e., the rows are first partitioned according to a range of values, and then subdivided using a hash function.

2.9 Data Warehouse Architecture

Figure 2.12 shows a typical architecture of a data warehouse system, which consist of several tiers:

- The **back-end tier** is composed of **extraction-transformation-loading (ETL) tools**, used to feed data in from operational databases and other **data sources**, which can be **internal** or **external**, and a **data staging area**, which is an intermediate database where all the data integration and transformation processes are run prior to the loading of the data into the data warehouse.
- The **data warehouse tier** is composed of an **enterprise data warehouse** and/or several **data marts**, and a **metadata repository** storing information about the data warehouse and its contents.
- The **OLAP tier** is an **OLAP server** that supports multidimensional data and operations.

- The **front-end tier** deals with data analysis and visualization. It contains **client tools** such as **OLAP tools**, **reporting tools**, **statistical tools**, and **data-mining tools**.

We describe the various components of the above architecture in detail next.

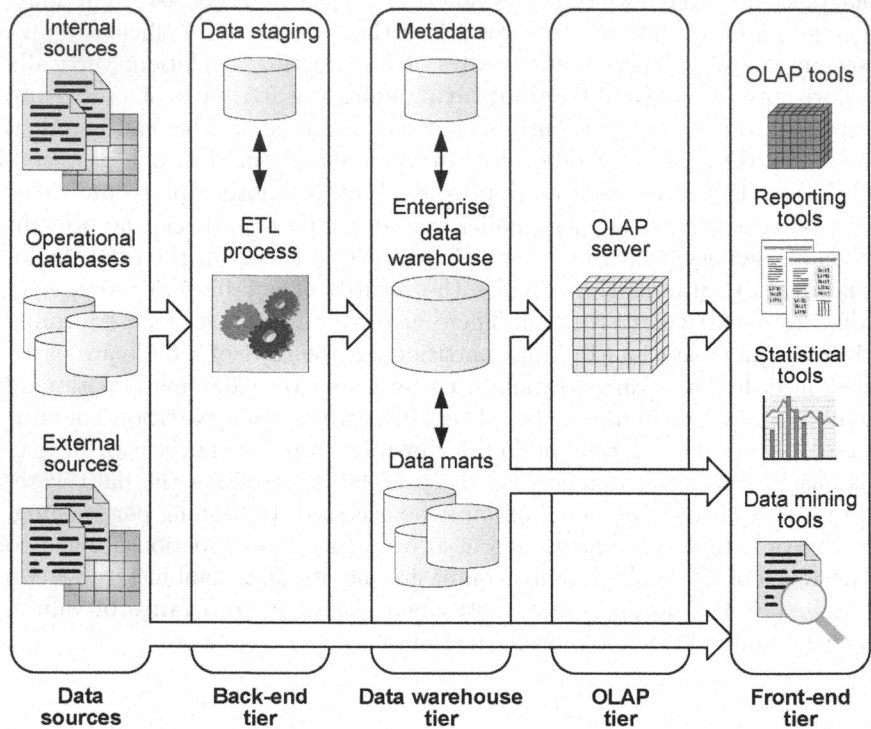

Fig. 2.12. Typical data warehouse architecture

2.9.1 Back-End Tier

In the back-end tier, the process commonly known as **extraction-transformation-loading** is performed. As the name indicates, it is a three-step process as follows.

- **Extraction** gathers data from multiple, heterogeneous data sources. These sources may be operational databases but may also be files in various formats; they may be **internal** to the organization or **external** to it. In order to solve interoperability problems, data is extracted whenever possible using application program interfaces such as ODBC (Open Database Connection), OLEDB (Open Linking and Embedding for Databases), and JDBC (Java Database Connectivity).

- **Transformation** modifies the data from the format of the data sources to the warehouse format. This includes several aspects: **cleaning**, which removes errors and inconsistencies in the data and converts it into a standardized format; **integration**, which reconciles data from different data sources, both at the schema and at the data level; and **aggregation**, which summarizes the data obtained from data sources according to the level of detail, or **granularity**, of the data warehouse.
- **Loading** feeds the data warehouse with the transformed data. This also includes **refreshing** the data warehouse, i.e., propagating updates from the data sources to the data warehouse at a specified frequency in order to provide up-to-date data for the decision-making process. Depending on organizational policies, the refresh frequency may vary from monthly to several times a day, or even to nearly in real time.

The ETL process usually requires a data staging area, i.e., a database in which the data extracted from the sources undergoes successive modifications to eventually be ready to be loaded into the data warehouse.

2.9.2 Data Warehouse Tier

The data warehouse tier in Fig. 2.12 depicts an enterprise data warehouse and several data marts. An **enterprise data warehouse** is a centralized data warehouse that encompasses all functional or departmental areas in an organization. On the other hand, a **data mart** is a specialized data warehouse targeted toward a particular functional area or user group in an organization. A data mart can be seen as a small, local data warehouse. The data in a data mart can be either derived from an enterprise data warehouse or collected directly from data sources.

Another component of the data warehouse tier is the metadata repository. **Metadata** can be defined as "data about data". Metadata has been traditionally classified into technical and business metadata. **Business metadata** describes the meaning (or semantics) of the data, and organizational rules, policies, and constraints related to the data. On the other hand, **technical metadata** describes how data is structured and stored in a computer system, and the applications and processes that manipulate the data.

In the data warehouse context, technical metadata can be of various natures, describing the data warehouse system, the source systems, and the ETL process. In particular, the metadata repository may contain information such as the following.

- Metadata describing the structure of the data warehouse and the data marts, both at the logical level (which includes the facts, dimensions, hierarchies, derived data definitions, etc.) and at the physical level (such as indexes, partitions, and replication). In addition, it contains security information (user authorization and access control) and monitoring information (such as usage statistics, error reports, and audit trails).

- Metadata describing the data sources, including their schemas (at the conceptual, logical, and/or physical levels), and descriptive information such as ownership, update frequencies, legal limitations, and access methods.
- Metadata describing the ETL process, including data lineage (i.e., tracing warehouse data back to the source data from which it was derived), data extraction, cleaning, transformation rules and defaults, data refresh and purging rules, algorithms for summarization, etc.

2.9.3 OLAP Tier

The OLAP tier in the architecture of Fig. 2.12 is composed of an OLAP server that presents business users with multidimensional data from data warehouses or data marts. As already said, there are several types of OLAP servers depending on the underlying implementation model: ROLAP, MOLAP, and HOLAP.

Most commercial database products provide OLAP extensions and related tools allowing the construction and querying of cubes, as well as navigation, analysis and reporting. Later on this section we describe two representative OLAP tools, Microsoft SQL Server Analysis Services 2005 and Oracle OLAP 10g. Unfortunately, there is not yet a standardized language for querying data cubes, and the underlying technology differs between the two. For example, Oracle has favored the Java programming language and developed the query language OLAP DML (Data Manipulation Language), while SQL Server uses the .NET framework and the query language MDX (MultiDimensional eXpressions). However, XML for Analysis (XMLA) aims at providing a common language for exchanging multidimensional data between client applications and OLAP servers working over the Web.

2.9.4 Front-End Tier

The front-end tier in Fig. 2.12 contains the client tools that allow users to exploit the contents of the data warehouse. Typical client tools include the following.

- **OLAP tools** allow interactive exploration and manipulation of the warehouse data in order to find patterns or trends of importance to the organization. They facilitate the formulation of complex queries that may involve large amounts of data. These queries are called **ad hoc queries**, since the system has no prior knowledge about them.
- **Reporting tools** enable the production, delivery, and management of reports, which can be paper-based reports or interactive-, Web-based reports. Reports use **predefined queries**, i.e., queries asking for specific information in a specific format that are performed on a regular basis.
- **Statistical tools** are used to analyze and visualize the cube data using statistical methods.

- Finally, **data-mining tools** allow users to analyze data in order to discover valuable knowledge such as patterns and trends; they also allow predictions to be made on the basis of current data.

2.9.5 Variations of the Architecture

Some of the components illustrated in Fig. 2.12 can be missing in a real environment.

In some situations there may be only an enterprise data warehouse without data marts or, alternatively, an enterprise data warehouse may not exist. Building an enterprise data warehouse is a complex task that is very costly in time and resources. In contrast, a data mart is typically easier to build than an enterprise warehouse. However, this advantage may turn into a problem when several data marts that were independently created need to be integrated into a data warehouse for the entire enterprise.

In some other situations, an OLAP server does not exist and the client tools directly access the data warehouse. This is indicated by the arrow connecting the data warehouse tier to the front-end tier. An extreme situation is where there is neither a data warehouse nor an OLAP server. This is called a **virtual data warehouse**, which defines a set of views over operational databases that are materialized for efficient access. The arrow connecting the data sources to the front-end tier depicts this situation. While a virtual data warehouse is easy to build, it does not provide a real data warehouse solution, since it does not contain historical data, does not contain centralized metadata, and does not have the ability to clean and transform the data. Furthermore, a virtual data warehouse can severely impact the performance of operational databases.

Finally, a data staging area may not be needed when the data in the source systems conforms very closely to the data in the warehouse. This situation typically arises when there is one data source (or only a few) having high-quality data. However, this is rarely the case in real-world situations.

2.10 Analysis Services 2005

Microsoft SQL Server 2005 provides an integrated platform for building analytical applications. It is composed of three main components, described below.

- **Integration Services** supports the ETL processes, which are used for loading and refreshing data warehouses on a periodic basis. Integration Services is used to extract data from a variety of data sources, to combine, clean, and summarize this data, and, finally, to populate a data warehouse with the resulting data.

- **Analysis Services** provides analytical and data-mining capabilities. It allows one to define, query, update, and manage OLAP databases. The MDX (Multi-Dimensional eXpressions) language is used to retrieve data. Users may work with OLAP data via client tools (Excel or other OLAP clients) that interact with Analysis Services' server component. Analysis Services is also used to build data-mining applications. It provides several data-mining algorithms, and uses the DMX (Data Mining eXtensions) language for creating and querying data-mining models and obtaining predictions.
- **Reporting Services** is used to define, generate, store, and manage reports. Reports can be built from various types of data sources, including data warehouses and OLAP cubes. Reports can be personalized and delivered in a variety of formats, with many interactivity and printing options. Users can view reports with a variety of clients, such as Web browsers or other reporting clients. Clients access reports via Reporting Services' server component.

SQL Server 2005 provides two tools for developing and managing these components. **Business Intelligence Development Studio** (BIDS) is an integrated development platform for system developers. BIDS supports Analysis Services, Reporting Services, and Integration Services projects. Finally, another tool, **SQL Server Management Studio**, provides integrated management of all SQL Server 2005 components.

We give next an introduction to the main features of SQL Server 2005 Analysis Services.

2.10.1 Defining an Analysis Services Database

The multidimensional model underlying Analysis Services is called the **Unified Dimensional Model** (UDM). It combines the characteristics of the relational and the multidimensional models. Typical multidimensional databases have a very simple data model that excludes complex schemas with many-to-many relationships, complex hierarchies, shared dimensions, etc. In addition, multidimensional models are typically dominated by hierarchies, where attributes are relegated to a secondary role. The UDM provides a relational view of dimensional structures. A cube is composed of one or several facts related to dimensions. A dimension contains attributes, which can be used for slicing and filtering queries, and dimensions can be combined into hierarchies. Dimensions can be shared between cubes.

There are two main approaches to building an Analysis Services database: on the basis of existing data sources, or on the basis of templates. With either approach, it is assumed that one or more source systems feed a relational data warehouse, which in turn is used to populate the Analysis Services database.

The first method of creating an Analysis Services database is to start from a source relational database, which can be either an operational database or a data warehouse. From the data source, the dimensions, cubes, and data-mining models can be defined. It is possible to build an Analysis Services database directly on an operational database, without first building a relational data warehouse. This method is applied if the source data needs minimal transformation, cleaning, and integration before it is useful for analytical purposes. However, in many cases it is necessary to use the classic data warehouse architecture, as described in Sect. 2.9, where there is a relational data warehouse, and the Analysis Services database acts as the query and analysis engine.

An alternative method of creating an Analysis Services application is to create a model without a data source. This approach generates a complete customizable application from a template: a Cube Wizard allows dimensional structures and analytic features to be defined, as well as an underlying relational data warehouse and Integration Services packages. It is also possible to combine the two approaches. A very common scenario is to create most of an Analysis Services database from an existing source, but to generate the time dimension from a template.

The Business Intelligence Development Studio is typically used as the development platform for designing Analysis Services databases. It is used to create a project in which a database is designed, and to send it to the Analysis Services server. BIDS is also used to directly connect to an Analysis Services database and make refinements to it. For building an Analysis Services database, it is necessary to define various kinds of objects, described in detail in the rest of this section.

2.10.2 Data Sources

Data warehouses must retrieve their data from one or several data stores. These data stores must be defined as **data sources**. Data sources are objects that contain details of a connection to a data store, which include the location of the server, a login and password, a method to retrieve the data, security permissions, etc. Analysis Services supports all data sources that have a connectivity interface through OLE DB or .NET Managed Provider. If the source is a relational database, then SQL is used by default to query the database.

In the simplest case there is only one data source, typically an operational data store, which is used to integrate data obtained from various heterogeneous sources, including OLTP systems. Another possibility is to take data directly from the OLTP system as an input to the decision support software.

2.10.3 Data Source Views

Once the data source objects have been created, a **data source view** (DSV) allows one to define the relational schema that will be used in the Analysis Services database. This schema is derived from the schemas of the various data sources. Cubes and dimensions are created from data source views rather than directly from data source objects.

It is common to initially create data source views using the DSV Wizard. When a data source view is being defined, the wizard connects to the relational database using the connection information contained in the data source object. The wizard then retrieves from the database all the tables and views, and their relationships, and shows them so that the user can select those needed for the Analysis Services database. The relationships between tables help the Cube Wizard and Dimension Wizard to identify fact and dimension tables, as well as hierarchies.

The DSV Designer can be used to make changes to the data source view by adding, deleting, or modifying tables and views, and to establish new relationships between tables. For example, it may be the case that some source systems do not specify the primary keys and the relationships between tables. Other common requirements are to select some columns from a table, to add a new derived column to a table, to restrict table rows on the basis of some specific criteria, and to merge columns from several tables into a single column. These operations can be done by creating views in the relational database. Analysis Services allows these operations to be performed in the DSV using named queries. A **named query** can be specified by filling in appropriate dialogs, or by using SQL directly. Creating a new column in a table can be accomplished by replacing the table with a named query and writing the appropriate SQL to create the additional column, or by defining a **named calculation** that allows one to add the new column, defined by an expression.

Analysis Services allows friendly names for tables and columns to be specified. Further, in order to facilitate visibility and navigation for large data warehouses, Analysis Services offers the possibility to define customizable views, called **diagrams**, within a DSV that show only certain tables. Also, the DSV Designer allows one to look at sample data in the tables for validation purposes.

2.10.4 Dimensions

Dimensions define the structure of a cube. A dimension must have at least one key attribute and other attributes that may define hierarchies. In Analysis Services, there are two types of hierarchies: attribute hierarchies and user-defined hierarchies. **Attribute hierarchies** correspond to a single column in a relational table, for instance an attribute Color for a dimension Product. **User-defined hierarchies**, or **multilevel hierarchies**, are derived from two or more attribute hierarchies, each attribute being a level in the hierarchy, for instance Product, Subcategory, and Category. An attribute can participate in more than one multilevel hierarchy, for instance a hierarchy Product, Model, and Line in addition to the previous one.

Analysis Services supports three types of multilevel hierarchies, depending on how the members of the hierarchy are related to each other: balanced, ragged, and parent-child hierarchies. We defined these kinds of hierarchies in

Sect. 2.6.1. Recall that in a ragged hierarchy, the parent of at least one member does not belong to the level immediately above the level of the member. In a table corresponding to a ragged dimension, the missing members can be represented in various ways. The table cells can contain nulls or empty strings, or they can contain the same value as their parent to serve as a placeholder. The HideMemberIf property of a level allows missing members to be hidden from end users in these cases.

The Dimension Wizard is typically used for defining dimensions. Dimensions must be defined either from a data source view that provides data for the dimension or from preexisting templates provided by Analysis Services. A typical example of the latter is the time dimension, which does not need to be defined from a data source. Dimensions can be built from one or more tables. In the case of a star schema, the dimension is created from a single table. Otherwise, if the dimension is created from a snowflake schema, it will contain several tables, one of which is the primary table of the dimension. Each of the columns of these tables results in an equivalent attribute in the dimension. The Dimension Wizard detects the key attributes and the hierarchies by taking a sample of the data from the data sources and scanning for relationships between the columns in the table. The wizard suggest keys and hierarchies for the one-to-many relationships detected in this sample.

Analysis Services has a set of predefined dimension types that are often used in business intelligence applications. Some such types are Organization, Products, Promotion, Rates, and Scenarios. These common dimension types have attributes associated with them that typically form levels of a hierarchy. When one of these predefined dimension types is chosen, the Dimension Wizard allows one to specify the columns in the user tables that correspond to the dimension attributes defining the hierarchies.

The Dimension Designer allows one to refine the hierarchies created by the Dimension Wizard. In particular, it allows one to specify relationships between the attributes of a dimension. An attribute can have **related attributes** (also called **member properties**), which are attributes of the same dimension that have a one-to-one or one-to-many relationship with the current attribute. Specifying such relationships between attributes helps to improve query performance, as well as changing the design of aggregation.

The Dimension Designer allows one to define **custom roll-ups**, which are user-defined aggregations across hierarchies. For parent-child hierarchies, a unary operator (such as $+$, $-$, or $\times$) may be used to state how the child members are aggregated into the parent members. For example, in a parent-child Account hierarchy, the members must be aggregated in different ways depending on the account category (for instance, incomes must be added and expenses must be subtracted). The unary operator for each member must be defined in a column of the data source. Similarly, if the value of a member is not derived from its children or if a complex calculation is needed for aggregating from children to their parent, then an associated MDX formula must be defined in a column of the data source. As an example, in the Account hierarchy above, depreciation can be calculated in more than one way according to the account category: it can be either an accelerated depreciation or a straight-line depreciation.

2.10.5 Cubes

In Analysis Services, a cube can be built using either a top-down or a bottom-up approach. The traditional way is to build cubes bottom-up from an existing relational database through a data source view. A cube can be built from one or several data source views. In the top-down approach, the cube is created and then a relational schema is generated based on the cube design.

A cube in Analysis Services consists of one or more **measure groups** from a fact table and one or more dimensions from dimension tables. Each measure group is formed by a set of **measures**. Measures are columns of the fact table that contain meaningful information for data analysis. Usually, measures are of numeric type and can be aggregated or summarized along hierarchies of a dimension. Users can specify the type of aggregation that needs to be applied for each measure; the most often used functions are sum, count, and distinct count. Measures have a property indicating whether the fact data will be stored in Analysis Services (i.e., using MOLAP), in the relational source (i.e., using ROLAP) or in both (i.e., using HOLAP).

To build a cube, it is necessary to specify the fact and dimension tables to be used. Each cube must contain at least one fact table, which determines the contents of the cube. The facts stored in the fact table are mapped as measures in a cube. Analysis Services allows multiple fact tables in a single cube. Each of these fact tables can be from a different data source. In this case the cube typically contains multiple measure groups, one from each fact table.

Analysis Services allows several types of dimensions, depending on the type of the relationship between measure groups and dimensions.

- **Regular:** There is a direct link between a fact table and a dimension table, as in a traditional star schema.
- **Reference:** The dimension is not directly related to the fact table, but is related indirectly through another dimension, as in a snowflake schema. An example is a Sales fact with a Customer dimension, where the latter is related to a Geography dimension. In this case Geography may be defined as a reference dimension for the Sales fact. Reference dimensions can be chained together; for instance, one can define another reference dimension from the Geography dimension.
- **Fact:** A table is used both as a fact and a dimension table. Fact dimensions, also referred to as **degenerate dimensions**, are similar to regular dimensions, although they have a one-to-one relationship with the facts. For example, a Sales fact table may contain information not only about each line of an order but about the order itself, such as the carrier's tracking number. Constructing a dimension with this attribute would allow one

to determine the total sales amount for all the orders bundled together under a single tracking number. Fact dimensions are frequently used to support drill-through actions.

- **Role-playing:** A single fact table is related to a dimension table more than once. Thus, the dimension has several roles, depending on the context. For example, the time dimension may be reused for both the order date and the shipping date of sale facts. A role-playing dimension is stored once and used multiple times.

- **Many-to-many:** A fact is joined to multiple dimension members. In these dimensions, also called **multiple-valued dimensions**, the dimension member has a one-to-many relationship with various facts and a single fact is associated with multiple dimension members. For example, when a bank account is owned by several customers, a single account transaction relates to multiple members of the Customer dimension.

- **Data-mining**: A data-mining dimension provides support for dimensions that result from data-mining models.

Analysis Services supports **semiadditive measures**, i.e., measures that can be aggregated in some dimensions but not in others. Recall that we defined such measures in Sect. 2.6.2. As an example, although the inventory level can be summed across the Product dimension, it cannot be meaningfully summed across the Time dimension: the inventory level for the month is not the sum of the inventory levels for each day but instead is the inventory level for the last day in the month. Analysis Services provide several functions for semiadditive measures.

Analysis Services allows one to define **perspectives**, which are views of a cube (in the traditional relational sense) showing a subset of it. This is useful for simplifying design and deployment when there are a large number of dimensions, measure groups, and measures. Further, **translations** provide the possibility of storing analytic data and presenting it to users in different languages, for instance in French and in English.

The Cube Wizard allows one to define an Analysis Services database, dimensions, hierarchies, attributes, and cubes. After that, it is possible to manually edit the design generated by the wizard using the Cube Designer. The Cube Wizard examines the database and the cardinality relationships in the data source view and characterizes tables as fact tables, dimension tables, or dimension-to-fact bridge tables that resolve a many-to-many relationship. Then it detects all potential measures and groups them into a measure group. Finally, the wizard scans all the dimension tables to identify hierarchies within them, if this has not been previously done.

2.11 Oracle 10g with the OLAP Option

Oracle provides a series of business intelligence (BI) products, including a platform, tools, and application components [212], which we briefly describe next.

- The **BI platform** is used to create and administer the data repository supporting analytical applications. Oracle 10g with the OLAP option allows one to store and query data in relational data warehouses, relational OLAP, and/or multidimensional OLAP. Oracle's relational data warehouse includes several features that improve access to data, query performance, data warehouse maintenance, and data-loading processes, among others. We shall refer to these features in more detail in Sect. 6.8. The OLAP option provides the ability to present data using a dimensional model and implement it using either relational data types or multidimensional data types. Two administrative tools can be used for modeling, creating, and managing multidimensional data sets: **Analytic Workspace Manager** and **Warehouse Builder**.
- **BI development tools** facilitate the development of customized solutions and allow expert users to query and analyze data. For example, software developers can use SQL and the OLAP API for Java to access and manipulate multidimensional data. Alternatively, they can use the **OLAP DML** (Data Manipulation Language), which is a procedural programming language that directly manipulates multidimensional data. The OLAP DML allows implementers to extend the analytical capabilities of SQL and the OLAP API for Java. Oracle also provides **BI Beans**, which are Java application components that can be used to build dedicated business intelligence applications or to embed business intelligence features into portals or other enterprise applications. The BI Beans include user interface components such as a query builder, a calculation builder, graphs, and cross tabs, as well as other components such as a catalog, metadata, and data managers. Using BI Beans, applications can be created with little coding. Several Oracle BI client tools, described next, were built using Oracle BI Beans.
- **BI client tools** are used to formulate ad hoc queries and calculations, to design reports and graphs, and to interactively navigate data, among other things. These tools include **OracleBI Discoverer**, which is a reporting tool that comes in two versions: **OracleBI Discoverer Plus**, which supports the relational data model, and **OracleBI Discoverer Plus OLAP**, dedicated to the multidimensional model. Another tool, **Oracle Spreadsheet Add-In**, is designed for users who prefer to use Microsoft Excel as the user interface to data managed by the OLAP option.
- **BI applications** are prebuilt business intelligence solutions that support decisions and business processes. As an example, Enterprise Planning and Budgeting can be used to develop analysis and planning processes. Oracle

groups BI applications into various categories, such as Financial Analytics, HR Analytics, and Marketing Analytics.

Oracle's BI products can be used either as an integrated platform or separately. For example, the Oracle BI Suite Enterprise Edition integrates several products in order to access data from various operational or data warehouse systems and present it to users for various kinds of analysis and reporting. On the other hand, Oracle 10g with the OLAP option can be used as a separate component for other third-party tools.

The Oracle BI solutions include many other products not mentioned above that can be used for supporting the decision-making process. However, we shall focus next on those products that are closely related to data warehouse and OLAP applications and rely on a multidimensional model.

2.11.1 Multidimensional Model

Oracle's multidimensional model is supported by the OLAP option of Oracle 10g. The dimensional model includes constructs such as dimensions, hierarchies, levels, attributes, cubes, and measures. A **dimension** provides context and structure for other embedded elements, i.e., levels, hierarchies, and attributes. **Levels** group members having the same characteristics, while **hierarchies** organize these levels at different granularities. **Attributes** are used to define nonhierarchical characteristics of dimension members. **Cubes** provide a means of organizing measures having the same dimensions. **Measures** supports **derived measures**; these measures are created by performing calculations on stored measures. In addition to dimensions, the definition of a cube includes aggregation rules for its measures. These aggregation rules can be simple, such as the sum, or more sophisticated ones, such as a hierarchical average that adds the data values and then divides the sum by the number of children.

The multidimensional model can be implemented as either relational OLAP and multidimensional OLAP.

The **relational OLAP** implementation uses relational data types and stores data as star schemas, snowflake schemas, or parent-child tables. The ROLAP implementation is appropriate for exploratory reporting applications that need to provide ad hoc query and calculation support, but whose usage patterns are somewhat predictable and whose calculation requirements are not extensive. Queries about relational data types can be optimized through techniques such as star queries, precalculation, and summary tables or materialized views.

In the **multidimensional OLAP** implementation, data is stored using multidimensional data types, represented as LOBs (Large OBjects) in relational tables. These tables are handled as array-like structures where each cell in the cube is identified by a unique combination of dimension members, one member from each dimension. Dimension members act as an index into the cube. This kind of storage is called an **analytic workspace** and is embedded within the Oracle database. Analytic workspaces are usually organized according to the subjects of analysis. For example, one analytic workspace might contain sales data and another might contain financial data. The multidimensional data types provided by the OLAP option are used for those applications where usage patterns are unpredictable, calculation requirements are more extensive, query performance requirements are high, and OLAP data sets must be managed efficiently.

2.11.2 Multidimensional Database Design

Oracle provides two tools for designing and managing multidimensional databases: Analytic Workspace Manager and Oracle Warehouse Builder. While both tools support the dimensional model in the ROLAP and MOLAP implementations, each tool is targeted toward a certain type of users and usage pattern. Further, users may combine the usage of both tools; for example, they may use Analytic Workspace Manager to define dimensions and cubes, and then use Warehouse Builder to define the ETL process and maintenance tasks for the analytic workspace.

Analytic Workspace Manager

Analytic Workspace Manager is designed for OLAP specialists who work interactively with analytic workspaces as part of the process of creating, developing, and maintaining an analytic application. This is typically the case during prototyping and application development, as well as with departmental implementations. Since Analytic Workspace Manager is not an ETL tool, it assumes that the data is available within the Oracle database, mainly as relational tables represented as star or snowflake schemas.

Analytic Workspace Manager provides several functionalities as follows:

- Defining a logical multidimensional model: dimensions, levels, attributes, cubes, and measures can be defined from the users' perspective independently of the underlying relational sources.
- Implementing a physical storage model: users are requested to provide information concerning physical implementation. On the basis of this information, Analytic Workspace Manager creates the physical design, considering data compression, partitioning, and preaggregation.
- Mapping relational sources: once the logical and storage methods have been defined, the data can be loaded into the analytic workspace. This requires that the logical model be mapped to source data represented as relational tables or views. If the relational data requires transformation, an ETL tool such as Oracle Warehouse Builder can be used. On the other hand, views might be used if the required transformations are minor. Note that an analytic workspace can provide the sole storage of dimensional

data, i.e., it is not necessary for relational data in the source systems to persist after it has been loaded into the analytic workspace.

- Managing the life cycle: users can indicate how maintenance should be done for the entire analytic workspace or for individual components. Analytic Workspace Manager can control the initial loading, refreshing, and aggregation procedures.

Oracle Warehouse Builder

Oracle Warehouse Builder is a tool for data and metadata management that enables users to create ETL processes, define a multidimensional schema (in a relational database or an OLAP system), deploy it in a relational database or in an analytic workspace, and populate it with data from multiple data sources. This tool is typically used by organizations that integrate dimensional modeling and deployment into an overall data warehouse process.

Warehouse Builder provides two methods for multidimensional modeling. A wizard allows one to quickly define a cube or a dimension, while most settings and options are defined by default. Alternatively, a manual configuration can be done using the **Data Object Editor**, which provides full control over the design and implementation options.

Warehouse Builder can be used to deploy a multidimensional model using either the ROLAP or the MOLAP implementation. For both options, the steps described above for Analytic Workspace Manager can be applied, with the difference that the mapping can be done from various kinds of source systems. The ROLAP option provides several different alternatives for storage, such as star schemas, snowflake schemas, or customized schemas. The process of assigning a specific method of storage for multidimensional objects is called **binding**. An additional feature included for ROLAP storage is the possibility to define slowly changing dimensions.[2] Three types of implementations are provided for these dimensions: type 1, which does not keep the history, type 2, which stores the complete change history, and type 3, which stores only the previous value.

During all the steps of creating a multidimensional model (whether relational or OLAP), Warehouse Builder creates metadata that can be used later in other tools, for example Oracle Discoverer.

2.11.3 Data Source Management

Multidimensional databases must be fed with data from source systems. These source systems may be Oracle databases or third-party databases such as Sybase, Informix, SQL Server, and DB2. The data from non-Oracle sources

---

[2] We refer to slowly changing dimensions in Sect. 5.1.

can be obtained either through transparent gateways or via an ODBC connection. In addition, Warehouse Builder provides direct support for SAP and PeopleSoft metadata and applications.

Warehouse Builder provides various features to support the ETL process. For example, the content and structure of data sources can be visualized, including constraints, indexing, storage options, and partitioning, among others. Users can filter source data before loading it into a multidimensional database. For developing the ETL process, users can employ a graphical user interface that allows them to define the mappings and required transformations between the various elements of the source and target systems, including analytic workspaces. Warehouse Builder generates code for these operations, and this code can be debugged.

Warehouse Builder includes other features that help to improve the quality of data obtained from source systems. For example, a data-profiling component allows users to inspect data in a source system to elicit its meaning, i.e., to validate assumptions about the data or to discover new characteristics. A wizard allows users to select which elements should be profiled. Using the profiling results, Warehouse Builder allows implementers to derive data rules that can include actions related to data cleaning and correction. These rules can be executed automatically during the ETL process. Warehouse Builder introduces data auditors to ensure that data quality is monitored while the ETL process is running on a scheduled basis. These auditors are process flow activities that evaluate one or more data rules and notify the users whether data fails or passes these rules.

There are other components related to improving the quality of data before loading it into a multidimensional database. These can be used to facilitate the implementation of the necessary transformations during the ETL process. For example, name and address cleansing includes several transformations that allow implementers to parse names, and standardize and correct them; a matching process identifies which records refer to the same logical data; and a merging process consolidates the data from a matched set into single records.

2.11.4 Dimensions

A dimension may include one or more hierarchies, levels, and attributes. All these elements should be defined separately within the scope of a dimension, and mapping rules should also be specified for each element.

Oracle distinguishes two types of hierarchies: level-based and value-based.

- **Level-based hierarchies** are the usual hierarchies, consisting of two or more levels related by one-to-many relationships. There are three types of level-based hierarchies:

  - **Normal hierarchies** correspond to the balanced hierarchies that we saw in Sect. 2.6.1, i.e., there is the same number of levels from every member to the root of the hierarchy.

– **Ragged hierarchies** contain leaf nodes at more than one level. They
correspond to the unbalanced hierarchies that we saw in Sect. 2.6.1.
– **Skip-level hierarchies** include at least one member whose parent
is more than one level above it in the hierarchical structure. They
correspond to the ragged hierarchies that we saw in Sect. 2.6.1.

• **Value-based hierarchies** are used to represent parent-child relationships
between members belonging to the same level. We have already referred
to this kind of hierarchy in Sect. 2.6.1.

If a dimension has more than one hierarchy, one of them can be set as the
default hierarchy.

Dimension attributes provide information about individual members. A
dimension attribute is applicable to one or more levels in a dimension. For
example, a dimension Product could have a hierarchy formed by the levels
Product, Category, and Department. This dimension could include an attribute
Description that is applicable to all levels. The list of dimension attributes
must include all the attributes that are needed for any of the levels in the
dimension. In addition to user-defined attributes, other dimension attributes
are created automatically, such as long and short descriptions, which are used
by OLAP client applications to manage display labels within reports.

Oracle distinguishes time dimensions from regular dimensions. A time dimension requires the definition of two attributes: the end date and the time
span. These are used to identify the last day and the number of days in
each time period. These attributes support time-series analysis, such as comparisons with earlier time periods. If this information is not available, time
dimensions may be defined as regular dimensions, although in this case they
will not support built-in time-based analysis.

Depending on the data available in the sources, a dimension can use either
natural keys or surrogate keys for its members. However, when natural keys are
used, it is necessary to ensure that every dimension member is unique across
all levels. For example, if New York is used as both a city name and a state
name, the second value will overwrite the first one. By default, Warehouse
Builder uses surrogate keys; however, this can be overridden in the Data
Object Editor.

Dimensions and mappings for source data can be created in advance and
stored as templates. In this way, they can be reused for other cubes. Further, Warehouse Builder allows one to define user-friendly object names that
are more understandable than physical names. It also allows one to provide
multilingual translations of object names.

2.11.5 Cubes

A cube consists of a set of ordered dimensions and measures. A cube may
have one or several dimensions.

Measures are defined in a similar way to dimension attributes. Calculated measures can be created either using a template-driven wizard or manually, using an editor for defining custom expressions. Calculated measures are not stored, although from the users' perspective they appear as stored measures and can be used in a nested fashion.

Aggregation methods may vary by dimension. This provides support for semiadditive and nonadditive measures, i.e., measures that cannot be aggregated using the sum operation. For example, the aggregation of a Stock quantity measure might use a sum over a Store dimension, while it might use an average over a Time dimension.

In addition to support many different types of calculation, Oracle 10g OLAP provides many predictive analytic functions, including forecasting. Using the **Expert option** that is integrated into Warehouse Builder, users can define a forecast by choosing the time-span information and the forecast method. Forecasting methods include nonlinear regression, and single exponential smoothing, among others. This allows one to create a new cube containing the forecast measures. The forecasting system also supports features such as seasonality of data.

A cube definition refers to several physical features that affect the performance of analytic workspaces:

- The sparsity of data represents the extent to which cells within a cube contain null values instead of data. By default, Warehouse Builder defines all dimensions, except the time dimension, as sparse. When this is done, a special compression method is applied that helps to better manage storage, aggregation, and query performance.
- Compressed storage can be used for measures that are extremely sparse, for example when there are many dimensions, dimensions with many members, and/or dimensions with deep hierarchies. Compressed storage uses less space and results in faster aggregation than does normal sparse storage.
- The order in which dimensions are listed for a cube affects performance, since it determines the way the data is stored on a disk. Performance is optimized when values that are accessed together are stored together. For regular storage, the dimensions are ordered from the largest to the smallest, the opposite of the case for compressed storage. In many applications, the time dimension is defined as the first one, to facilitate data maintenance.
- Cubes can be physically partitioned in a similar way to relational tables. For example, implementers can choose the Quarter level of the Time dimension as a partitioning criterion. Each quarter and its descendants are stored in a separate partition. Partitioning gives several benefits, such as parallelizing data loading and aggregation, working with smaller data sets, which facilitates I/O operations, and simplifying the removal of old data from storage.

- The Data Object Editor allows one to select the levels in a dimension along which the measure data is precomputed. It is possible to precompute all levels for each dimension. However, by default Warehouse Builder alternates the precomputed levels, starting with leaf nodes. This balances the need for query performance against the amount of disk space consumed.

2.12 Conclusion

This chapter has introduced the background concepts needed for the rest of the book.

We started by describing database systems and the usual steps followed for designing them, i.e., requirements specification, conceptual design, logical design, and physical design. We then described in more detail the phases of conceptual, logical, and physical design and the models used in these phases. We presented the entity-relationship model, a well-known conceptual model, followed by two common logical models, the relational and object-relational models. We finished this introduction to database systems by describing several issues related to physical database design.

We continued by describing data warehouse systems, highlighting their differences with respect to traditional database systems. We then covered the multidimensional model that is the basis for data warehouse systems, and typical logical solutions for implementing data warehouses, i.e., the star, snowflake, and constellation schemas. Physical considerations related to implementing data warehouses were also discussed. As data warehouse systems include many different components, we discussed the basic architecture of data warehouse systems and several variants of it that may be considered. We finished this overview by introducing two common data warehouse tools: Microsoft Analysis Services 2005, and Oracle 10g with the OLAP option.

3

Conventional Data Warehouses

The advantages of using conceptual models for designing applications are well known. In particular, conceptual models facilitate communication between users and designers, since they do not require knowledge about specific features of the underlying implementation platform. Further, schemas developed using conceptual models can be mapped to various logical models, such as relational, object-relational, or object-oriented models, thus simplifying responses to changes in the technology used. Moreover, conceptual models facilitate the maintenance and evolution of applications, since they focus on users' requirements; as a consequence, they provide better support for subsequent changes in the logical and implementation schemas.

In this chapter, we present the MultiDim model, a conceptual multidimensional model that allows one to represent data requirements for data warehouse and OLAP applications. The definition of the model is given in Sect. 3.1. Since hierarchies are essential for exploiting data warehouse and OLAP systems to their full capabilities, in Sect. 3.2 we consider various kinds of hierarchies that exist in real-world situations. We classify these hierarchies, giving a graphical representation of them and emphasizing the differences between some kinds of hierarchies that are currently ignored in the literature. Section 3.3 discusses some advanced aspects of modeling, namely complex hierarchies, and role-playing, fact, and multivalued dimensions. The metamodel of our conceptual multidimensional model is given in Sect. 3.4.

We then give a mapping of our conceptual model to the relational and object-relational models, thus showing the feasibility of implementing our model in current database management systems. We first describe in Sect. 3.5 some general mapping rules for the conceptual model. Then, we discuss in Sect. 3.6 specific mappings for the various kinds of hierarchies, comparing alternative implementations for some of them. Section 3.7 refers to the implementation of hierarchies in commercial platforms, using as examples Microsoft Analysis Services 2005 and Oracle OLAP 10g. Finally, Sect. 3.8 refers to work related to multidimensional conceptual models and hierarchies, and Sect. 3.9 summarizes this chapter.

3.1 MultiDim: A Conceptual Multidimensional Model

In this section, we define the MultiDim model, a conceptual multidimensional model. Our model allows one to represent at the conceptual level all elements required in data warehouse and OLAP applications, i.e., dimensions, hierarchies, and facts with associated measures. The graphical notation of the Multi-Dim model is shown in Fig. 3.1;[1] it is similar to that of the entity-relationship (ER) model.

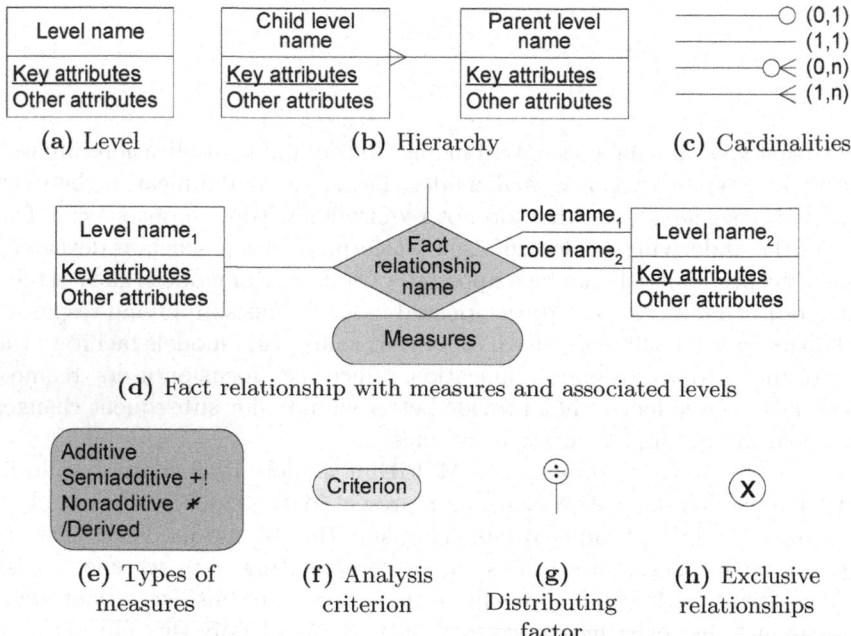

Fig. 3.1. Notation of the MultiDim model

In order to give a general overview of the model, we shall use the example in Fig. 3.2, which illustrates the conceptual schema of a sales data warehouse. This figure includes several types of hierarchies, which will be presented in more detail in the subsequent sections.

A **schema** is composed of a set of dimensions and a set of fact relationships.

A **dimension** is an abstract concept that groups data that shares a common semantic meaning within the domain being modeled. A dimension is composed of a set of hierarchies (see later), and a hierarchy is in turn composed of a set of levels.

---

[1] A more detailed description of our notation is given in Appendix B.

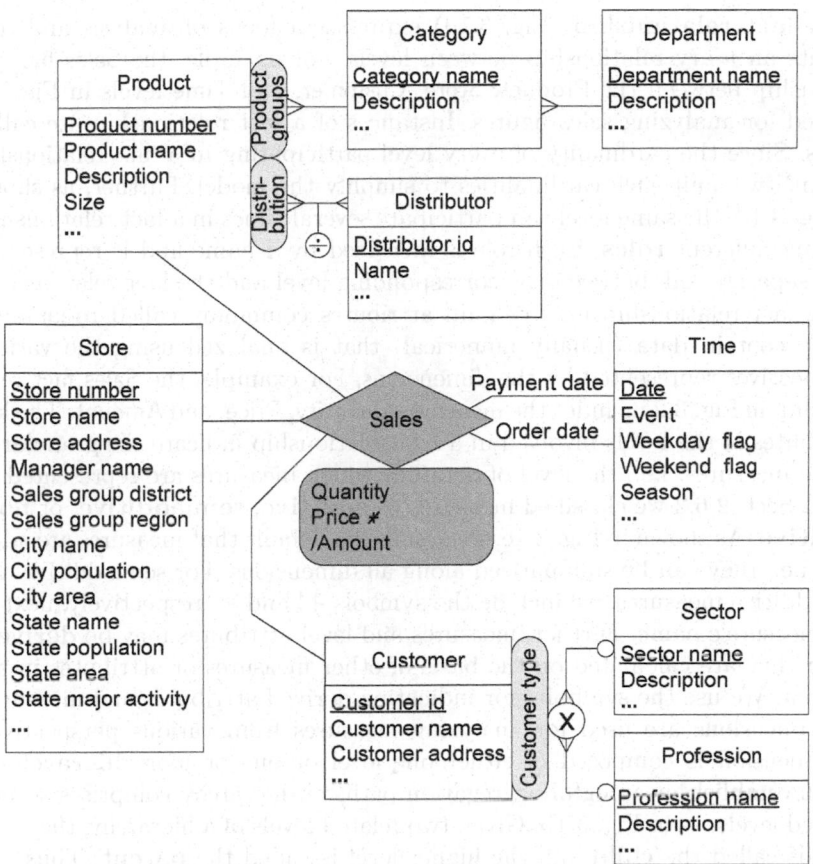

Fig. 3.2. A conceptual multidimensional schema of a sales data warehouse

A **level** corresponds to an entity type in the ER model. It describes a set of real-world concepts that, from the application's perspective, have similar characteristics. For example, Product, Category, and Department are some of the levels in Fig. 3.2. Instances of a level are called **members**. As shown in Fig. 3.1a, a level has a set of **attributes** that describe the characteristics of their members. In addition, a level has one or several **keys** that identify uniquely the members of a level, each key being composed of one or several attributes. For example, in Fig. 3.2, Category name is a key attribute of the Category level. Each attribute of a level has a type, i.e., a domain for its values. Typical value domains are integer, real, and string. For brevity, we do not include type information for attributes in the graphical representation of our conceptual schemas. This can be done if necessary, and it is done in the textual representation of our model, which is shown in Appendix A.

A **fact relationship** (Fig. 3.1d) expresses a focus of analysis and represents an n-ary relationship between levels. For example, the Sales fact relationship between the Product, Store, Customer, and Time levels in Fig. 3.2 is used for analyzing sales figures. Instances of a fact relationship are called **facts**. Since the cardinality of every level participating in a fact relationship is (0,n), we omit such cardinalities to simplify the model. Further, as shown in Fig. 3.1d, the same level can participate several times in a fact relationship, playing different **roles**. Each role is identified by a name and is represented by a separate link between the corresponding level and the fact relationship.

A fact relationship may contain attributes commonly called **measures**. These contain data (usually numerical) that is analyzed using the various perspectives represented by the dimensions. For example, the Sales fact relationship in Fig. 3.2 includes the measures Quantity, Price, and Amount. The key attributes of the levels involved in a fact relationship indicate the granularity of the measures, i.e., the level of detail at which measures are represented.

In Sect. 2.6.2 we classified measures as **additive, semiadditive,** or **nonadditive**. As shown in Fig. 3.1e, we assume by default that measures are additive, i.e., they can be summarized along all dimensions. For semiadditive and nonadditive measures, we include the symbols $+!$ and $\nparallel$, respectively, next to the measure's name. Further, measures and level attributes may be **derived**, when they are calculated on the basis of other measures or attributes in the schema. We use the symbol $/$ for indicating derived attributes and measures.

Dimensions are used for analyzing measures from various perspectives. A dimension is composed of either one level or one or more **hierarchies**, which establish meaningful aggregation paths. A hierarchy comprises several related levels, as in Fig. 3.1b. Given two related levels of a hierarchy, the lower level is called the **child** and the higher level is called the **parent**. Thus, the relationships composing hierarchies are called **parent-child relationships**. Since these relationships are used only for traversing from one level to the next, they are simply represented by a line to simplify the notation.

Parent-child relationships are characterized by **cardinalities**, shown in Fig. 3.1c, indicating the minimum and the maximum number of members in one level that can be related to a member in another level. For example, in Fig. 3.2 the child level Product is related to the parent level Category with a one-to-many cardinality, which means that every product belongs to only one category and that each category can have many products.

The hierarchies in a dimension may express various structures used for analysis purposes; thus, we include an **analysis criterion** (Fig. 3.1f) to differentiate them. For example, the Product dimension in Fig. 3.2 includes two hierarchies: Product groups and Distribution. The former hierarchy comprises the levels Product, Category, and Department, while the latter hierarchy includes the levels Product and Distributor. Single-level hierarchies, such as Time and Store in Fig. 3.2, indicate that even though these levels contain attributes that may form a hierarchy, such as City name and State name in the Store dimension, the user is not interested in using them for aggregation purposes.

The levels in a hierarchy are used to analyze data at various **granularities**, i.e., levels of detail. For example, the Product level contains specific information about products, while the Category level may be used to see these products from the more general perspective of the categories to which they belong. The level in a hierarchy that contains the most detailed data is called the **leaf level**; it must be the same for all hierarchies included in a dimension. The leaf-level name is used for defining the dimension's name. The last level in a hierarchy, representing the most general data, is called the **root level**. If several hierarchies are included in a dimension, their root levels may be different. For example, both hierarchies in the Product dimension in Fig. 3.2 contain the same leaf level, Product, while they have different root levels, the Department and the Distributor levels.

In some publications, the root of a hierarchy is represented using a level called All. We leave to designers the decision of including this level in multidimensional schemas. In this book, we do not present the All level for the various hierarchies, since we consider that it is meaningless in conceptual schemas and, in addition, it adds unnecessary complexity to them.

The key attributes of a parent level define how child members are grouped. For example, in Fig. 3.2, Department name in the Department level is a key attribute; it is used for grouping different category members during the roll-up operation from the Category to the Department level. However, in the case of many-to-many parent-child relationships, it is also needed to determine how to distribute the measures from a child to its parent members. For example, in Fig. 3.2 the relationship between Product and Distributor is many-to-many, i.e., the same product can be distributed by several distributors. The notation in Fig. 3.1g indicates that a distributing factor is used to divide the measures associated with a product between its categories.

Finally, it is sometimes the case that two or more parent-child relationships are **exclusive**. This is represented using the symbol in Fig. 3.1h. An example is given in Fig. 3.2, where customers can be either persons or organizations. Thus, according to their type, customers participate in only one of the relationships departing from the Customer level: persons are related to the Profession level, while organizations are related to the Sector level.

3.2 Data Warehouse Hierarchies

Hierarchies are fundamental in analytical applications, since they provide the means to represent the parameters under analysis at different abstraction levels. In real-world situations, users must deal with complex hierarchies of various kinds. However, the logical models of current data warehouse and OLAP systems allow only a limited set of kinds of hierarchies. Therefore, users are often unable to capture the essential semantics of multidimensional applications and must limit their analysis to considering only the predefined set of hierarchies provided by the tools in use. Further, current approaches for representing hierarchies at the conceptual level focus mainly on aggregation paths represented at the schema level, i.e., they focus on establishing sequences of levels that should be traversed during roll-up and drill-down operations. However, as we shall see in this section, the distinction between the various kinds of hierarchies should also be made at the instance level, i.e., we should consider the cardinalities in parent-child relationships.

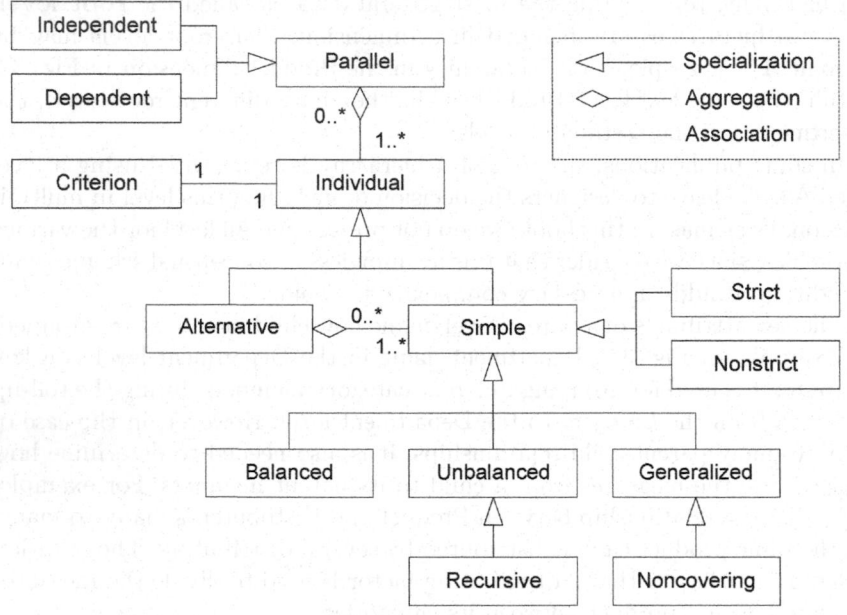

Fig. 3.3. Hierarchy classification

In this section, we discuss various kinds of hierarchies that can be represented in the MultiDim model. To give a more general view of our proposal for the various kinds of hierarchies and the relationships between them, we first present a classification of them using the metamodel shown in Fig. 3.3. As can be seen in the figure, parallel hierarchies can be specialized into independent and dependent hierarchies, depending on whether they share common levels. A parallel hierarchy is an aggregation of individual hierarchies. Each individual hierarchy is associated with one analysis criterion. An individual hierarchy can be simple or alternative, where the latter is composed of one or several simple hierarchies associated with the same analysis criterion. The simple hierarchies include further kinds: balanced, unbalanced, and generalized hierarchies. Also, recursive and noncovering hierarchies are a special case of unbalanced and generalized hierarchies, respectively. For each of these simple hierarchies, another specialization can be applied, depending on whether the cardinalities between the parent and child levels are one-to-many or many-to-many. The former are called strict hierarchies, while the latter are called nonstrict hierarchies.

In the following, we refer in more detail to the various kinds of hierarchies, presenting them at the conceptual and at the logical level. Their implementation at the physical level, using as examples Microsoft Analysis Services 2005 and Oracle OLAP 10g, is covered in Sect. 3.7.

3.2.1 Simple Hierarchies

Simple hierarchies are those hierarchies where, if all its component parent-child relationships are one-to-many, the relationship between their members can be represented as a tree. Further, these hierarchies use only one criterion for analysis. Simple hierarchies are divided into balanced, unbalanced, and generalized hierarchies.

Balanced Hierarchies

A **balanced hierarchy** has only one path at the schema level (Fig. 3.4a). At the instance level, the members form a tree where all the branches have the same length (Fig. 3.4b). As implied by the cardinalities, all parent members have at least one child member, and a child member belongs to only one parent member. For example, in Fig. 3.4 each category has assigned to it at least one product, and a product belongs to only one category.

Unbalanced Hierarchies

An **unbalanced hierarchy**[2] has only one path at the schema level. However, as implied by the cardinalities, at the instance level some parent members may not have associated child members. Figure 3.5a shows a hierarchy in which a bank is composed of several branches: some of them have agencies with ATMs, some have only agencies, and small branches do not have any organizational division. As a consequence, at the instance level the members represent an unbalanced tree (Fig. 3.5b), i.e., the branches of the tree have different lengths, since some parent members do not have associated child members. As in the case of balanced hierarchies, the cardinalities imply that every child member should belong to at most one parent member. For example, in Fig. 3.5 every agency belongs to one branch.

Unbalanced hierarchies include a special case that we call **recursive hierarchies**.[3] In this kind of hierarchy the same level is linked by the two roles of a parent-child relationship. An example is given in Fig. 3.6a, which represents an organizational chart in terms of the employee-supervisor relationship. The

---

[2] These hierarchies are also called heterogeneous [106] and non-onto [231] hierarchies.

[3] These are also called parent-child hierarchies [156, 206].

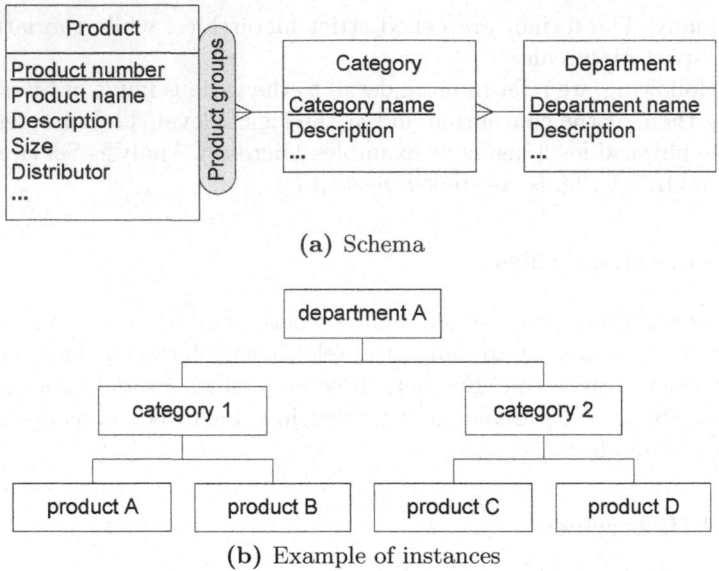

(a) Schema

(b) Example of instances

Fig. 3.4. A balanced hierarchy

subordinate and supervisor roles of the parent-child relationship are linked to the Employee level. Recursive hierarchies are mostly used when all hierarchy levels express the same semantics, i.e., when the characteristics of the children and parents are similar (or the same), as in Fig. 3.6a, where an employee has a supervisor who is also an employee.

As shown in Fig. 3.6b, a recursive hierarchy could be used to represent the unbalanced hierarchy shown in Fig. 3.5a. However, in this representation the semantics of the hierarchy is somewhat lost, since levels corresponding to concepts at different granularities are represented together, for instance ATM, agency, and branch. Note that an additional discriminating attribute, Entity type, is needed to identify the level to which the members belong. Further, the structure of the hierarchy is not clear and it is necessary to retrieve all members to reconstruct it, i.e., we must move to the instance level to recover information about the schema. In addition, the representation of this hierarchy shown in Fig. 3.6b is less expressive than that in Fig. 3.5a, since the grouping of attributes according to the levels to which they belong is lost.

Generalized Hierarchies

Sometimes a dimension includes subtypes that can be represented by a generalization/specialization relationship [5, 8, 160, 169]. Moreover, the specialized subtypes can include their own hierarchy. In the example given in Fig. 3.7, the Customer supertype is specialized into the Person and the Company subtypes

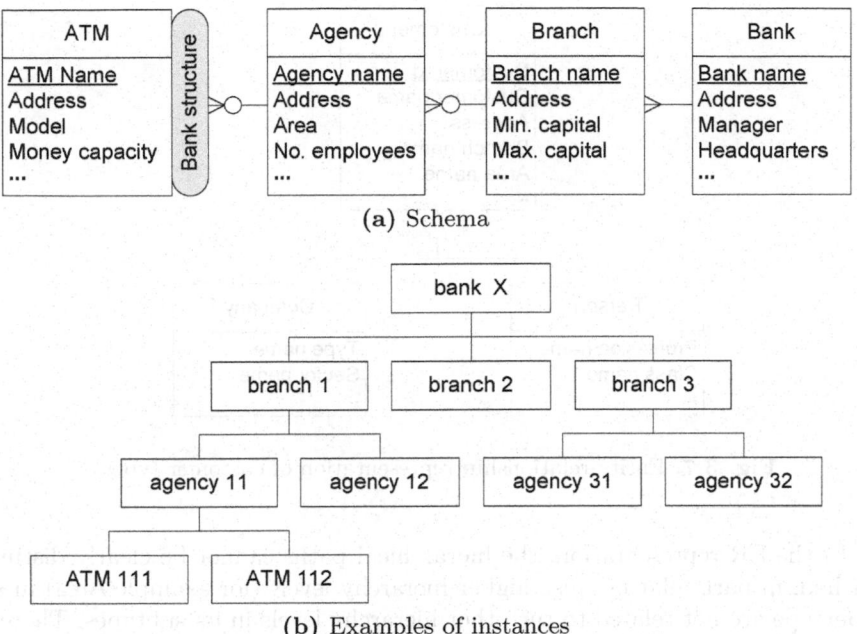

(a) Schema

(b) Examples of instances

Fig. 3.5. An unbalanced hierarchy

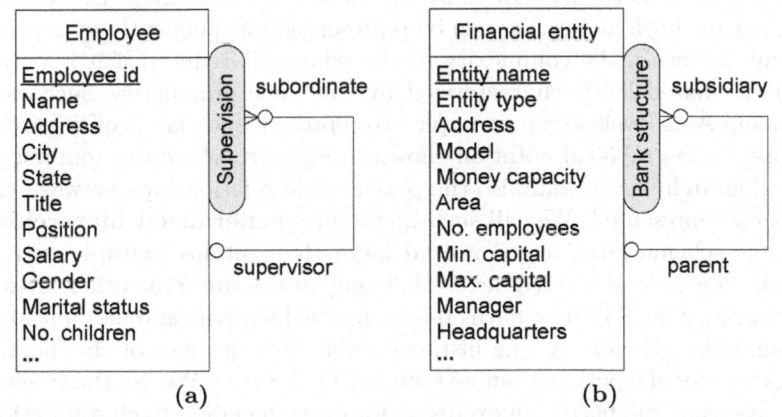

Fig. 3.6. Examples of recursive hierarchies

to distinguish the two types of customers. Further, the subtypes have hierarchies with common levels (for example Branch and Area) and specific levels (for example Profession and Class for Person, and Type and Sector for Company).

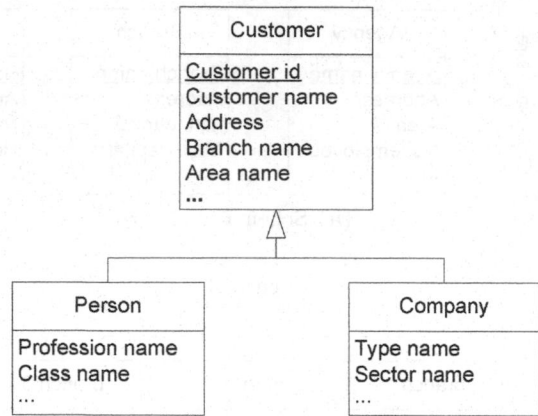

Fig. 3.7. Entity-relationship representation of customer types

In the ER representation, the hierarchical paths cannot be clearly distinguished, in particular because higher hierarchy levels (for example Area) in a supertype are not related to the other hierarchy levels in its subtypes. Therefore, the information about the levels forming a hierarchy and the parent-child relationships between them cannot be retrieved. For example, in Fig. 3.7 it is not clear that measures related to a customer that is a person can be aggregated using a hierarchy formed by the levels Profession–Class–Branch–Area. Further, while both hierarchies can be represented independently by repeating the common levels, the complexity of the schema is reduced if it is possible to include shared levels characterized by the same granularity, such as the Branch and Area levels in our example. To represent such kinds of hierarchies, we propose the graphical notation shown in Fig. 3.8, where the common and specific hierarchy levels and also the parent-child relationships between them are clearly represented. We call such hierarchies **generalized hierarchies**.

At the schema level, a generalized hierarchy contains multiple exclusive paths sharing at least the leaf level; they may also share some other levels, as shown in Fig. 3.8a. All these paths represent one hierarchy and account for the same analysis criterion. At the instance level, each member of the hierarchy belongs to only one path, as can be seen in Fig. 3.8 (b).[4] We use the symbol $\otimes$ to indicate that the paths are exclusive for every member. Such a notation is equivalent to the **xor** annotation used in UML [29].[5] The levels at which the alternative paths split and join are called the **splitting** and **joining levels**, respectively.

---

[4] Note that, as indicated by the ellipses, not all members are represented in the figure.

[5] This notation is also used by Hurtado et al. [105] to denote the "splitting constraints".

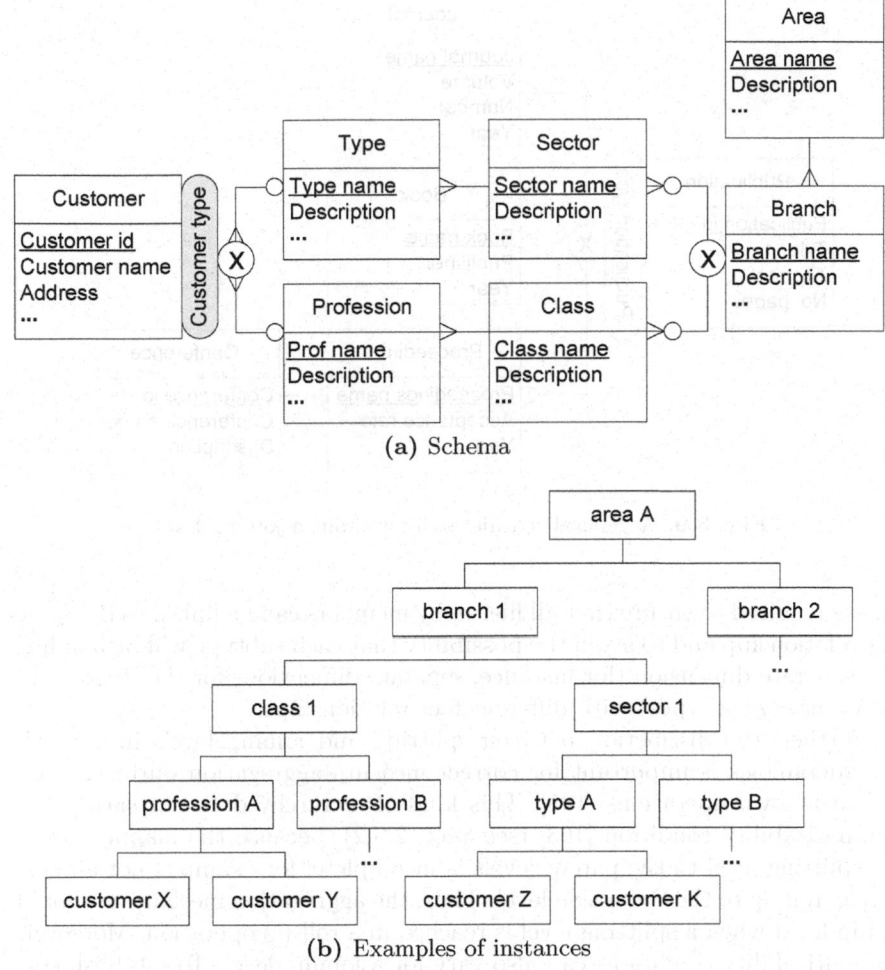

(a) Schema

(b) Examples of instances

Fig. 3.8. A generalized hierarchy

In the example in Fig. 3.8a, the paths between the two ⊗ symbols refer to attributes that can be used for aggregation purposes for the specialized subtypes shown in Fig. 3.7. The lower path (i.e., the path containing the Profession and Class levels) corresponds to the Person subtype, while the upper path (i.e., that containing the Type and Sector levels) indicates the Company subtype.

Note that the supertype included in the generalization/specialization relationship in Fig. 3.7 is used in generalized hierarchies for representing a leaf level. However, it will include only those attributes that represent concepts at the lowest granularity (for example Customer id, Customer name, and Address).

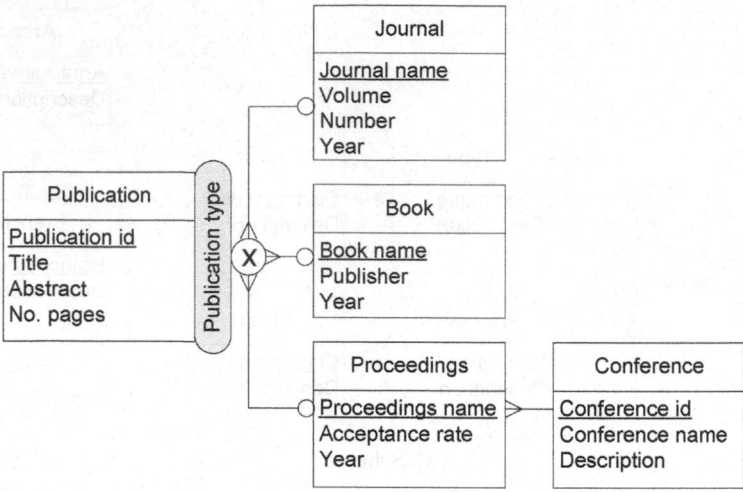

Fig. 3.9. A generalized hierarchy without a joining level

This is required to ensure that all hierarchy members can be linked to the same fact relationship and to avoid the possibility that each subtype will be handled as a separate dimension (for instance, separate dimensions for the Person and the Company subtypes) with different fact relationships.

Further, the distinction between splitting and joining levels in generalized hierarchies is important for correct measure aggregation during roll-up and drill-down operations [105]. This kind of hierarchy does not satisfy the summarizability condition [163] (see Sect. 2.6.2), because the mapping from the splitting level to the parent levels is incomplete; for example, not all customers roll up to the Profession level. Thus, the aggregation mechanism should be modified when a splitting level is reached in a roll-up operation. Moreover, summarizability conditions can also vary for a joining level. To establish correct aggregation procedures, the approach proposed by Hurtado et al. [105] for heterogeneous hierarchies could be applied. On the other hand, the traditional approach can be used for aggregating measures for common hierarchy levels.

In generalized hierarchies, it is not necessary that splitting levels must be joined. For example, Fig. 3.9 shows a generalized hierarchy used for analyzing international publications. Three kinds of publications are considered, i.e., journals, books, and conference proceedings. The latter can be aggregated to the conference level. However, there is not a common joining level for all paths.

Generalized hierarchies include a special case commonly referred to as **noncovering hierarchies** or **ragged hierarchies** [156, 231]. The example given in Fig. 3.10 represents a distribution company that has warehouses in several different states. However, the geographical division in these states may vary; for example, the division into counties may be skipped for some of them. A noncovering hierarchy is a generalized hierarchy with the additional restriction that the alternative paths are obtained by skipping one or several intermediate levels. At the instance level, every child member has only one parent member, although the path length from the leaves to the same parent level can be different for different members.

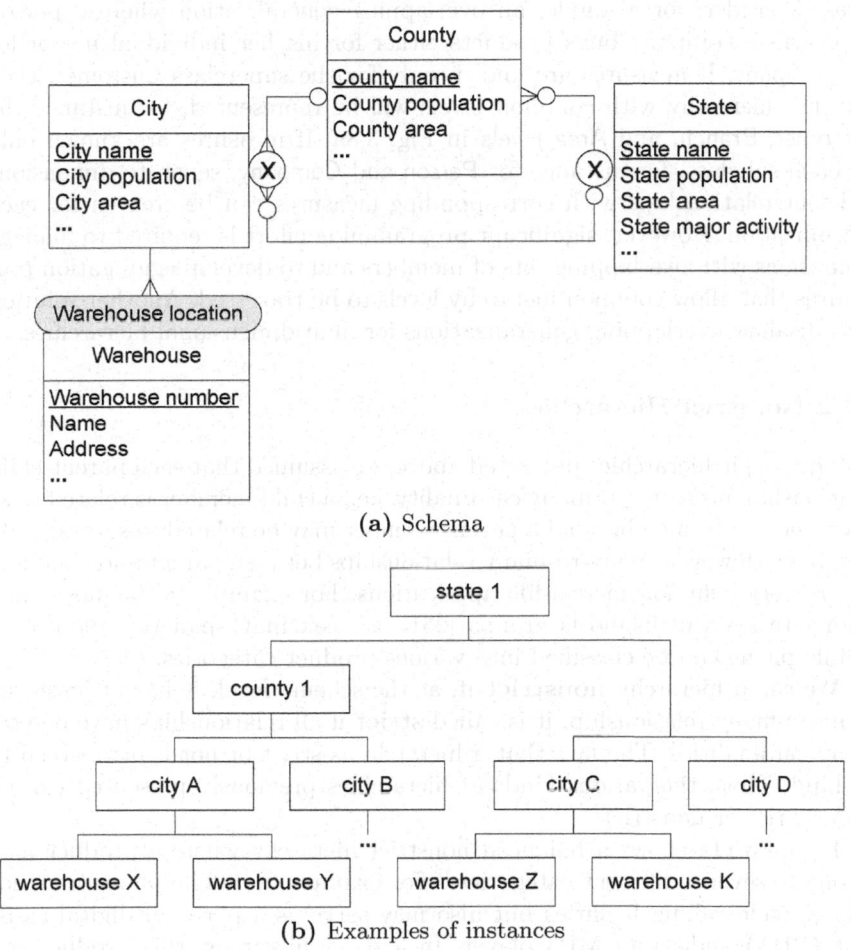

(a) Schema

(b) Examples of instances

Fig. 3.10. A noncovering hierarchy

Note that not all generalization/specialization hierarchies can be represented by generalized hierarchies in a multidimensional model. Recall that a generalization/specialization can be (a) total or partial, and (b) disjoint or overlapping [66] (see Sect. 2.2). Partial specializations induce an additional path in the generalized hierarchy that relates the common levels. For example, if the generalization in Fig. 3.7 were partial, indicating that there were some customers that were considered to be neither persons nor companies, then an additional path between the Customer and Branch levels would be needed in the hierarchy of Fig. 3.8a.

On the other hand, for overlapping generalizations, various options are possible, according to the users' requirements and the availability of measures. Consider, for example, an overlapping generalization where a person who owns a company buys products either for his/her individual use or for the company. If measures are known only for the superclass Customer, then only the hierarchy with common levels will be represented, for instance the Customer, Branch, and Area levels in Fig. 3.8a. If measures are known only for each subclass, for instance for Person and Company, separate dimensions and fact relationships with corresponding measures can be created for each specialization. However, significant programming effort is required to manage dimensions with overlapping sets of members and to develop aggregation procedures that allow common hierarchy levels to be traversed. Another solution is to disallow overlapping generalizations for multidimensional hierarchies.

3.2.2 Nonstrict Hierarchies

For the simple hierarchies presented above, we assumed that each parent-child relationship has a one-to-many cardinality, i.e., a child member is related to at most one parent member and a parent member may be related to several child members. However, many-to-many relationships between parent and child levels are very common in real-life applications. For example, a diagnosis may belong to several diagnosis groups [231], a week may span two months, a mobile phone can be classified into various product categories, etc.

We call a hierarchy **nonstrict** if, at the schema level, it has at least one many-to-many relationship; it is called **strict** if all relationships have one-to-many cardinalities. The fact that a hierarchy is strict or not is orthogonal to its kind. Thus, the various kinds of hierarchies previously presented can be either strict or nonstrict.

Figure 3.11a shows a balanced nonstrict hierarchy where a product may belong to several different categories.[6] For example, a mobile phone may not only provide calling facilities but also may serve as a personal digital assistant (PDA) and as an MP3 player. In a strict hierarchy, this product can only belong to one category, i.e., "phone". In a nonstrict hierarchy it may be considered as belonging to three categories, i.e., "phone", "PDA", and "MP3 player". Therefore, since at the instance level a child member may have more than one parent member, the members of the hierarchy form an acyclic graph (Fig. 3.11b).

---

[6] This example is inspired by [103].

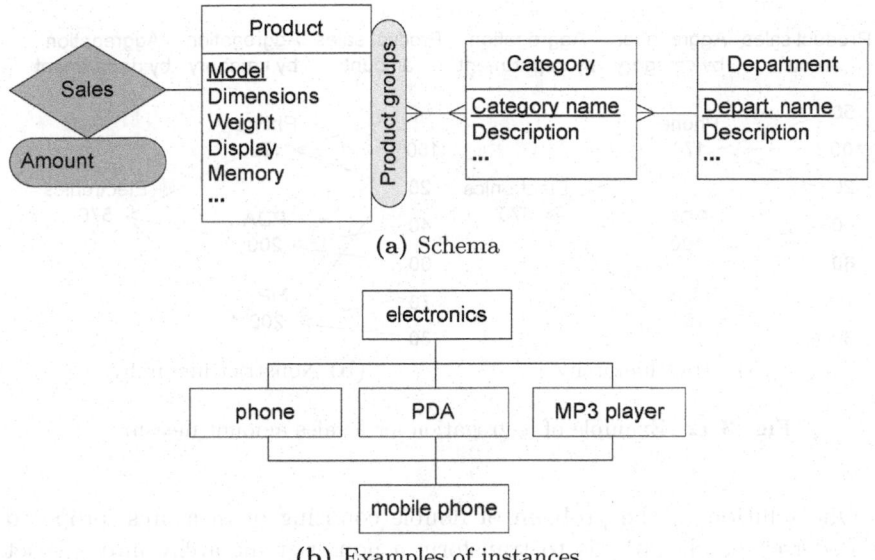

(a) Schema

(b) Examples of instances

Fig. 3.11. A balanced nonstrict hierarchy

Note the slight abuse of terminology. We use the term "nonstrict hierarchy" to denote an acyclic classification graph [103]. We use this term for several reasons. Firstly, the term "hierarchy" conveys the notion that users need to analyze measures at different levels of detail, which is less clear with the term "acyclic classification graph". Secondly, the term "hierarchy" is already used by practitioners and, as we shall see in the next section, some tools, in particular SQL Server Analysis Services 2005, allow many-to-many parent-child relationships. Finally, the term " hierarchy" is also used by several researchers (e.g., [5, 169, 231, 298]).

Nonstrict hierarchies induce the problem of **double counting** of measures when a roll-up operation reaches a many-to-many relationship. Let us consider the example in Fig. 3.12. This illustrates sales of products with aggregations according to the Category and Department levels. Suppose that the product with sales equal to 100 is a mobile phone that also includes PDA and MP3 player functionalities. In a strict hierarchy, this mobile phone can belong to only one category. Therefore, as shown in Fig. 3.12a, the sum of sales by category and by department can be calculated straightforwardly. However, it might be interesting also to consider this mobile phone when calculating the sum of sales for PDAs and MP3 players. Nevertheless, as can be seen in Fig. 3.12b, this approach causes incorrect aggregated results, since the phone's sales are counted three times instead of only once. Note that incorrect results will propagate when higher hierarchy levels are traversed, for example Department in Fig. 3.11a.

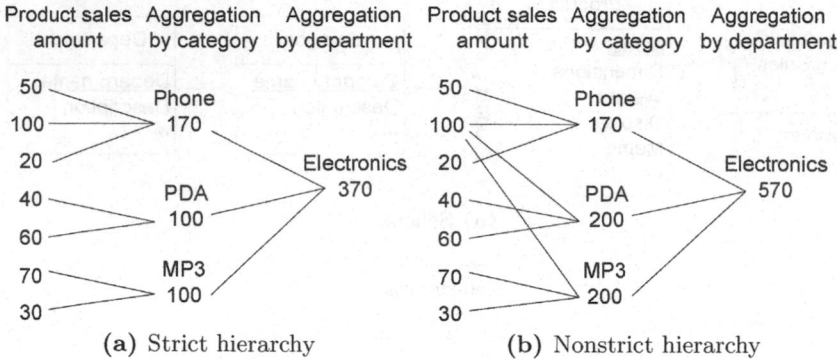

(a) Strict hierarchy   (b) Nonstrict hierarchy

Fig. 3.12. Example of aggregation for a sales amount measure

One solution to the problem of double counting of measures, proposed by Pedersen et al. [231], is to transform a nonstrict hierarchy into a strict hierarchy by creating a new member that joins into one group the parent members participating in a many-to-many relationship. In our mobile phone example, a new category member will be created that represents the three categories together: mobile phone, PDA, and MP3 player. Another solution is to ignore the existence of several parent members and to choose one of them as the primary member [201]; for example, we may choose only the mobile phone category. However, neither of these solutions may correspond to the users' analysis requirements, since in the former, artificial categories are introduced, and in the latter, some pertinent analysis scenarios are ignored.

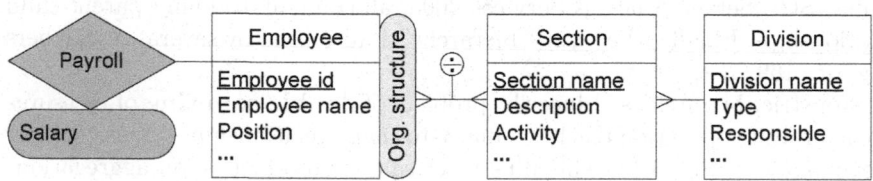

Fig. 3.13. A nonstrict hierarchy with a distributing factor

An alternative approach to the double-counting problem can be taken by indicating how measures are distributed between several parent members for many-to-many relationships. For example, Fig. 3.13 shows a nonstrict hierarchy where employees may work in several sections. The schema includes a measure that represents an employee's overall salary, i.e., the sum of the salaries paid in each section. It may be the case that the percentage of time for which an employee works in each section is known. Therefore, in this case we include an additional symbol $\ominus$, indicating that a **distributing factor** determines how measures should be divided between several parent members in a many-to-many relationship. The choice of an appropriate distributing factor is important in order to avoid approximate results. For example, suppose that the distributing factor represents the percentage of time for which an employee works in a specific section. If an employee has a higher position in one section, then even though he/she works less time in that section, he/she may earn a higher salary. Therefore, applying the percentage of time as the distributing factor for measures representing an employee's overall salary may not give an exact result.

Note that in the mobile phone example, distribution of the sales measure between three categories according to previously specified percentages, for example 70% to "phone", 20% to "PDA", and 10% to "MP3", may not give meaningful information to users. In other situations, this distribution is impossible to specify. In the latter case, if a user does not require exact values of measures, this distributing factor can be calculated by considering the total number of parent members with which the child member is associated [280]. In our mobile phone example, since we have three categories with which this phone is associated, 1/3 of the value of the measure will be aggregated for each category.

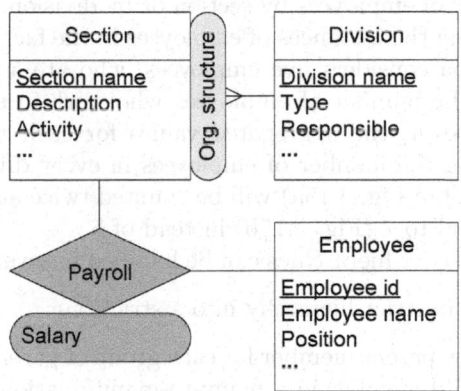

Fig. 3.14. Transforming a nonstrict hierarchy into a strict hierarchy with an additional dimension

Another solution could be to transform a nonstrict hierarchy into independent dimensions as shown in Fig. 3.14.[7] This solution corresponds to a different conceptual schema, where the focus of analysis has been changed from employee's salaries to employee's salaries by section. Note that this solution can only be applied when the exact distribution of the measures is known, for instance, when the amounts of salary paid for working in the different

---

[7] For simplicity, we omit other dimensions, such as the time dimension.

sections are known. It cannot be applied for nonstrict hierarchies without a distributing factor, as in Fig. 3.11a.

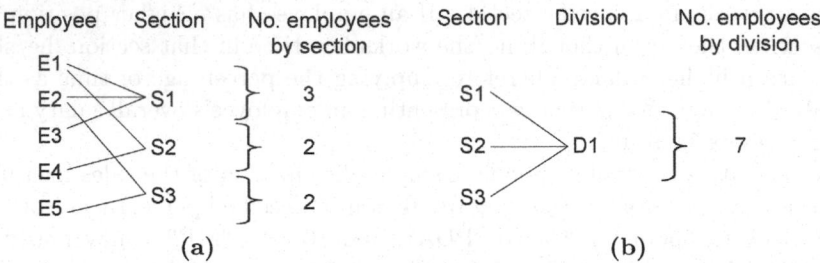

Fig. 3.15. Double-counting problem for a nonstrict hierarchy

Nevertheless, although the solution in Fig. 3.14 allows one to aggregate correctly the salary measure when applying the roll-up operation from the Section to the Division level, the problem of double counting of the same employee will occur. Suppose that we want to use the schema in Fig. 3.14 to calculate the number of employees by section or by division; this value can be calculated by counting the instances of employees in the fact relationship. The example in Fig. 3.15a considers five employees, who are assigned to various sections. Counting the number of employees who work in each section gives correct results. However, the aggregated values for each section cannot be reused for calculating the number of employees in every division, since some employees (E1 and E2 in Fig. 3.15a) will be counted twice and the total result will give a value equal to 7 (Fig. 3.15b) instead of 5.

In summary, nonstrict hierarchies can be handled in several different ways:

- Transforming a nonstrict hierarchy into a strict one:

  - Creating a new parent member for each group of parent members linked to a single child member in a many-to-many relationship [231].
  - Choosing one parent member as the primary member and ignoring the existence of other parent members [201].
  - Splitting the hierarchy in two at the many-to-many relationship, where the levels from the parent level and beyond become a new dimension.

- Including a distributing factor [146].
- Calculating approximate values of a distributing factor [280].

Since each solution has its advantages and disadvantages and requires special aggregation procedures, the designer must select the appropriate solution according to the situation at hand and the user's requirements.

3.2.3 Alternative Hierarchies

Alternative hierarchies represent the situation where, at the schema level, there are several nonexclusive simple hierarchies that share at least the leaf level and account for the same analysis criterion. An example is given in Fig. 3.16a, where the Time dimension includes two hierarchies, corresponding to different subdivisions of years. As seen in Fig. 3.16b,[8] at the instance level such hierarchies form a graph, since a child member can be associated with more than one parent member, and these parent members may belong to different levels.

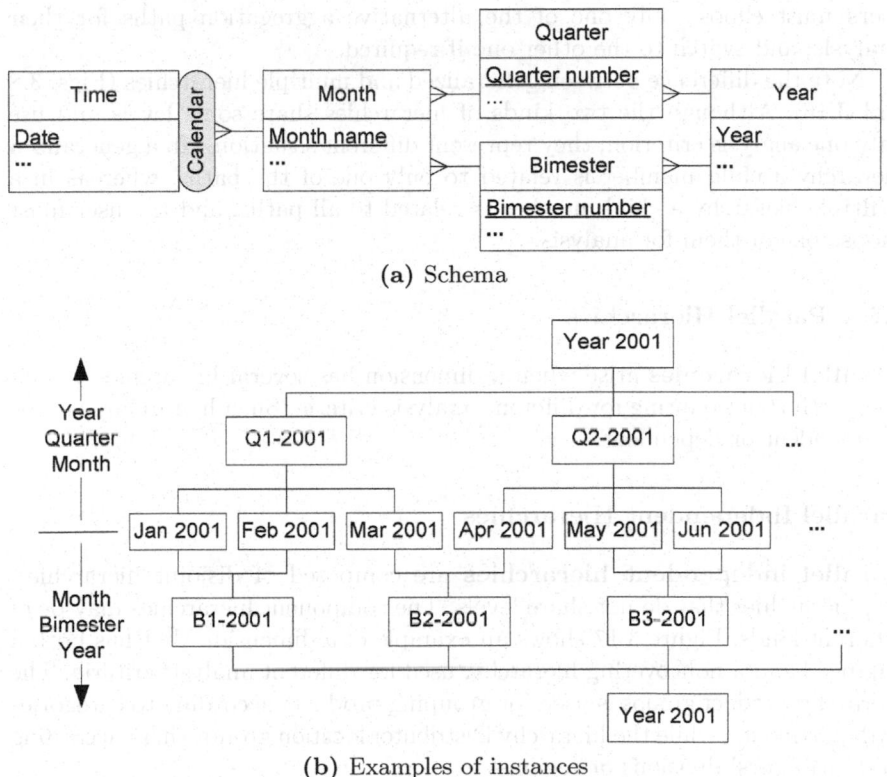

(a) Schema

(b) Examples of instances

Fig. 3.16. Alternative hierarchies

Alternative hierarchies are needed when the user needs to analyze measures from a unique perspective (for example time) using alternative aggregations. Since the measures in the fact relationship will participate totally in each component hierarchy, measure aggregation can be performed as for simple

---

[8] Note that in the figure, the members of the Time level are not shown.

hierarchies. Aggregated measures for a common hierarchy level can be reused when the various component hierarchies are traversed.

However, in the case of alternative hierarchies it is not semantically correct to simultaneously combine different component hierarchies, since this would result in meaningless intersections [292], i.e., a combination of members that do not have values for aggregated measures. In the present example, this would be the case when combining the members B1-2001 and Q2-2001 or Q2-2001 and Jan-2001. This situation induces problems in the interpretation of query results, since the lack of an aggregated value for measures, for example sales, can be perceived by a user as if there were not sales during the corresponding period, instead of realizing that the combination of levels is invalid. Therefore, users must choose only one of the alternative aggregation paths for their analysis, and switch to the other one if required.

Note the difference between generalized and multiple hierarchies (Figs. 3.8 and 3.16). Although the two kinds of hierarchies share some levels and use only one analysis criterion, they represent different situations. In a generalized hierarchy a child member is related to only one of the paths, whereas in a multiple hierarchy a child member is related to all paths, and the user must choose one of them for analysis.

3.2.4 Parallel Hierarchies

Parallel hierarchies arise when a dimension has several hierarchies associated with it, accounting for different analysis criteria. Such hierarchies can be independent or dependent.

Parallel Independent Hierarchies

Parallel independent hierarchies are composed of disjoint hierarchies, i.e., hierarchies that do not share levels. The component hierarchies may be of different kinds. Figure 3.17 shows an example of a dimension that has both a balanced and a noncovering hierarchy, used for different analysis criteria. The hierarchy Product groups is used for grouping products according to categories or departments, while the hierarchy Distributor location groups them according to distributors' divisions or regions.

Parallel Dependent Hierarchies

Parallel dependent hierarchies are composed of several hierarchies that account for different analysis criteria and share some levels. The example given in Fig. 3.18 represents an international company that requires sales analysis for stores located in several countries. The Store dimension contains two balanced hierarchies. The hierarchy Store location represents the geographic division of the store address and includes the levels Store, City, State, and Country. The

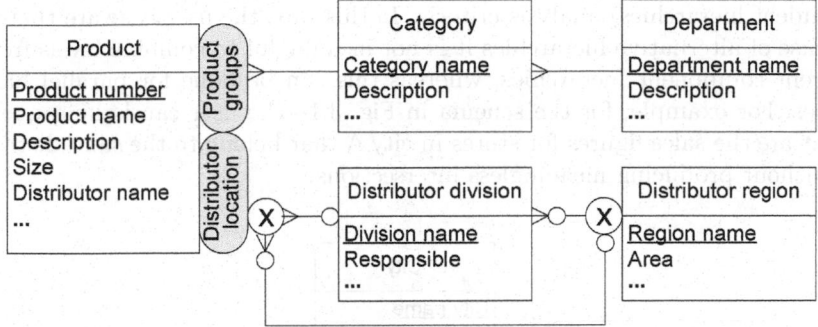

Fig. 3.17. Parallel independent hierarchies, composed of a balanced and a
noncovering hierarchy

hierarchy Sales organization represents the organizational division of the company and includes the levels Store, Sales district, State, and Sales region. Since
the two hierarchies share the State level, this level plays different roles according to hierarchy chosen for the analysis. Sharing levels in a conceptual schema
allows the designer to reduce the number of its elements, i.e., to improve its
readability, without loosing its semantics. Note that in order to unambigu-
ously define the levels composing the various hierarchies, the analysis criteria
must be included in the sharing level for hierarchies that continue beyond that
level.

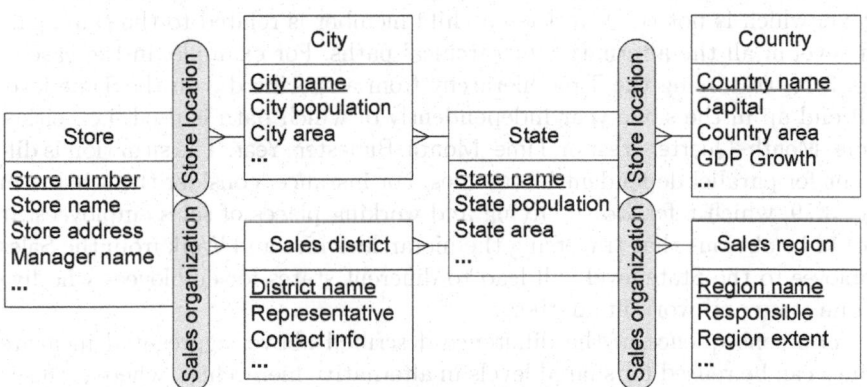

Fig. 3.18. Parallel dependent hierarchies, composed of two balanced hierarchies

Note that even though both multiple and parallel hierarchies share some
levels and may include several simple hierarchies, they represent different situations and should be clearly distinguishable at the conceptual level. This is
done by including only one (for alternative hierarchies) or several (for parallel

dependent hierarchies) analysis criteria. In this way, the user is aware that in the case of alternative hierarchies it is not meaningful to combine levels from different component hierarchies, whereas this can be done for parallel hierarchies. For example, for the schema in Fig. 3.18 the user can issue a query "what are the sales figures for stores in city A that belong to the sales district X" without producing meaningless intersections.

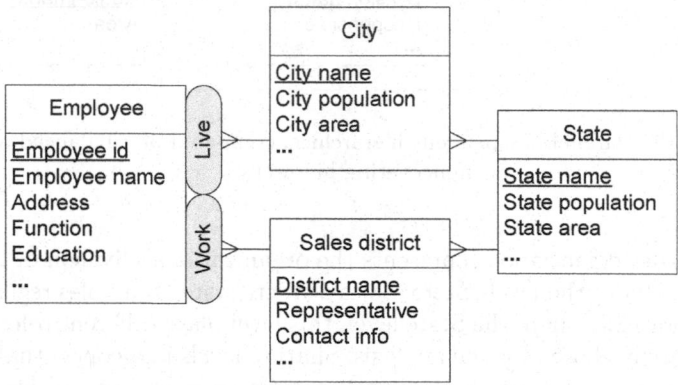

Fig. 3.19. Parallel dependent hierarchies leading to different parent members of the shared level

Further, if the hierarchies composing a set of alternative hierarchies share a level which is not the leaf, then a child member is related to the same parent level in all the alternative hierarchical paths. For example, in the case of Fig. 3.16, traversing the Time hierarchy from a specific day in the Time level will end up in the same year independently of which path is used, i.e., either Time–Month–Quarter–Year or Time–Month–Bimester–Year. The situation is different for parallel dependent hierarchies. For instance, consider the schema in Fig. 3.19, which refers to the living and working places of sales employees. It should be obvious that traversing the hierarchies Live and Work from the Sales employee to the State level will lead to different states for employees who live in one state and work in another.

As a consequence of the difference described above, aggregated measure values can be reused for shared levels in alternative hierarchies, whereas this is not the case for parallel dependent hierarchies. For example, suppose that the amount of sales generated by employees E1, E2, and E3 are €50, €100, and €150, respectively. If all employees live in state A, but only E1 and E2 work in this state, aggregating the sales of all employees to the State level following the Live hierarchy gives a total amount of €300, whereas the corresponding value will be equal to €150 when the Work hierarchy is traversed. Note that both results are correct, since the two hierarchies represent different analysis criteria.

3.3 Advanced Modeling Aspects

In this section, we discuss some particular features of the MultiDim model in order to show its applicability for representing complex multidimensional schemas. We begin by describing some complex hierarchies that can be represented by the model. We then discuss role-playing dimensions, fact dimensions, and multivalued dimensions.

3.3.1 Modeling of Complex Hierarchies

The example in Fig. 3.20 includes all the kinds of hierarchies described in the previous sections. This example shows three fact relationships, indicating that the focus of analysis is related to sales, the payroll, and sales incentive programs. In the figure only the names of the levels are shown; attribute names are not included.

As shown in the figure, the MultiDim model allows us to represent shared levels and shared dimensions. Sharing levels allows existing data to be reused, while sharing dimensions opens the possibility of analyzing measures obtained from different fact relationships using the drill-across operation (see Sect. 2.6.3) [146]. For example, the level Month is shared between two hierarchies (Sales time and Payroll time), while the level State is shared between three hierarchies (Sales organization, Store location, and Customer location). The Store dimension, with two hierarchies Sales organization and Store location, is shared between the Sales and the Payroll fact relationships. Thus, the measure Amount (from the Sales fact relationship) can be analyzed together with Base salary (from the Payroll fact relationship) for different stores and for different periods of time.

Further, the MultiDim model allows one to define several parent-child relationships between the same levels, for example the relationships between Employee and Section for the Works and Affiliated hierarchies. Each relationship corresponds to a different analysis criterion and has an associated set of instances.

Another feature of the MultiDim model is the possibility to represent complex **generalized hierarchies**, which, to our knowledge, are not considered in other models with graphical notations (for example, the model in [109]). The example shown in Fig. 3.21a (inspired by [105]) models the situation where stores belong to several different provinces, but only some provinces are grouped into states, as can be seen at the instance level in Fig. 3.21b. The cardinality (0,1) in Fig. 3.21a indicates that not all members of the Province level will have assigned to them a parent in the State level. This is shown in Fig. 3.21b, where the member all is used to represent the maximum aggregation level that exists for this hierarchy. Figure 3.22 (inspired by [104]) shows another example of a complex generalized hierarchy. This indicates three different aggregation paths that exist for products: Product–Brand, Product–Brand–Category–Department, and Product–Category–Department.

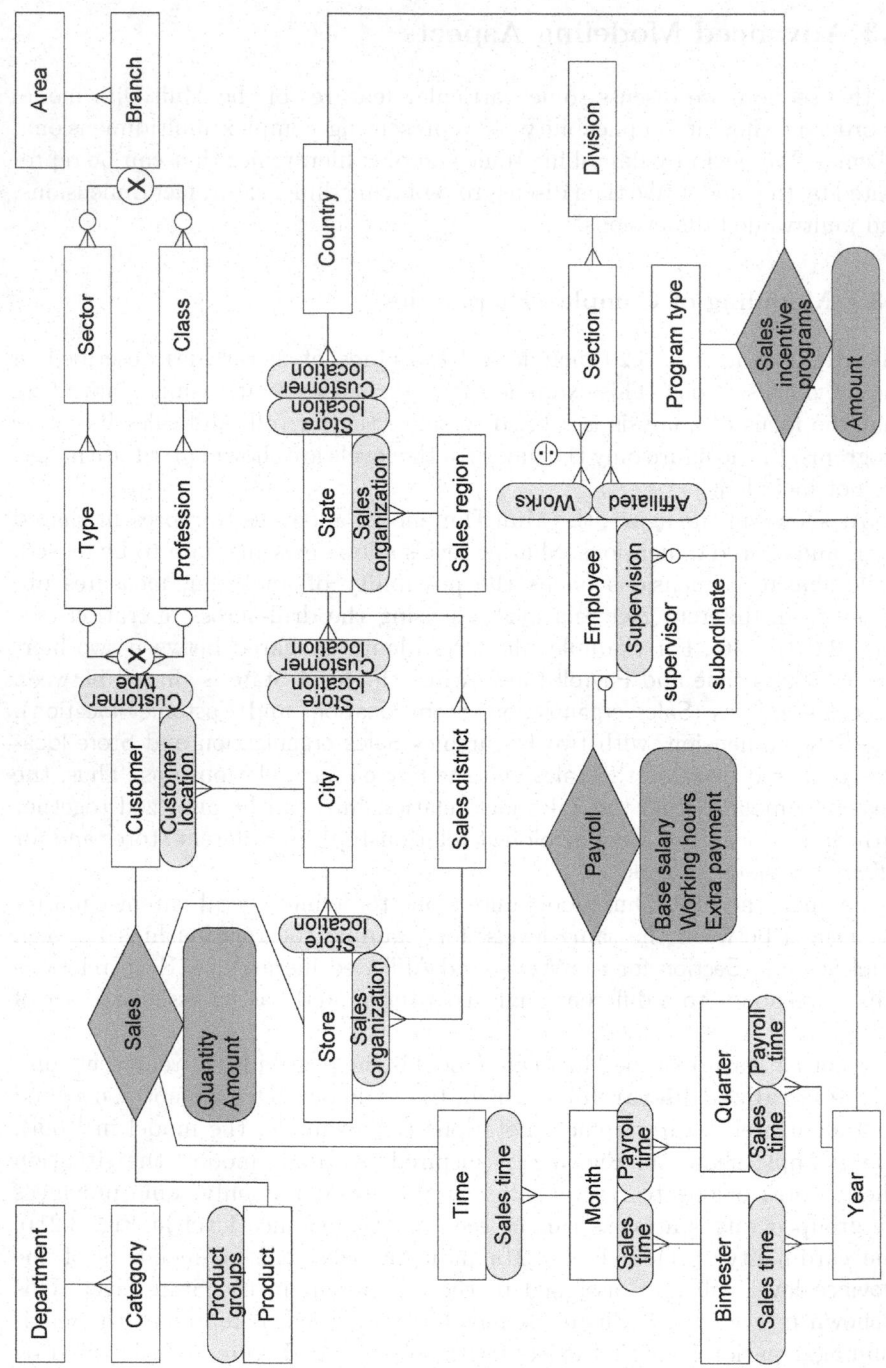

Fig. 3.20. A multidimensional schema containing several kinds of hierarchies

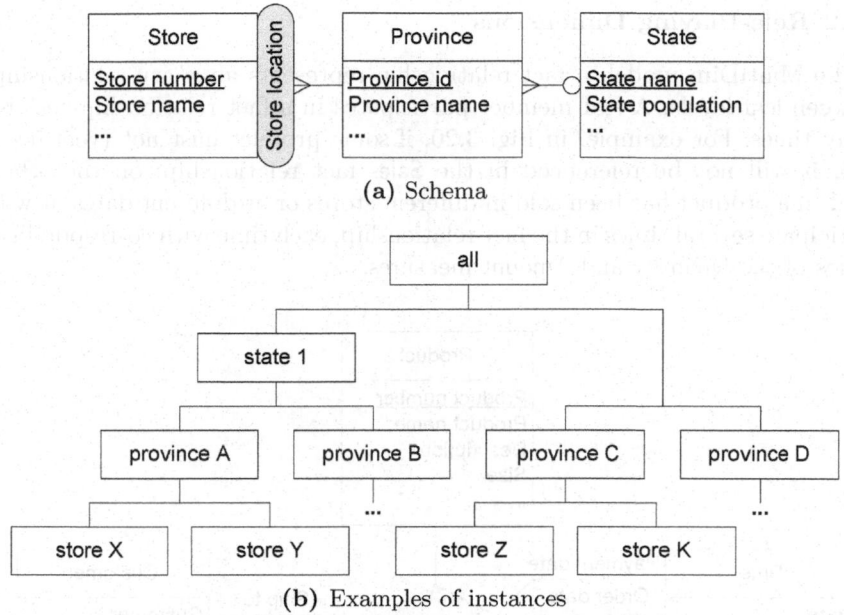

(a) Schema

(b) Examples of instances

Fig. 3.21. A generalized hierarchy with different root levels

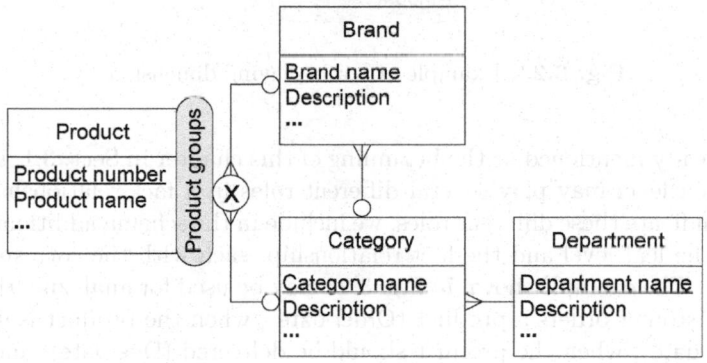

Fig. 3.22. A generalized hierarchy with three aggregation paths and different root levels

3.3.2 Role-Playing Dimensions

In the MultiDim model, a fact relationship represents an n-ary relationship between leaf levels. A leaf member participates in a fact relationship zero to many times. For example, in Fig. 3.20, if some product hast not (yet) been sold, it will not be referenced in the Sales fact relationship; on the other hand, if a product has been sold in different stores or at different dates, it will participate several times in the fact relationship, each time with corresponding values of the Quantity and Amount measures.

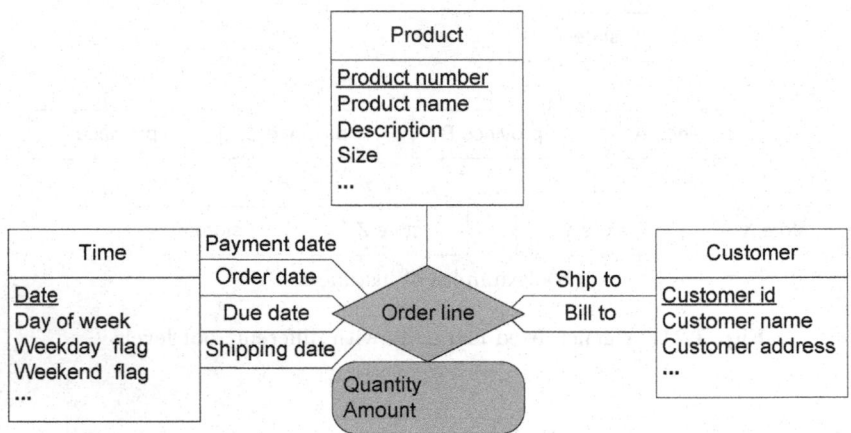

Fig. 3.23. Example of a role-playing dimension

As already mentioned at the beginning of this chapter in Sect. 3.1, in some situations a level may play several different roles in a fact relationship [100, 169]. To indicate these different roles, we include in the schema additional links between the leaf level and the fact relationship, each with the corresponding role name. The example shown in Fig. 3.23 may be used for analyzing the time when a customer orders a product (Order date), when the product is shipped (Shipping date), when the product should be delivered (Due date), and when it is actually paid for (Payment date). Furthermore, since a product can be shipped to one customer and be paid for by another, the Customer dimension also plays two different roles, i.e., Ship to and Bill to.

The schema in Fig. 3.23 could include four distinct time dimensions indicating the order, shipping, due, and payment dates, and two distinct dimensions referring to the clients to whom the products are shipped and the clients who pay for them. However, as can be seen in the figure, the complexity of the schema is reduced if the different roles that can be played by the dimensions are used instead of duplicating the dimensions.

Note that role-playing dimensions are supported by Microsoft Analysis Services 2005 [100, 156].

3.3.3 Fact Dimensions

We refer now to **fact dimensions** [156], or **degenerate dimensions** [146, 169]. A typical example is shown in Fig. 3.24, where users need to analyze product sales at the lowest granularity, which is represented by the transaction (or order) line. Note that fact relationship in Fig. 3.24 differs from the fact relationship Sales in Fig. 3.20, since in the latter the daily sales of products are summarized prior to their inclusion in the fact relationship.

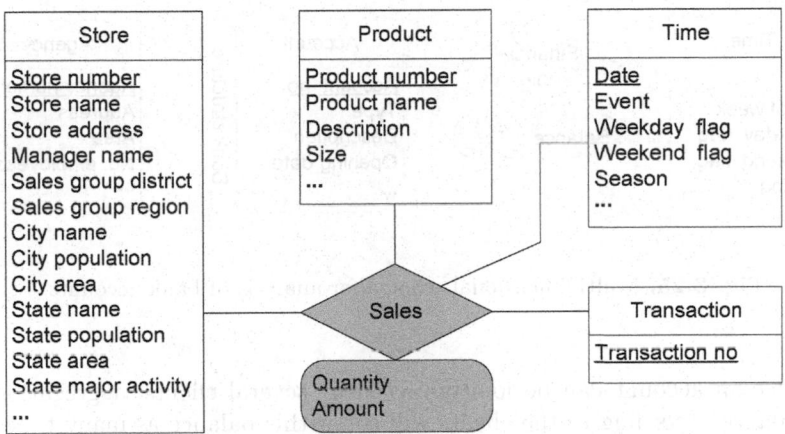

Fig. 3.24. A schema containing a fact dimension

As can be seen in Fig. 3.24, the Transaction dimension includes only the transaction number, since other data, such as the transaction date or the store number, are already included in the other dimensions. In logical representations of this schema, this dimension is typically converted into an attribute of the fact relationship. As a consequence, this dimension is called either a degenerate dimension, since it does not have other attributes describing it, or a fact dimension, since it is included as part of the fact relationship while playing the role of a dimension, i.e., it is used for grouping purposes.

We consider that transforming this schema by including the dimension attribute in the fact relationship must be done at the logical or implementation level. The conceptual schema should keep this dimension to indicate that users may require grouping according to dimension members, i.e., according to the transaction number.

3.3.4 Multivalued Dimensions

The terms "multivalued dimensions" [146] and "many-to-many relationships between facts and dimensions" [169, 231, 280] have recently been proposed for representing the situation where several members of a dimension participate in the same instance of a fact relationship. A common example used to represent this situation is an analysis of clients' balances in bank accounts, as shown in Fig. 3.25.

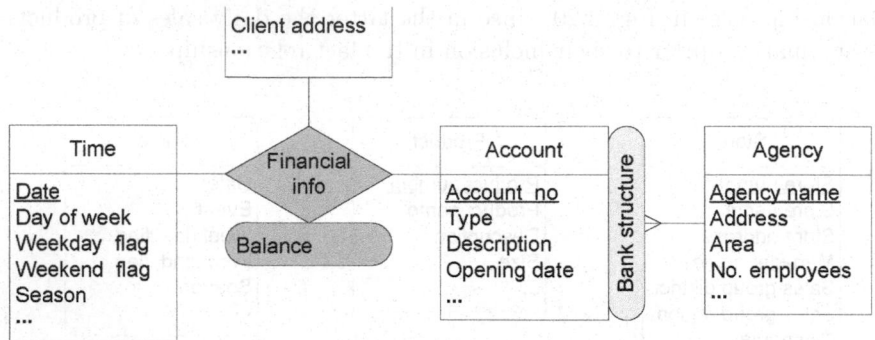

Fig. 3.25. Multidimensional schema for analysis of bank accounts

Since an account can be jointly owned by several clients, aggregation of the balance according to the clients will count this balance as many times as the number of account holders. Let us consider the example in Fig. 3.26. At some point in time T1, we have two accounts A1 and A2 with balances of, respectively, 100 and 500. As shown in the figure, both accounts are shared between several clients: account A1 between C1, C2, and C3, and account A2 between C1 and C2. The total balance of the two accounts is equal to 600; however, aggregation (for example, according to the Time or the Client dimension) gives a value equal to 1300.

Time	Account	Client	Balance
T1	A1	C1	100
T1	A1	C2	100
T1	A1	C3	100
T1	A2	C1	500
T1	A2	C2	500

Fig. 3.26. Example of double-counting problem for a multivalued dimension

The problem of **double counting** can be analyzed by considering the **multidimensional normal forms** (MNFs) defined by Lehner et al. [162] and Lechtenbörger and Vossen [160]. MNFs focus on summarizability (see Sect. 2.6.2) and determine conditions that ensure correct measure aggregation in the presence of generalized hierarchies (see Sect. 3.2.1). MNFs are defined using a generalization of functional dependencies that takes null values into account. The first multidimensional normal form (1MNF) establishes four conditions. The first three conditions aim at ensuring that the functional dependencies implied by a schema reflect those in the application domain. The last condition requires that each measure is uniquely identified by the set of leaf levels. The 1MNF is the basis for correct schema design and is used to define the remaining MNFs. The second and third multidimensional normal forms focus on establishing conditions for generalized hierarchies. We omit their descriptions here since our model is more general than that of Lechtenbörger and Vossen [160], i.e., the MultiDim model includes nonstrict (Sect. 3.2.2) and parallel dependent hierarchies (Sect. 3.2.4) and also allows one to share levels among dimensions that participate in the same fact relationship (as in Fig. 3.20). In addition, the MNFs do not take multivalued dependencies into account, and these dependencies are important to ensure correctness of multidimensional schemas.

Let us analyze the schema in Fig. 3.25 in terms of the 1MNF: we need to establish the functional dependencies that exist between the leaf levels and the measures. Since the balance depends on the specific account and the time when it is considered, we have the situation that the account and the time determine the balance. Therefore, the schema in Fig. 3.25 does not satisfy the 1MNF, since the measure is not determined by all leaf levels, and thus the fact relationship must be decomposed.

If we consider the multivalued dependencies, there are two possible ways in which the Account holders fact relationship in Fig. 3.25 can be decomposed. In the first case, the same joint account may have different clients assigned to it during different periods of time, and thus the time and the account multidetermine the clients. This situation leads to the the solution shown in Fig. 3.27a, where the original fact relationship is decomposed into two fact relationships, i.e., Account holders and Financial info. On the other hand, the second case arises when the client composition of joint accounts does not change over time, and thus only the account multidetermines the clients. In this case, the link relating in Fig. 3.27a the Time level and the Account holders fact relationship should be eliminated. Alternatively, this situation can be modeled with a non-strict hierarchy as shown in Fig. 3.27b.

Even though the solutions proposed in Fig. 3.27 eliminate the double-counting problem, the two schemas in Fig. 3.27 require programming effort for queries that ask for information about individual clients. The difference lies in the fact that in Fig. 3.27a a drill-across operation (see Sect. 2.6.3) between the two fact relationships is needed, while in Fig. 3.27b special procedures for aggregation in nonstrict hierarchies must be applied.In the case of Fig. 3.27a,

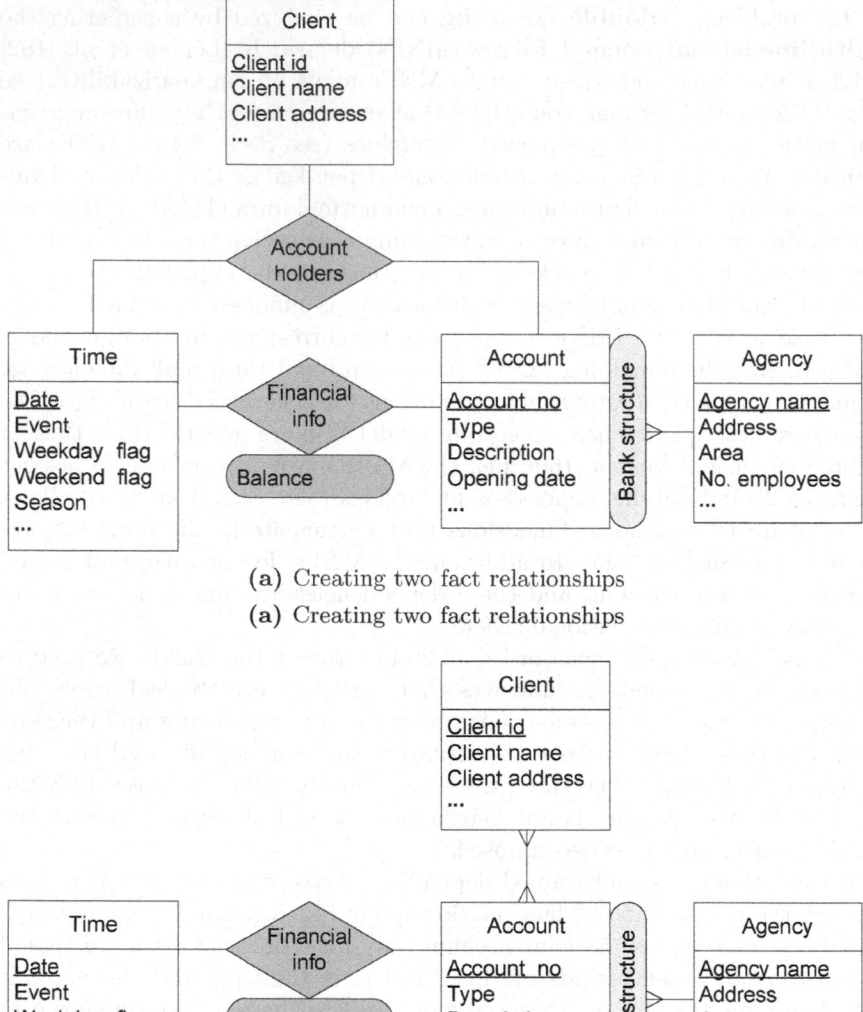

(a) Creating two fact relationships

(a) Creating two fact relationships

(b) Including a nonstrict hierarchy

(b) Including a nonstrict hierarchy

Fig. 3.27. Decomposition of the fact relationship in Fig. 3.25

since the two fact relationships represent different granularities, queries with drill-across operations are complex, demanding a conversion either from a finer to a coarser granularity (for example, grouping clients to know who holds a specific balance in an account) or vice versa (for example, distributing a balance between different account holders). Note also that the two schemas in Fig. 3.27 could represent the information about the percentage of ownership of accounts by customers (if this is known). This could be represented by a measure in the Account holders fact relationship in Fig. 3.27a, and by a distributing factor in the many-to-many relationship in Fig. 3.27b.

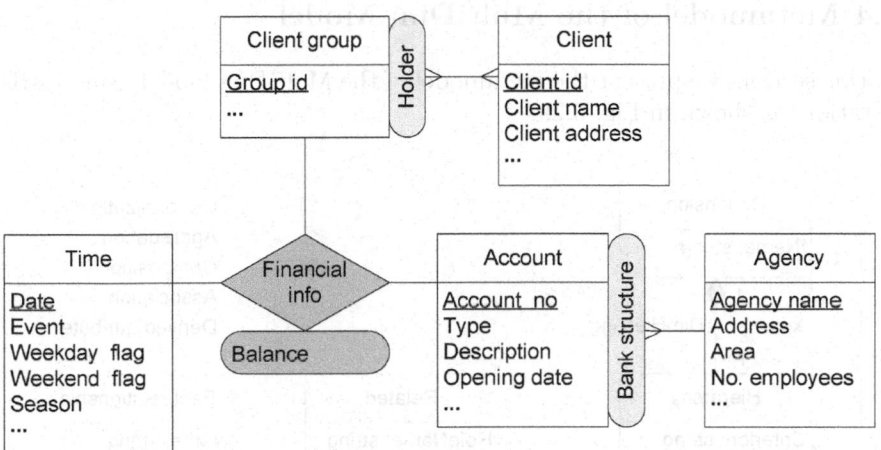

Fig. 3.28. Alternative representation of the schema shown in Fig. 3.25

Another solution to this problem, implicitly presented by Pedersen [230] and Song et al. [280], is shown in Fig. 3.28. In this solution, an additional level is created, which represents the groups of clients participating in joint accounts. For our example in Fig. 3.26, we create two groups: one that includes clients C1, C2, and C3, and another with clients C1 and C2. The difference between Fig. 3.28 and Pedersen's solution consists in the fact that the latter includes the groups of clients together with individual clients, i.e., the same level will contain members with different data granularities. However, the schema in Fig. 3.28 is not in the 1MNF, since the measure Balance is not determined by all leaf levels, i.e., it is only determined by Time and Account. Therefore, the schema must be decomposed leading to schemas similar to those in Fig. 3.27, with the difference that in this case the Client level in the two schemas in Fig. 3.27 is replaced by a non-strict hierarchy composed of the Client group and the Client levels.

Another solution for avoiding multivalued dimensions is to choose one client as the primary account owner and ignore the other clients. In this way, only one client will be related to a specific balance, and the schema in Fig. 3.25 can be used without any problems related to double counting of measures. However, this solution may not represent the real-world situation, and may exclude from the analysis the other clients of joint accounts.

In summary, we can avoid multivalued dimensions in multidimensional schemas by using one of the solutions presented in Fig. 3.27. The choice between these alternatives depends on the functional and multivalued dependencies existing in the fact relationship, the kinds of hierarchies that are allowed in the schema, and the complexity of the implementation.

3.4 Metamodel of the MultiDim Model

In this section, we present the metamodel of the MultiDim model using UML notation, as shown in Fig. 3.29.

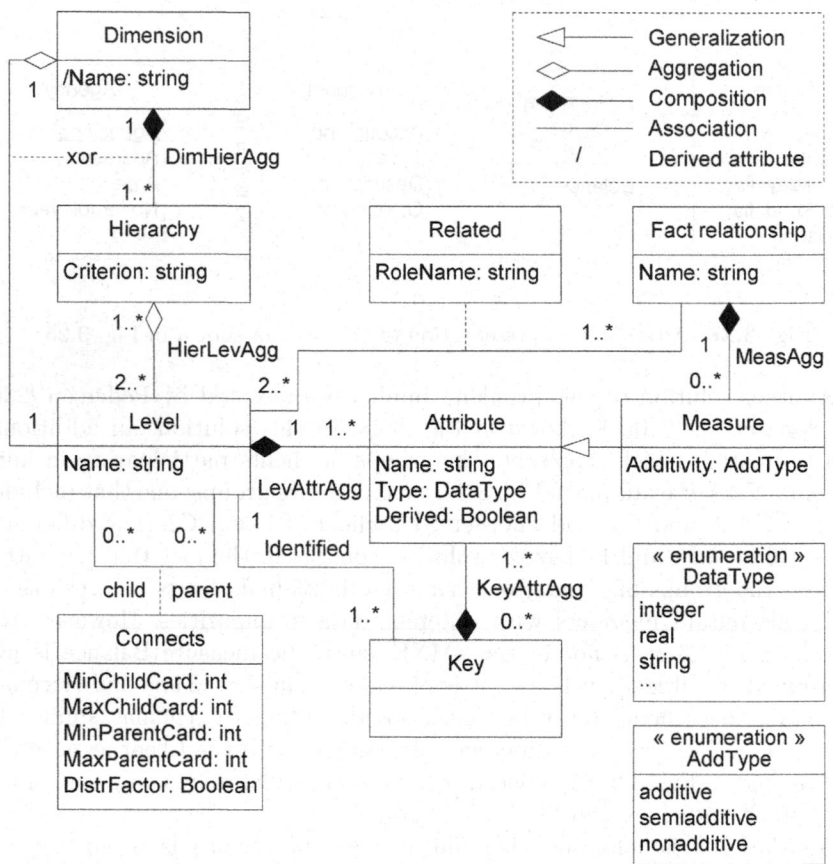

Fig. 3.29. Metamodel of the MultiDim model

A dimension is composed of either one level or one or more hierarchies, and each hierarchy belongs to only one dimension. A hierarchy contains two or more related levels that can be shared between different hierarchies. A criterion name identifies a hierarchy and each level has a unique name. The leaf-level name is used to derive a dimension's name: this is represented by the derived attribute Name in Dimension. Levels include attributes, some of which are key attributes used for aggregation purposes, while others are descriptive attributes. Attributes have a name and a type, which must be a data type, i.e., integer, real, string, etc. Further, an attribute may be derived.

The levels forming hierarchies are related through the Connects association class. This association class is characterized by the minimum and maximum cardinalities of the parent and child roles, and the distributing factor, if any. A constraint not shown in the diagram specifies that a parent-child relationship has a distributing factor only if the maximum cardinalities of the child and the parent are equal to "many".

A fact relationship represents an n-ary association between leaf levels with $n > 1$. Since leaf levels can play different roles in this association, the role name is included in the Related association class. A fact relationship may contain attributes, which are commonly called measures; they may be additive, semiadditive, or nonadditive.

3.5 Mapping to the Relational and Object-Relational Models

3.5.1 Rationale

Conceptual models can be implemented using various logical models. In this section, we present the mapping rules that allow one to implement conceptual schemas designed using the MultiDim model into the relational and object-relational models.

As already explained in Sect. 2.7, data warehouses are usually implemented in relational databases as star or snowflake schemas. There are several reasons for using a relational implementation for multidimensional models.

- Relational databases use well-known strategies for data storage, and are well standardized, tool independent, and readily available. In contrast, there is no consensus in the research community about defining a logical level for OLAP systems [8]. Furthermore, multidimensional OLAP (MOLAP) systems differ greatly in the structures used for data storage.
- Much research has been done in the field of relational databases on improving query optimization, indexing, join operations, and view materialization, considering the particularities of data warehouse applications.
- Even though MOLAP systems offer better performance during roll-up operations, many commercial and prototype systems use ROLAP (relational OLAP) or HOLAP (Hybrid OLAP, a combination of ROLAP and MOLAP) storage for large volumes of data, as described in Sect. 2.7.
- Commercial relational database systems [13, 100, 214] include extensions to represent and manage a multidimensional view of data.

- Relational databases are an accepted resource layer in the **Common
  Warehouse Model** (CWM) [207] from the Object Management Group
  (OMG).[9]

The object-relational model is increasingly being used in conventional
database applications. One reason for this is that it preserves the foundations
of the relational model while organizing data using an object model [154] (see
Sect. 2.3.2). In addition, the object-relational model is supported by many
commercial DBMSs. Since data warehouses are specific databases devoted to
analytical purposes, they can also be implemented using the object-relational
model.

We chose as implementation models, the surrogate-based relational model
[89] and the the object-relational model. In these models, each row in a table
has an associated surrogate, i.e., a system-generated artificial key. Surrogate
values cannot be seen or modified by users. The reason for using surrogates
instead of relying on user-defined keys is that data warehouses usually use
system-generated keys to ensure both better performance during join operations and independence from transactional systems. Further, surrogates do
not vary over time, so that two entities that have identical surrogates represent
the same entity, thus allowing one to include historical data in an unambiguous way. In addition, in some situations it is necessary to store information
about an entity either before it has been assigned a user-controlled key value
(for example, to introduce a student into a database before an ID number has
been assigned to him/her) or after it has ceased to have one (for example,
when an employee leaves an organization). Further, the SQL:2003 standard
[192, 193] and some current DBMSs (for example, Oracle [213]) support tables with rows that have automatically created surrogates. Nevertheless, the
relational or object-relational models without surrogates can also be used in
our mapping for those DBMSs that do not allow surrogates. In this case, implementers are responsible for creating artificial keys independent from those
used in the source systems.

3.5.2 Mapping Rules

Since the MultiDim model is based on the ER model, its mapping to the
relational model is based on well-known rules, such as those described in
Sect. 2.3.1:

Rule 1: A level corresponds to an entity type in the ER model. It is mapped
to a table that contains all attributes of the level and includes an additional attribute for a surrogate key.

Rule 2: A parent-child relationship corresponds to a binary relationship type
in the ER model. Recall that in order to have meaningful hierarchies, the
maximum cardinality of the parent role must be n. Thus, two different
mappings exist, depending on the cardinality of the child role:

---

[9] See the description of the CWM in Sect. 3.8.

Rule 2a: If the cardinality of the child role is (0,1) or (1,1), i.e., the parent-child relationship is one-to-many, the table corresponding to the child level is extended with the surrogate key of the corresponding parent level, i.e., there is a foreign key in the child table pointing to its parent table.

Rule 2b: If the cardinality of the child role is (0,n) or (1,n), i.e., the parent-child relationship is many-to-many, a new table is created that contains as attributes the surrogate keys of the parent and child levels. If the parent-child relationship has a distributing factor, an additional attribute is added to the table to store this information.

Rule 3: A fact relationship type corresponds to an n-ary relationship type in the ER model. A fact relationship involving leaf levels $L_1, \ldots, L_n$ is mapped to a table that includes as attributes the surrogate keys of the participating levels, one for each role that each level is playing. In addition, every attribute (i.e., measure) of the fact relationship is mapped to an attribute of the table.

Applying the above rules to the example given at the beginning of this chapter (Fig. 3.2) gives the tables shown in Fig. 3.30. The Sales table includes five foreign keys, i.e., one for each participating level, and two attributes representing the measures, i.e., Quantity and Amount. Note that when a level plays several roles, as is the case for Time, each role is represented by a foreign key, i.e., Order date and Payment date are both foreign keys to the Time dimension. Note also that the many-to-many parent-child relationship between Product and Distributor is mapped to the table ProdDist, containing two foreign keys.

A set of SQL DDL commands defining an excerpt from the relational schema of Fig. 3.30 is as follows.

```
create table Sales as (
      ProductFkey integer not null,
      OrderTimeFkey integer not null,
      PaymentTimeFkey integer not null,
      StoreFkey integer not null,
      CustomerFkey integer not null,
      Quantity integer not null,
      Amount decimal(7,2) not null,
      primary key (ProductFkey, OrderTimeFkey, PaymentTimeFkey,
          StoreFkey, CustomerFkey),
      foreign key ProductFkey references Product(ProductKey),
      foreign key OrderTimeFkey references Time(TimeKey),
      foreign key PaymentTimeFkey references Time(TimeKey),
      foreign key StoreFkey references Store(StoreKey),
      foreign key CustomerFkey references Customer(CustomerKey) );
create table Product as (
      ProductKey integer primary key not null,
      ProductNumber character varying(9) not null,
      ProductName character varying(30) not null,
```

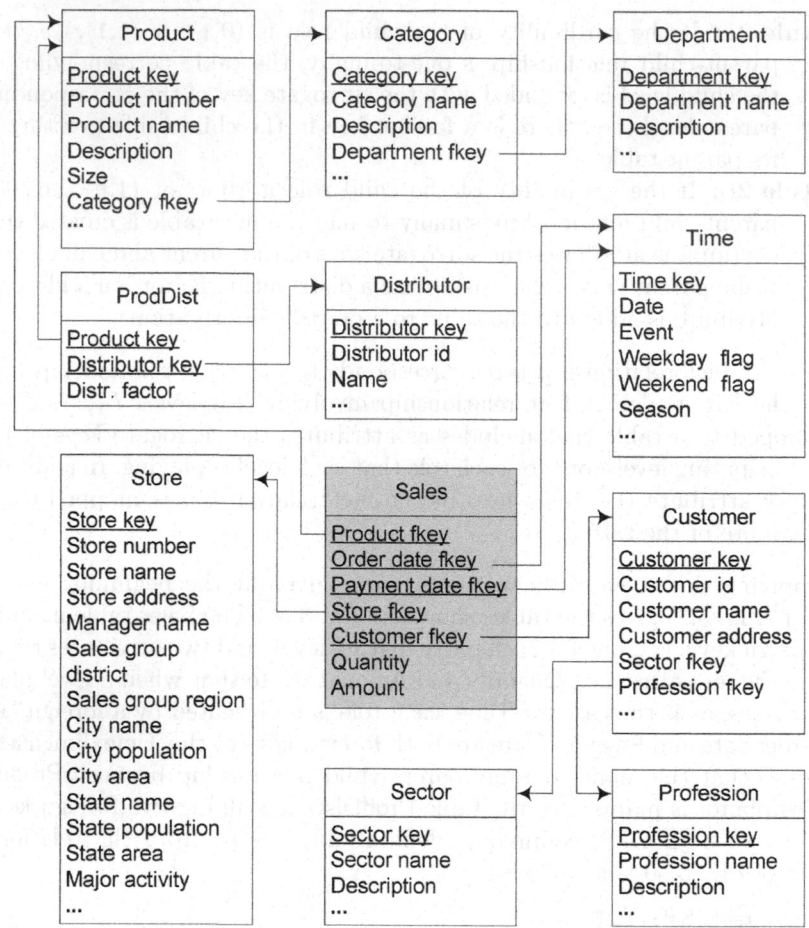

Fig. 3.30. Relational representation of the example in Fig. 3.2

```
        Description text not null,
        Size integer not null,
        CategoryFkey integer not null,
        ...
        foreign key CategoryFkey references Product(ProductKey) );
create table Category as (
        CategoryKey integer primary key not null,
        CategoryName character varying(20) not null,
        Description text not null,
        DepartmentFkey integer not null,
        ...
        foreign key DepartmentFkey references Department(DepartmentKey) );
create table Department as (
```

```
DepartmentKey integer primary key,
DepartmentName character varying (30) not null,
Description text not null,
... );
```

In the small excerpt given above, we have defined only the Sales fact table and the tables for the hierarchy composed of the levels Product–Category–Department. The definitions for the other tables are similar.

The transformation of the MultiDim model to the object-relational model is based on rules similar to those specified above for the mapping to the relational model.[10] The difference is that, as explained in Sect. 2.3.2, the object-relational model provides a ref type, which can be used instead of foreign keys in the above rules.

The set of SQL:2003 commands defining an object-relational schema corresponding to the relational schema of Fig. 3.30 is as follows.

```
create type SalesType as (
    ProductRef ref(ProductType) scope Product references are checked,
    OrderTimeRef ref(TimeType) scope Time references are checked,
    PaymentTimeRef ref(TimeType) scope Time references are checked,
    StoreRef ref(StoreType) scope Store references are checked,
    CustomerRef ref(CustomerType) scope Customer references are checked,
    Quantity integer,
    Amount decimal(7,2) )
    ref is system generated;
create table Sales of SalesType (
    constraint SalesPK primary key (ProductRef, OrderTimeRef,
        PaymentTimeRef, StoreRef, CustomerRef),
    ProductRef with options constraint ProductRefNN ProductRef not null,
    /* not null constraints for all other attributes */
    ref is oid system generated );
create type ProductType as (
    ProductKey integer,
    ProductNumber character varying(9),
    ProductName character varying(30),
    Description text,
    Size integer,
    CategoryRef ref(CategoryType) scope Category references are checked,
    ... )
    ref is system generated;
create table Product of ProductType (
    constraint ProductPK primary key (ProductKey),
    ref is oid system generated );
create type CategoryType as (
    CategoryKey integer,
    CategoryName character varying(20),
```

---

[10] As explained in Sect. 3.6.2, there is one exception to this, since the relational and the object-relational representation of nonstrict hierarchies may differ.

```
          Description text,
          DepartmentRef ref(DepartmentType) scope Department
              references are checked,
          ... )
          ref is system generated;
    create table Category of CategoryType (
          constraint CategoryPK primary key (CategoryKey),
          ref is oid system generated );
    create type DepartmentType as (
          DepartmentKey integer,
          DepartmentName character varying (30),
          Description text,
          ... )
          ref is system generated;
    create table Department of DepartmentType (
          constraint DepartmentPK primary key (DepartmentKey),
          ref is oid system generated );
```

Note that we have implemented the above schema using typed tables. As already said above, typed tables allow the system-generated identifiers to be used for linking rows between tables. Typed tables also allow methods to be defined for the types associated with the tables. However, since we do not refer to any aspects of manipulation in the following, we have not included the specification of methods for object-relational schemas. Note also that constraints associated with the types (for instance, primary key constraints) are specified in the corresponding table declarations. For brevity, in the above SQL commands we have not specified many constraints, in particular not null constraints for mandatory attributes. An example of such a constraint is given for the attribute ProductRef in the Sales table.

3.6 Logical Representation of Hierarchies

The general mapping rules given in the previous section do not capture the specific semantics of all of the kinds of hierarchies described in Sect. 3.2. In addition, for some kinds of hierarchies, alternative logical representations exist. In this section, we consider in detail the logical representation of the various kinds of hierarchies. In what follows we discuss the relational representation, and we mention the object-relational representation when it differs from the relational one.

3.6.1 Simple Hierarchies

Balanced Hierarchies

Applying the mapping rules described in Sect. 3.5.2 to balanced hierarchies leads to the well-known relational representations of data warehouses:

- **Normalized tables** or **snowflake structure:** each level is represented as a separate table that includes a key and the descriptive attributes of the level. For example, using Rules 1 and 2a of Sect. 3.5.2 for the Product groups hierarchy in Fig. 3.4 gives a snowflake structure with tables Product, Category, and Department, as shown in Fig. 3.31a.
- **Denormalized** or **flat tables:** the key and descriptive attributes of all levels forming a hierarchy are included in one table. This structure can be obtained in two ways: (1) denormalizing the tables that represent several hierarchy levels (for example, including in one table all attributes of the Product, Category, and Department tables shown in Fig. 3.31a), or (2) mapping a dimension that includes a one-level hierarchy according to Rule 1 (for example, the Store dimension in Fig. 3.2 may be represented as shown in Fig. 3.31b).

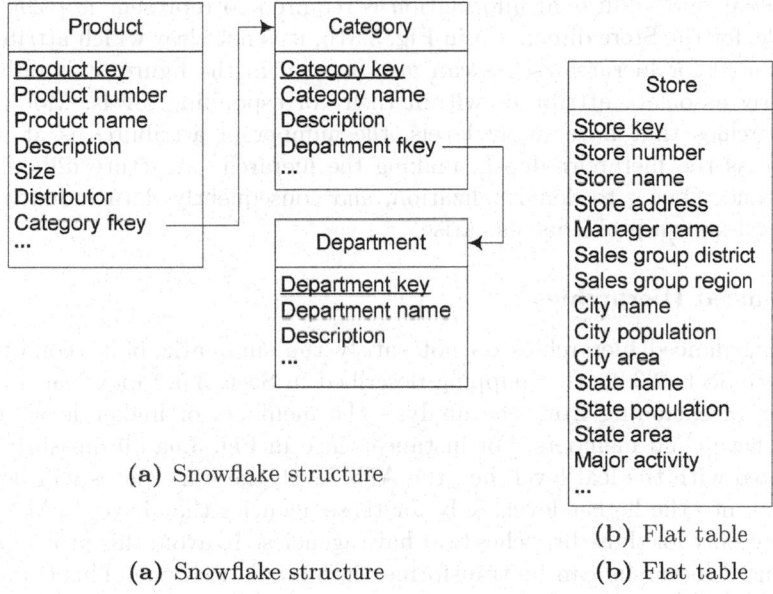

(a) Snowflake structure

(b) Flat table

(a) Snowflake structure   (b) Flat table

Fig. 3.31. Relations for a balanced hierarchy

Normalized tables are used in **snowflake schemas** (see Sect. 2.7). They represent hierarchical structures better, since every level can be easily distinguished and, further, levels can be reused between different hierarchies. Additionally, this representation can easily manage heterogeneity across levels [129], i.e., it allows different levels of a hierarchy to include specific attributes. For example, the Product, Category, and Department tables in Fig. 3.31a have specific attributes. This data structure allows measures to be aggregated using, for example, the SQL group by, roll-up, or cube operators (see Sect. 2.6.3). Further, in some applications, snowflake schemas can improve system performance in spite of requiring join operations between the relations representing the levels [188]. However, snowflake schemas have some limitations; for example, they cannot be used without modification if there is heterogeneity within a level [129], i.e., different characteristics within a group of members of a level. This situation may occur when different products have their own specific attributes, as in the case of generalized hierarchies (see Sect. 3.2.1).

On the other hand, the inclusion of flat tables in the schema, leading to **star schemas** (see Sect. 2.7), has several advantages: the schema is easy to understand for the purpose of query formulation, it can be easily accessed, and few joins are needed for expressing queries, owing to the high level of denormalization. Additionally, much research has been done to improve system performance during processing of star queries (e.g., [82, 141]). Finally, as for the normalized approach, the aggregations of measures are straightforward.

Some disadvantages of the star schema representation are as follows: it does not allow hierarchies to be modeled adequately, since the hierarchy structure is not clear and additional information is required to represent it [129]. For example, for the Store dimension in Fig. 3.31b, it is not clear which attributes can be used for hierarchies. As can also be seen in the figure, it is difficult to clearly associate attributes within their corresponding levels. Moreover, in hierarchies that have many levels, the number of attributes is at least as large as the hierarchy depth, making the hierarchy structure difficult to understand. Owing to denormalization, and consequently data redundancy, other well-known problems also arise.

Unbalanced Hierarchies

Since unbalanced hierarchies do not satisfy the summarizability conditions [163] (see Sect. 2.6.2), the mapping described in Sect. 3.5.2 may lead to the problem of excluding from the analysis the members of higher levels that do not have child members. For instance, since in Fig. 3.5a all measures are associated with the leaf level, i.e., the ATM level, these measures will be aggregated into the higher levels only for those agencies that have ATMs and, similarly, only for those branches that have agencies. To avoid this problem, an unbalanced hierarchy can be transformed into a balanced one. This transformation is performed using placeholders (marked PH in Fig. 3.32) [231] or null values [156] in missing levels. Afterwards, the logical mapping for balanced hierarchies may be applied.

Nevertheless, this representation has the following shortcomings:

1. A fact table must include common measures belonging to different hierarchy levels (in our example, ATMs, agencies, and branches).
2. The common measures used for all levels have different data granularities. For example, measures for the ATM level and for the Agency level for those members that do not have an ATM are included in the same fact table.
3. Special placeholders must be created and managed for aggregation purposes. For example, in Fig. 3.32 the same measure value must be repeated for branch 2, while using two placeholders for the missing levels.

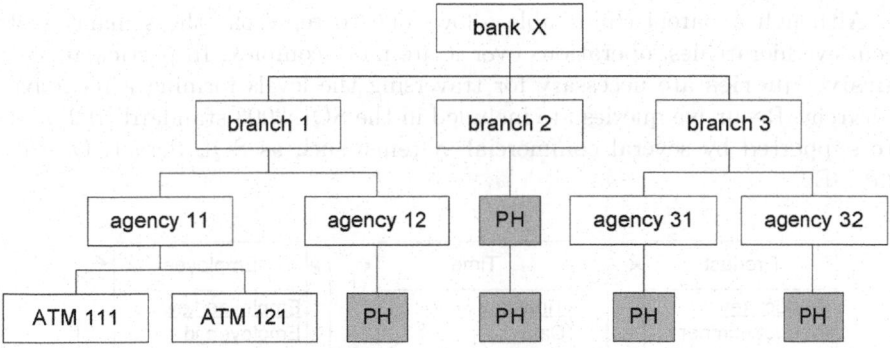

Fig. 3.32. Transformation of the unbalanced hierarchy shown in Fig. 3.5b into a balanced one using placeholders

4. The unnecessary introduction of meaningless values requires more storage space.
5. A special interface needs to be implemented to hide placeholders from users.

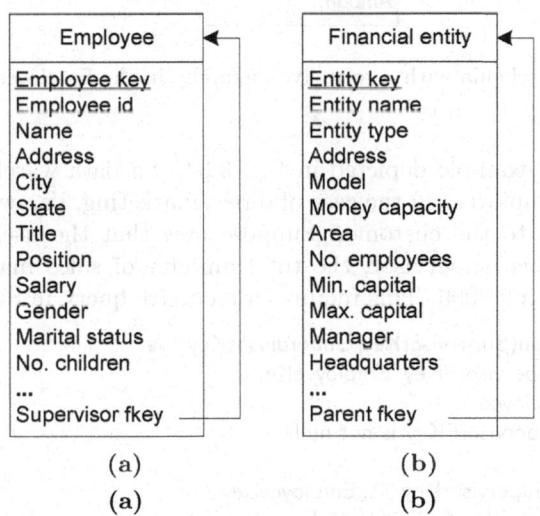

Fig. 3.33. Relational implementation of the recursive hierarchies shown in Fig. 3.6

Recall from Sect. 3.2.1 that **recursive hierarchies** are a special case of unbalanced hierarchies. Mapping recursive hierarchies to the relational model yields **parent-child tables** containing all attributes of a level, and an additional foreign key relating child members to their corresponding parent. Figure 3.33a shows the table representing the recursive hierarchy shown in Fig. 3.6a.

Although a parent-child table allows one to represent the semantics of recursive hierarchies, operations over it are more complex. In particular, **recursive queries** are necessary for traversing the levels forming a recursive hierarchy. Recursive queries are included in the SQL:2003 standard [192] and are supported by several commercial systems, such as SQL Server, Oracle, and DB2.

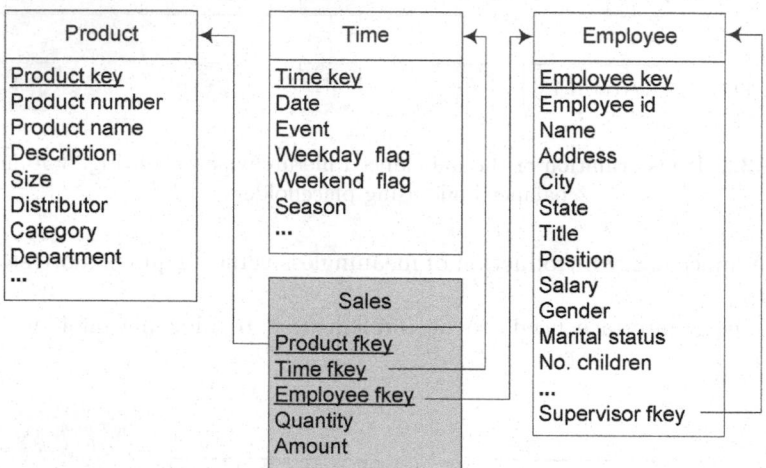

Fig. 3.34. A schema with a recursive hierarchy in the Employee dimension

Consider the example depicted in Fig. 3.34 of a data warehouse for analyzing sales by employees in the case of direct marketing, i.e., when the goods are sold directly to the customer. Suppose now that the user is interested in finding, for each supervisor, the total amount of sales made by his/her subordinates during 2006. This requires a recursive query in SQL as follows:

```
with Supervision(SupervisorKey, SubordinateKey) as
(     select SupervisorFKey, EmployeeKey
      from Employee
      where SupervisorFKey is not null
union all
      select S.SupervisorKey , E.EmployeeKey
      from Supervision S, Employee E
      where S.SubordinateKey = E.SupervisorFKey
)
select SupervisorKey, sum(S1.Amount)
from Sales S, Supervision U, Time T
where S.EmployeeFKey=U.SubordinateKey
and S1.TimeFkey=T.TimeKey
and T.Date >= '1/1/2006' and T.Date =< '31/12/2006'
group by SupervisorKey
```

In the above query, the clause with defines a temporary table Supervision that computes recursively all direct and indirect subordinates of an employee. For example, if E1 supervises E2, and E2 supervises E3, the Supervision table will contain in addition one line stating that E1 supervises E3. Then, the select statement joins the tables Supervision, Sales, and Time and groups the result by SupervisorKey in order to sum all sales of the subordinates.

Financial entity											
Entity key	Entity name	Entity type	Addr.	Model	Money capacity	Area	No. emp.	Min. capital	Max. capital	...	Parent fkey
1	ATM111	ATM	...	T1	100000	null	null	null	null	...	3
2	ATM112	ATM	...	T2	150000	null	null	null	null	...	3
3	agency11	Agency	...	null	null	150	12	null	null	...	5
4	agency12	Agency	...	null	null	135	9	null	null	...	5
5	branch1	Branch	...	null	null	null	null	10000	99000	...	10
6	branch2	Branch	...	null	null	null	null	1500	55000	...	10
7	agency31	Agency	...	null	null	230	15	null	null	...	9
8	agency32	Agency	...	null	null	185	12	null	null	...	9
9	branch3	Branch	...	null	null	null	null	7500	98500	...	10
10	bankX	Bank	...	null	null	null	null	null	null	...	null

Fig. 3.35. Parent-child relational schema of an unbalanced hierarchy with instances from Fig. 3.5b

One solution for implementing unbalanced hierarchies is to transform them into recursive hierarchies, which are then transformed at the logical level into **parent-child tables**. For example, such a transformation of the recursive hierarchy in Fig. 3.6b will give the table shown in Fig. 3.33b. However, the problem of such tables is that they represent semantically different levels, each one having particular attributes. As can be seen in Fig. 3.35, since a parent-child table contains the attributes of all hierarchy levels, a member of a level will contain null values for all attributes belonging to the other levels.

For this reason, **integrity constraints** must be implemented to ensure that a member has values for the attributes of its level. For example, the following check constraint states that members of the ATM level must have values for the attributes Address, Model, and Money capacity and null values for the other attributes:

```
alter table FinancialEntity
    add constraint ATMAttributes
    check ( (EntityType != 'ATM') or ( Address is not null and
    Model is not null and MoneyCapacity is not null and
    Area is null and NoEmployees is null and ... ) )
```

Note that in the above constraint the logical implication (EntityType = 'ATM') $\Rightarrow$ (...) is encoded as (EntityType != 'ATM') $\vee$ (...). Similar check constraints are needed for the other levels of the hierarchy, i.e., for the Agency, Bank, and Branch levels.

Other triggers are needed to ensure that the links between levels are correct. For example, the following trigger, implemented in Microsoft SQL Server, enforces the constraint that ATM members are related to Agency members:

```
create trigger ATM_Agency on FinancialEntity
after insert, update as
if exists ( select * from Inserted I, FinancialEntity F
    where I.ParentFkey = F.EntityKey and
    I.EntityType = 'ATM' and F.EntityType != 'Agency' )
begin
    raiserror 13000 'An ATM member must be related to an Agency member'
end
```

Generalized Hierarchies

Several approaches have been proposed for the logical representation of generalized hierarchies: (1) representing heterogeneity within levels by creating separate tables for each level of the hierarchy and keeping the information about which tables compose the different paths [129]; (2) using a flat representation with null values for attributes that do not describe specific members [162] (for example, companies will have null values in attributes corresponding to persons); (3) using separate star schemas, i.e., separate fact and dimension tables for each path [15, 94, 146]; and (4) creating one table for the common levels and another table for the specific ones [15, 272].

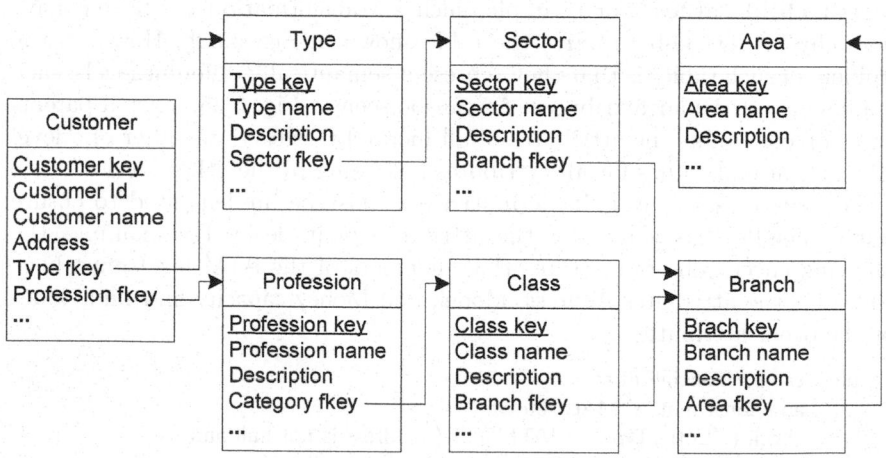

Fig. 3.36. Relations for the generalized hierarchy shown in Fig. 3.8

One disadvantage of the first three approaches above is that the common levels of the hierarchy cannot be easily distinguished and managed. Further, the inclusion of null values requires one to specify additional constraints to ensure correct queries (for example, to avoid grouping Type with Profession in Fig. 3.8). Even though these difficulties do not exist for solution (4) above, in that case an additional attribute must be created in the table representing the common levels of the hierarchy to specify the name of the table for the specific levels. In consequence, queries that manage metadata and data are required to access the tables of the specific levels, which is not an easy task to accomplish in SQL. The traditional mapping of generalization/specialization to relational tables (e.g., Rule 7 in Sect. 2.3.1) also gives problems owing to the inclusion of null values and the loss of the hierarchical structure.

Applying the mapping described in Sect. 3.5.2 to the generalized hierarchy given in Fig. 3.8 yields the relations shown in Fig. 3.36. Even though this mapping clearly represents the hierarchical structure, it does not allow one to traverse only the common levels of the hierarchy. To ensure heterogeneity within levels and the possibility of traversing either specific or common levels, we propose in addition the following mapping rule:

Rule 4: A table corresponding to a splitting level in a generalized hierarchy may have an additional attribute, which is a foreign key of the next joining level, provided this level exists. The table may also include a discriminating attribute that indicates the specific aggregation path of each member.

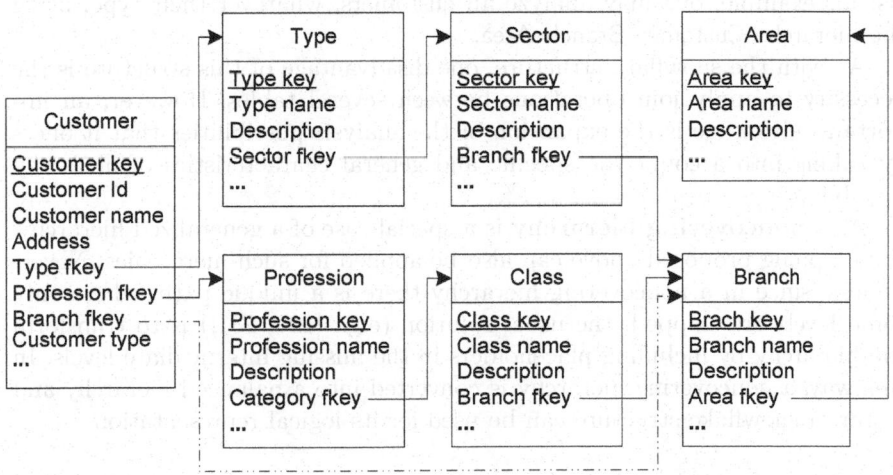

Fig. 3.37. Improved relational representation for the generalized hierarchy shown in Fig. 3.8

An example of the relations for the hierarchy in Fig. 3.8 is given in Fig. 3.37. The table Customer includes two kinds of foreign keys: one that indicates the next specialized hierarchy level (for example, Profession fkey), which is obtained by applying Rules 1 and 2a in Sect. 3.5.2, and another one corresponding to the next joining level (for example, Branch fkey), which is obtained by applying the above Rule 4. The discriminating attribute Customer type, with values Person and Company, indicates the specific aggregation path of members to facilitate aggregations. Finally, integrity constraints must be specified to ensure that only one of the foreign keys for the specialized levels may have a value, according to the value of the discriminating attribute:

```
alter table Customer
    add constraint CustomerTypeCK
    check ( CustomerType in ('Person', 'Company') )
alter table Customer
    add constraint CustomerPersonFK
    check ( (CustomerType != 'Person') or
    ( ProfessionFkey is not null and TypeFkey is null ) )
alter table Customer
    add constraint CustomerCompanyFK
    check ( (CustomerType != 'Company') or
    ( ProfessionFkey is null and TypeFkey is not null ) )
```

The structure in Fig. 3.37 allows one to choose alternative paths for analysis. One possibility is to use paths including the specific levels, for example Profession or Type. Another possibility is to access the levels that are common to all members, i.e., to ignore the levels between the splitting and joining levels; for example, one may analyze all customers, whatever their type, using the hierarchy Customer–Branch–Area.

As with the snowflake structure, one disadvantage of this structure is the necessity to apply join operations between several tables. However, an important advantage is the expansion of the analysis possibilities that it offers by taking into account the specific and general characteristics of hierarchy members.

As a **noncovering hierarchy** is a special case of a generalized hierarchy, the mapping proposed above can also be applied for such hierarchies. Nevertheless, since in a noncovering hierarchy there is a unique path, where only some levels are skipped, the usual solution (e.g., [100, 231]) is to transform the hierarchy by including placeholders in the missing intermediate levels. In this way, a noncovering hierarchy is converted into a balanced hierarchy and a star or snowflake structure can be used for its logical representation.

3.6.2 Nonstrict Hierarchies

The traditional mapping of nonstrict hierarchies to the relational model, as specified in Sect. 3.5.2, creates relations for representing the levels and an additional relation for representing the many-to-many relationship between them.

An example for the case of the hierarchy in Fig. 3.11a is given in Fig. 3.38. The table representing the many-to-many relationship (EmplSection in the figure) is usually called a **bridge table** [146].

Note that if the parent-child relationship has a distributing factor, the bridge table will include an additional attribute for storing the values required for measure distribution.[11] For example, for an employee X, the table EmplSection could include three rows: (1) employee X, section 1, 30%; (2) employee X, section 2, 50%; (3) employee X, section 3, 20%. In order to aggregate measures correctly, a special aggregation procedure that uses this distributing factor must be implemented.

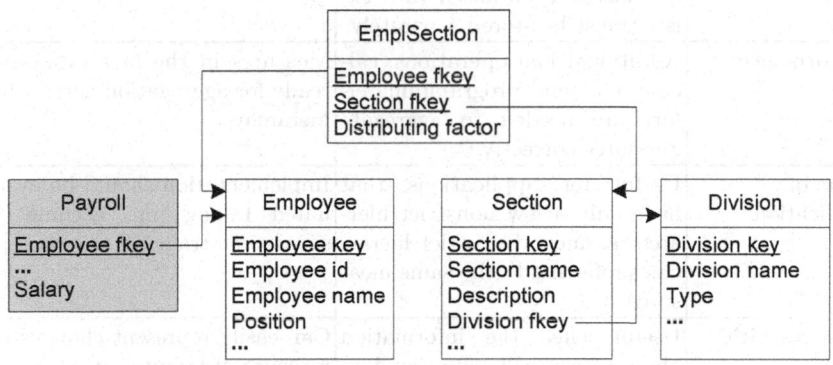

Fig. 3.38. Relational tables for the nonstrict hierarchy of Fig. 3.11

Recall from Sect. 3.2.2 that another solution is to transform a nonstrict hierarchy into a strict one by including an additional dimension in the fact relationship, as shown in Fig. 3.14. Then, the corresponding mapping for a strict hierarchy can be applied. The choice between the two solutions may depend on various factors, such as those mentioned in Table 3.1. Still another option is to convert the many-to-many relationship into a one-to-many relationship by defining a "primary" relationship [201], i.e., to convert the nonstrict hierarchy into a strict one, to which the corresponding mapping for a simple hierarchy is applied.

There are two possibilities for representing nonstrict hierarchies in the object-relational model. We can use a solution similar to the bridge table above, with the difference that references are used instead of foreign keys. The other solution extends the table representing a child level with an additional multivalued attribute (of an array or multiset type) that includes references to the parent members. If the schema contains a distributing factor, the additional multivalued attribute is composed of two attributes: one referencing a

---

[11] This is called the weighting factor in [146].

Table 3.1. Comparison of implementations of nonstrict hierarchies using a bridge table and an additional dimension

Factor	Bridge table	Additional dimension
Table size	The fact table is smaller.	The fact table is bigger if child members are related to many parent members. The additional foreign key in the fact table will also increase the space required.
Additional structure	Additional information about the parent-child relationship and distribution factor (if it exists) must be stored separately.	Measures are only included in the fact table.
Performance	Additional join operations, calculations, and programming effort are needed to aggregate measures correctly.	Measures in the fact table are ready for aggregation across the hierarchy.
Type of application	Useful for applications that have only a few nonstrict hierarchies, and other strict hierarchies referring to the same measures.	Implementation should be evaluated taking into account a space-time trade-off.
Changes with time	Useful when the information about measure distribution does not change with time.	Can easily represent changes in measure distribution that occur with time.
Current OLAP implementation	Current tools can manage to some extent this type of hierarchy.	Current tools can manage this type, but in some cases the problems of double counting may occur.

parent member and another one specifying the value of the distributing factor for this particular parent member. This is shown in Fig. 3.39.

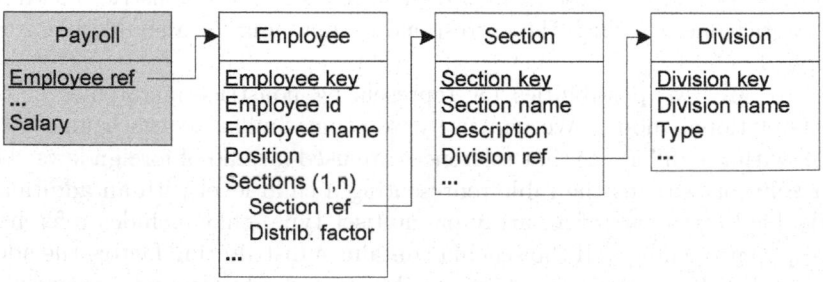

Fig. 3.39. Object-relational representation of the nonstrict hierarchy of Fig. 3.13

Even though nonstrict hierarchies are very common in real-world situations, not much research has been done to them. The search for solutions that allow better management of these hierarchies is considered to be of fundamental importance [103]. In this section, we have analyzed only the case of nonstrict balanced hierarchies. When one has a nonstrict hierarchy of one of the other types presented earlier, a combination of the proposed mapping for the specific hierarchy and the mapping for a nonstrict hierarchy should be applied.

3.6.3 Alternative Hierarchies

For alternative hierarchies, the traditional mapping to relational tables can be applied. This is shown in Fig. 3.40 for the conceptual schema in Fig. 3.16. Note that even though generalized and multiple hierarchies can be easily distinguished at the conceptual level (see Figs. 3.8a and 3.16), this distinction cannot be made at the logical level (compare Figs. 3.36 and 3.40).

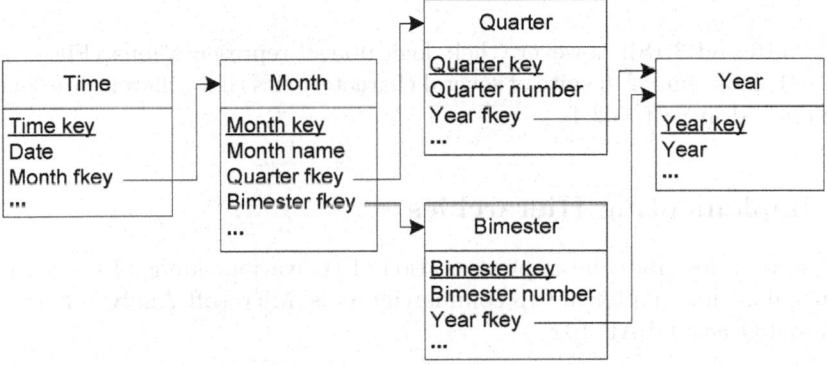

Fig. 3.40. Relations for the alternative hierarchies in Fig. 3.16

3.6.4 Parallel Hierarchies

As parallel hierarchies are composed of either alternative or simple hierarchies, their logical mapping consists in combining the mappings for the specific types of hierarchies. For example, Fig. 3.41 shows the result of applying this mapping to the relational model for the schema shown in Fig. 3.18.

Note that shared levels in parallel dependent hierarchies are represented in one table. Since this level plays different roles in each hierarchy, during implementation it is necessary to create views to avoid unnecessary repetition of data and to facilitate queries. Note also that both multiple and parallel dependent hierarchies can be easily distinguished at the conceptual level

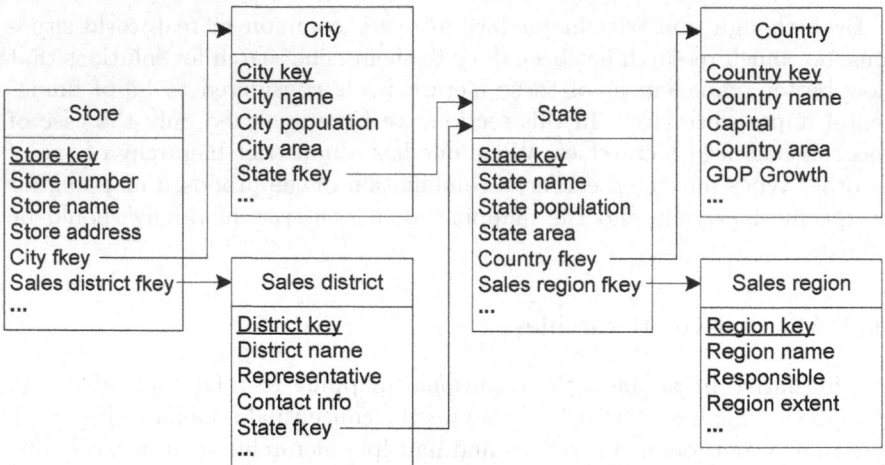

Fig. 3.41. Relational schema of a set of parallel dependent hierarchies composed of two balanced hierarchies

(Figs. 3.16 and 3.18); however, their logical-level representations (Figs. 3.40 and 3.41) look similar in spite of several characteristics that differentiate them, as explained in Sect. 3.2.4.

3.7 Implementing Hierarchies

This section describes the implementation of the various kinds of hierarchies discussed in this chapter in two commercial tools, Microsoft Analysis Services 2005 and Oracle OLAP 10g.

3.7.1 Hierarchies in Analysis Services 2005

Balanced Hierarchies

As already said in Sect. 2.10, both the snowflake and the star schema representation are supported in Analysis Services [100]. Recall that what we have called hierarchies are called in Analysis Services **user-defined hierarchies** or **multilevel hierarchies**. Multilevel hierarchies can be defined by means of all attributes in a dimension; these attributes may be stored in a single table or in several tables of a snowflake schema.

Unbalanced Hierarchies

Analysis Services proposes the implementation of unbalanced hierarchies using parent-child hierarchies. Nevertheless, this implies that the measures must be the same for all hierarchy levels.

In a parent-child hierarchy, nonleaf members may also have associated measures in a fact table, in addition to measures aggregated from their child members. For these nonleaf members, special system-generated child members are created that contain the underlying fact table data. These are referred to as **data members**, and are always included as child members for the purposes of aggregation. However, the normal aggregation behavior can be overridden using a **custom roll-up** formula.

The MembersWithData property is used to control the visibility of data members. Data members are shown when this property is set to NonLeaf-DataVisible in the parent attribute; data members are hidden when this property is set to NonLeafDataHidden. The MembersWithDataCaption property of the parent attribute allows one to define a naming template, used to generate member names for data members.

Generalized Hierarchies

Analysis Services does not support generalized hierarchies. If the members differ in attributes and in hierarchy structure, the common solution is to define one hierarchy for the common levels and another hierarchy for each of the exclusive paths containing the specific levels.

On the other hand, Analysis Services supports **noncovering hierarchies**, called **ragged hierarchies**. A ragged hierarchy is defined using all its levels (i.e., the longest path). Members that do not include a particular level in the hierarchy may contain null values or empty strings, or they can contain the same value as their parent to serve as a placeholder. The dimension property HideMemberIf is used to display these members appropriately. The possible values for the HideMemberIf property and their associated behaviors are as follows:

- Never: Level members are never hidden.
- OnlyChildWithNoName: A level member is hidden when it is the only child of its parent and its name is null or an empty string.
- OnlyChildWithParentName: A level member is hidden when it is the only child of its parent and its name is the same as the name of its parent.
- NoName: A level member is hidden when its name is empty.
- ParentName: A level member is hidden when its name is identical to that of its parent.

Nonstrict Hierarchies

Analysis Services 2005 allows one to define many-to-many relationships between a dimension and a fact relationship. This is done using a bridge table such as that given in Fig. 3.38. In the case of this table, using the terminology of Analysis Services, the Section dimension has a many-to-many relationship with the Payroll measure group, through the Employee intermediate dimension and the EmplSection measure group.

Alternative Hierarchies

Several hierarchies can be attached to a dimension, and they can share levels. For example, two hierarchies can be defined on the Time dimension of Fig. 3.16: a first one composed of Time–Month–Quarter–Year, and another one composed of Time–Month–Bimester–Year.

However, there is no notion of alternative hierarchies since hierarchies cannot be defined as exclusive. For example, the two hierarchies in Fig. 3.16 can be combined, giving meaningless intersections such as Q2-2001 and B1-2001. Even though the measures are manipulated adequately, giving an empty display for these meaningless intersections, the tool should switch automatically between the exclusive component hierarchies. For example, if a user is analyzing measures using the hierarchy Time–Month–Quarter–Year and requests the hierarchy Time–Month–Bimester–Year, the first hierarchy should be excluded from the current analysis.

Parallel Hierarchies

Analysis Services supports parallel hierarchies, whether dependent or independent. Levels can be shared among the various component hierarchies.

3.7.2 Hierarchies in Oracle OLAP 10g

The general steps for defining hierarchies in Oracle OLAP 10g [212] are to create first a dimension, then the levels, and then the hierarchies, indicating the levels that compose them. In the next step, the mapping of levels to source data is performed.

As already said in Sect. 2.11, Oracle OLAP provides two kinds of hierarchies: level-based and value-based hierarchies. The former require the definition of the component levels, which usually correspond to columns in a relational table (in a flat or normalized form). On the other hand, value-based hierarchies are formed by considering relationships between members that are defined in a parent-child table.

To ensure the uniqueness of the members forming a dimension, Oracle OLAP allows designers to use natural or surrogate keys. Natural keys are read from the relational source without modification. Surrogate keys are created before the members are loaded into the analytic workspace; they consist of the prefix of the name of the level added to each member. Since levels are not defined in value-based hierarchies, Oracle OLAP imposes the restriction that only natural keys and not surrogate keys can be used with value-based hierarchies.

Balanced Hierarchies

Oracle OLAP refers to balanced hierarchies as **normal** hierarchies. They are defined as level-based hierarchies, i.e., the hierarchy structure is formed in terms of levels and relationships between them. Oracle OLAP allows the mapping of this kind of hierarchies to both the star and the snowflake schema representation, i.e., to columns stored in a single table or in several tables.

Unbalanced Hierarchies

Unbalanced hierarchies are called **ragged hierarchies** in Oracle OLAP. Oracle OLAP provides two solutions for implementing them, depending on whether or not levels can be distinguished. One option is to define unbalanced hierarchies as level-based; this requires the inclusion of null values in the columns of the source table corresponding to "missing" child members. Another option is to define unbalanced hierarchies as value-based hierarchies, whose data comes from parent-child tables. Note that **recursive hierarchies** can be implemented only as value-based hierarchies.

Generalized Hierarchies

Currently it is not possible to manage generalized hierarchies in Oracle OLAP. They can be defined either as one hierarchy that includes the common levels, or as separate cubes, each cube having a dimension composed of a hierarchy specific to a group of members.

On the other hand, Oracle OLAP supports **noncovering hierarchies**, which are referred to as **skip-level hierarchies**. These can also be defined as either level-based or value-based hierarchies. Defining noncovering hierarchies as level-based hierarchies requires that the source table contains null values in the columns corresponding to "missing" levels. When a noncovering hierarchy is defined as value-based, a parent-child table will represent the relationships between members of the hierarchy.

Nonstrict Hierarchies

Oracle OLAP does not support nonstrict hierarchies.

Alternative Hierarchies

Oracle OLAP does not distinguish between multiple and parallel hierarchies. Therefore, even though developers can define several hierarchies within a dimension, it is the users' responsibility to know that the combination of the component hierarchies is meaningless and should not be applied.

Parallel Hierarchies

Since Oracle OLAP supports the definition of several hierarchies within the same dimension, parallel independent hierarchies can be implemented by defining each hierarchy separately and by indicating the levels composing it.

On the other hand, for parallel dependent hierarchies, each shared level must be defined separately for every hierarchy that includes it. Further, to ensure the uniqueness of members belonging to the same dimension, members of "repeated" levels should have different surrogates assigned to them.

3.8 Related Work

The advantages of conceptual modeling for database design have been acknowledged for several decades and have been studied in many publications (e.g., [29, 45]). However, the analysis presented in [259] shows the small interest of the research community in conceptual multidimensional modeling. A detailed description of multidimensional models can be found in [248, 295]. Some proposals provide graphical representations based on the ER model ([263, 298]), or on UML ([5, 297]), or propose new notations ([82, 109, 301]), while other proposals do not refer to a graphical representation ([106, 231, 242]). Other publications related to data warehouses and OLAP systems refer to the summarizability problem (e.g., [104, 105, 106, 163]), multidimensional normal forms (e.g., [160, 162]), or implementation considerations (e.g., [57, 141]), among other things.

The publications reviewed in this section fall into three categories. First, publications describing various kinds of hierarchies. Second, publications referring to logical representations of hierarchies. Finally, we briefly cover the commercial solutions for implementing hierarchies. For better comparison, in the description that follows, we use our own terminology for the various kinds of hierarchies.

There is great variation in the kinds of hierarchies supported by current multidimensional models. Some models (e.g., [82, 263, 295]) support only simple hierarchies. This situation is considered as a shortcoming of existing models for data warehouses [103]. Table 3.2 compares the multidimensional models that, to the best of our knowledge, cope with hierarchies. We have used three symbols in this table for this comparison: — when no reference to the specified kind of hierarchy exists, ± when only a description and/or definition of the hierarchy is presented, and ✓ when a description and a graphical representation are given. If a different name for a hierarchy is proposed, it is included in the table.

As can be seen in Table 3.2, the existing multidimensional models do not consider all of the kinds of hierarchies described in this book. Several models give only a description and/or a definition of some of these hierarchies, without a graphical representation (e.g., [104, 129, 206, 207, 231, 242]). Further, different names are used for the same kind of hierarchy (for example for unbalanced, generalized, and nonstrict hierarchies). Although some proposals refer to generalized hierarchies (e.g., [169]), they include the supertype and the subtypes as hierarchy levels, without considering the attributes that form a hierarchy as done in our model. All of the proposed models support strict

Table 3.2. Comparison of conceptual models that support various kinds of hierarchies

Model	Balanced	Unbalanced	Generalized	Non-covering	Nonstrict	Alternative or parallel dependent
[231]	± Onto	± Non-onto	—	± Non-covering	± Nonstrict	± Multiple
[109]	✓ Simple	—	✓ Multiple optional	—	—	✓ Alternative
[298]	✓	✓	—	—	✓ Nonstrict	✓ Multiple
[104, 106]	± Strictly homogeneous	± Homogeneous	± Heterogeneous	± Heterogeneous	—	± Heterogeneous
[242]	± Total classification	± Partial classification	± Multiple	± Multiple	—	± Multiple multiplicity
[129]	±	± Unbalanced	± Heterogeneous	± Heterogeneous	—	±
[15]	—	—	±	—	—	—
[5]	✓	± Non-onto	—	±	±	✓
[169]	✓	±	± Specialization	✓	✓	✓ Alternative
[301]	✓	—	—	✓	±	—
[82, 263] [295]	✓	—	—	—	—	✓
[206]	± Balanced	± Unbalanced	—	± Ragged	—	—
[207]	± Level based	± Value based	± Value based	± Value based	—	—

hierarchies explicitly or implicitly. None of the models takes into account the possibility of several different analysis criteria; in consequence, alternative and parallel hierarchies cannot be distinguished. As we have already explained in Sect. 3.2.4, this distinction is important and helps to avoid the assumption that every path in parallel hierarchies should lead to the same members of the shared level [159]. Very few models propose a graphical representation for the various hierarchies that facilitates their distinction at the schema and instance levels (e.g., [169, 263, 298]).

The Object Management Group (OMG) has proposed the Common Warehouse Model (CWM) [207] as a standard for representing data warehouse and OLAP systems. This model provides a framework for representing metadata about data sources, data targets, transformations and analysis, and processes and operations for the creation and management of warehouse data. This model is represented as a layered structure consisting of a number of submodels. One of these submodels, the resource layer, defines models that can be used for representing data in data warehouses and includes the relational model as one of these models. Further, the analysis layer presents a metamodel for OLAP, which includes the concepts of a dimension and a hierarchy. In the CWM, it is possible to represent all of the kinds of hierarchies specified in Table 3.2.

Current commercial OLAP tools do not allow conceptual modeling of hierarchies. They usually provide a logical-level representation limited to star or snowflake schemas. In the previous sections we have referred to publications that use the relational approach to represent the various kinds of hierarchies, where complex hierarchies are sometimes transformed into simpler ones. For example, we discussed the approaches taken by Pedersen et al. [231] for transforming unbalanced, noncovering, and nonstrict hierarchies, and the solutions proposed by Kimball and Ross [146] for managing balanced, generalized, and nonstrict hierarchies. We also discussed the advantages and disadvantages of the proposed solutions. Even though some of these solutions (e.g., [129]) provide a correct mapping that captures the semantics of the hierarchy, they produce a significant number of relations and require SQL extensions for their management. On the other hand, Song et al. [280] have proposed several logical solutions to the problem of managing many-to-many relationships between facts and dimension tables. Besides the solutions described in Sect. 3.2.2, these authors have proposed other solutions that count instances instead of aggregating the measures from a fact table.

Some commercial products, such as Microsoft Analysis Services 2005 and Oracle OLAP 10g, can cope with some of the hierarchies presented in this chapter, as we described in Sect. 3.7. On the other hand, IBM provides some other tools for implementing data warehouses and analytical applications. Alphablox Analytics [111] is used for multidimensional analysis of data represented in a relational DB2 server as a star schema, a snowflake schema, or combination of the two. Alphablox analytics allows the definition of balanced, unbalanced, and noncovering hierarchies, where the latter two hierarchies are implemented as a flat table with null values for missing members. Additionally, unbalanced hierarchies that represent recursive parent-child relationships can be stored in a parent-child table. Alphablox Analytics does not allow the definition of multiple and parallel hierarchies.

3.9 Summary

Data warehouses are defined using a multidimensional view of data, which is based on the concepts of facts, measures, dimensions, and hierarchies. OLAP systems allow users to interactively query warehouse data using operations such as drill-down and roll-up, and these operations require the definition of hierarchies for aggregating measures.

A hierarchy represents some organizational, geographic, or other type of structure that is important for analysis purposes. However, although many kinds of hierarchies can be found in real-world applications, current OLAP systems can manage only a limited number of them. This imposes important restrictions when users require complex multidimensional analysis including several kinds of hierarchies. Thus, designers must apply various implementation strategies to transform some kinds of hierarchies into simpler ones.

This chapter has presented the MultiDim model, a conceptual multidimensional model. The MultiDim model is based on the entity-relationship model and provides an intuitive graphical notation. It is well known that graphical representations facilitate the understanding of application requirements by users and designers. The model also has an associated textual notation, which is described in Appendix A. That appendix also provides a formalization of the model.

We have classified the hierarchies, taking into account their differences at the schema and at the instance level. We have distinguished between simple and multiple hierarchies. The latter are composed of one or more simple hierarchies accounting for the same analysis criterion. Moreover, simple hierarchies can be of various kinds: balanced, unbalanced, and generalized. Recursive and noncovering hierarchies are special cases of unbalanced and generalized hierarchies, respectively. For each of these simple hierarchies, another specialization can be applied, making them either strict or nonstrict. The latter allows many-to-many relationships between parent and child levels, defining graphs at the instance level. Finally, we have distinguished the situation where several hierarchies accounting for different analysis criteria may be attached to the same dimension. Depending on whether or not they share common levels, we have called these hierarchies parallel dependent and parallel independent hierarchies, respectively.

We have also provided a mapping of our conceptual model to the relational and object-relational models. For this mapping, we used well-known rules that transform entity-relationship models to logical models. We have proposed some modifications of these rules for the case of generalized and nonstrict hierarchies. We have also compared alternative relational approaches to implementing some kinds of hierarchies and given indications of when to use these alternative approaches. Finally, we have described how to implement the various kinds of hierarchies in commercial products, using as examples Microsoft Analysis Services 2005 and Oracle OLAP 10g.

The notation of the MultiDim model allows a clear distinction of each kind of hierarchy, which is not the case with most existing conceptual multidimensional models. It is important to capture this distinction at the conceptual level, since when a mapping to the logical level is performed it is no longer possible to distinguish between generalized, alternative, and parallel dependent hierarchies. Therefore, the MultiDim model allows us to capture in a better way the multidimensional semantics of data, which can then be exploited in data warehouse and OLAP systems.

4

Spatial Data Warehouses

It is estimated that about 80% of the data stored in databases has a spatial or location component [257]. Therefore, the location dimension has been widely used in data warehouse and OLAP systems. However, this dimension is usually represented in an alphanumeric, nonspatial manner (i.e., using solely the place name), since these systems are not able to manipulate spatial data. Nevertheless, it is well known that including spatial data in the analysis process can help to reveal patterns that are difficult to discover otherwise.

Taking into account the growing demand to incorporate spatial data into the decision-making process, we have extended the multidimensional model presented in the previous chapter by the inclusion of spatial data. Section 4.1 briefly introduces some concepts related to spatial databases, providing background information for the rest of the chapter. In Sect. 4.2 we present an example of a spatial multidimensional schema, which is used to facilitate the comprehension of the spatially extended MultiDim model.

In the following sections we refer to each element of the model. In Sect. 4.3 we define spatial levels, which may have spatial attributes. In Sect. 4.4 we refer to the previous classification of hierarchies given in Sect. 3.2 showing by means of examples its applicability to spatial hierarchies. We also classify the topological relationships that may exist between spatial levels forming a hierarchy according to the procedures required for measure aggregation. In Sect. 4.5 we introduce the concept of spatial fact relationships, which may relate more than one spatial dimension. Section 4.6 refers to spatial measures represented by a geometry and conventional measures calculated using spatial or topological operators. The metamodel of the spatially extended MultiDim model is presented in Sect. 4.7.

Section 4.8 presents our rationale for implementing spatial data warehouses in current DBMSs, and Sect. 4.9 proposes a mapping to the object-relational model and gives examples of implementation using Oracle Spatial 10g. This mapping also considers the integrity constraints needed to ensure semantic equivalence between the conceptual and logical schemas. After summarizing the mapping rules in Sect. 4.10, Sect. 4.11 covers related work, and Sect. 4.12 concludes this chapter.

4.1 Spatial Databases: General Concepts

Spatial databases have been used for several decades for storing and manipulating spatial data. They allow us to describe the spatial properties of real-world phenomena. There are two complementary ways of modeling spatial data in database applications. In the **object-based** approach, space is decomposed into identifiable objects and their shapes are described. This allows us, for example, to represent a road as a line or a state as a surface. The **field-based** approach is used to represent phenomena that vary over space, associating with each point in a relevant extent of space a value that characterizes a feature at that point. Typical examples are temperature, altitude, soil cover, and pollution level. In this book, we cover only object-based spatial data. The extension of our model to include field-based data is a topic for future work; this is described in Sect. 8.2.

4.1.1 Spatial Objects

A **spatial object** corresponds to a real-world entity for which an application needs to store spatial characteristics. Spatial objects consist of a descriptive (or conventional) component and a spatial component. The **descriptive component** is represented using conventional data types, such as integer, string, and date; it contains general characteristics of the spatial object. For example, a state object may be described by its name, population, area, and major activity. The **spatial component** includes the geometry, which can be of various spatial data types, such as point, line, or surface, to which we refer in more detail below.

4.1.2 Spatial Data Types

Several spatial data types can be used to represent the spatial extent of real-world objects. In this book, we use the spatial data types defined by the conceptual spatiotemporal model MADS [227]. These data types and their associated icons are shown in Fig. 4.1. As shown in the figure, these data types are organized in a hierarchy and provide support for two-dimensional features.

Point represents zero-dimensional geometries denoting a single location in space. A point can be used to represent, for instance, a village in a country.

Line represents one-dimensional geometries denoting a set of connected points defined by one or more linear (in)equations. A line can be used to represent, for instance, a road in a road network. A line is closed if it has

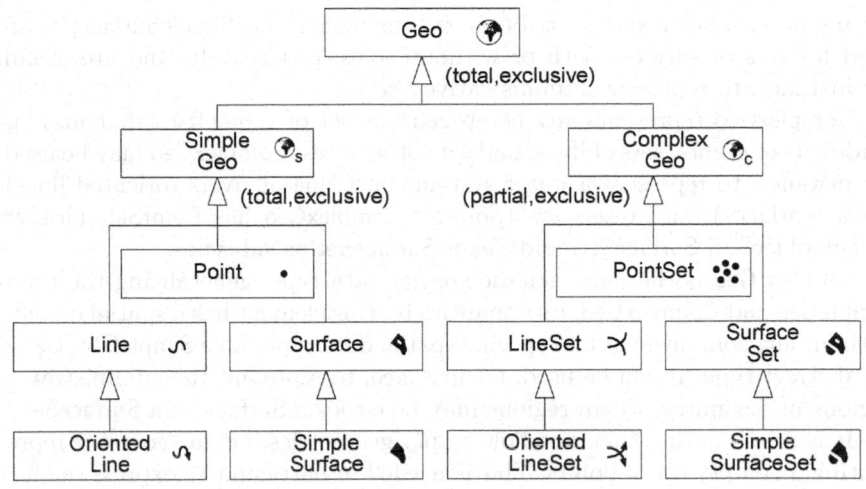

Fig. 4.1. Spatial data types

no identifiable extremities (in other words, its start point is equal to its end point).

OrientedLine represents lines whose extremities have the semantics of a start point and an end point (the line has a given direction from the start point to the end point). OrientedLine is a specialization of Line. An oriented line can be used to represent, for instance, a river in a hydrographic network.

Surface represents two-dimensional geometries denoting a set of connected points that lie inside a boundary formed by one or more disjoint closed lines. If the boundary consists of more than one closed line, one of the closed lines contains all the other ones, and the latter represent holes in the surface defined by the former line. In simpler words, a surface may have holes but no islands (no exterior islands and no islands within a hole).

SimpleSurface represents surfaces without holes. For example, the extent of a lake may be represented by a surface or a simple surface, depending on whether the lake has or not islands.

SimpleGeo is a generic spatial data type that generalizes the types Point, Line, and Surface. SimpleGeo is an abstract type, i.e., it is never instantiated as such: upon creation of a SimpleGeo value it is necessary to specify which of its subtypes characterizes the new element. A SimpleGeo value can be used, for instance, to generically represent cities, whereas a small city may be represented by a point and a bigger city by a simple surface.

Several spatial data types are used to describe spatially homogeneous sets. PointSet represents sets of points, which could be used to represent, for instance, the houses in a town. LineSet represents sets of lines, which could be used to represent, for instance, a road network. OrientedLineSet (a specialization of LineSet) represents a set of oriented lines, which could represent, for instance, a river and its tributaries. SurfaceSet and SimpleSurfaceSet are used for sets of surfaces with or without holes, respectively, and are useful, for instance, to represent administrative regions.

ComplexGeo represents any heterogeneous set of geometries that may include sets of points, sets of lines, and sets of surfaces. ComplexGeo may be used, for instance, to represent a water system consisting of rivers (oriented lines), lakes (surfaces), and reservoirs (points). ComplexGeo has PointSet, LineSet, OrientedLineSet, SurfaceSet, and SimpleSurfaceSet as subtypes.

Finally, Geo is the most generic spatial data type, generalizing the types SimpleGeo and ComplexGeo; its semantics is "this element has a spatial extent" without any commitment to a specific spatial data type. Like SimpleGeo, Geo is an abstract type. It can be used, for instance, to represent the administrative regions of a country, where regions may be either a Surface or a SurfaceSet.

It is worth noting that we allow empty geometries, i.e., a geometry representing an empty set of points. This is needed in particular to express the fact that the intersection of two disjoint geometries is also a geometry, although it may be an empty one.

4.1.3 Reference Systems

The locations in a given geometry are expressed with respect to some coordinates of a plane, i.e., a spatial reference system. The latter represents a function that associates real locations in space with geometries of coordinates defined in mathematical space and vice versa [127]. For example, projected coordinate systems give Cartesian coordinates that result from mapping a point on the Earth's surface to a plane. Various spatial reference systems can be used; see [127] for more details.

4.1.4 Topological Relationships

Topological relationships specify how two spatial values relate to each other. The are extremely used in spatial applications, since they allow one to test, for instance, whether two states have a common border, whether a highway crosses a state, or whether a city is located in a state.

The definitions of the topological relationships are based on the definitions of the boundary, the interior, and the exterior of spatial values. Intuitively, the **exterior** of a spatial value is composed of all the points of the underlying space that do not belong to the spatial value. The **interior** of a spatial value is composed of all its points that do not belong to the boundary. The **boundary** is defined for the different spatial data types as follows.

A point has an empty boundary, and its interior is equal to the point. The boundary of a line is given by its extreme points, provided that they can be distinguished (for example, a closed line has no boundary). The boundary of a surface is given by the enclosing closed line and the closed lines defining the holes. Finally, the boundary of a ComplexGeo is defined (recursively) by the spatial union of (1) the boundaries of its components that do not intersect other components, and (2) the intersecting boundaries that do not lie in the interior of their union. We give precise definitions of these concepts in Appendix A.

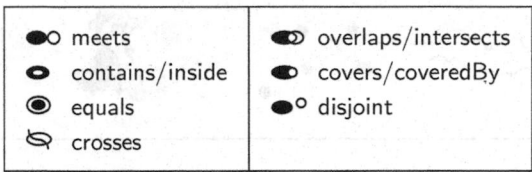

●○	meets	⬤	overlaps/intersects
●	contains/inside	◖●	covers/coveredBy
◉	equals	●○	disjoint
⤸	crosses		

Fig. 4.2. Icons for the various topological relationships

We describe next the topological relationships used in our model; the associated icons are given in Fig. 4.2 and examples are shown in Fig. 4.3.

meets: Two geometries meet if they intersect but their interiors do not. Note that two geometries may intersect in a point and not meet. The examples of the crosses relationship in Fig. 4.3 do not satisfy the meets relationship, since the intersection of the two geometries is located in the interior of the two of them. A similar situation occurs for the leftmost example of the overlaps relationship.

overlaps: Two geometries overlap if the interior of each one intersects both the interior and the exterior of the other one.

contains/inside: contains and inside are symmetric relationships: a contains b if and only if b inside a. A geometry contains another one if the inner object is located in the interior of the other object and the boundaries of the two objects do not intersect.

covers/coveredBy: covers and coveredBy are symmetric relationships: a covers b if and only if b coveredBy a. A geometry covers another one if it includes all points of the other, inner geometry. This means that the first geometry contains the inner one, as defined previously, but without the restriction that the boundaries of the geometries do not intersect. As a particular case, the two geometries may be equal.

disjoint/intersects: disjoint and intersects are inverse relationships: when one applies, the other does not. Two geometries are disjoint if the interior and the boundary of one object intersect only the exterior of the other object.

equals: A geometry equals another one if they share exactly the same set of points.

crosses: One geometry crosses another if they intersect and the dimension of this intersection is less than the greatest dimension of the geometries.

Some of the above relationships do not apply to all spatial data types. For example, the meets topological relationship does not apply to the point data types. We refer to [227] for details of all these restrictions.

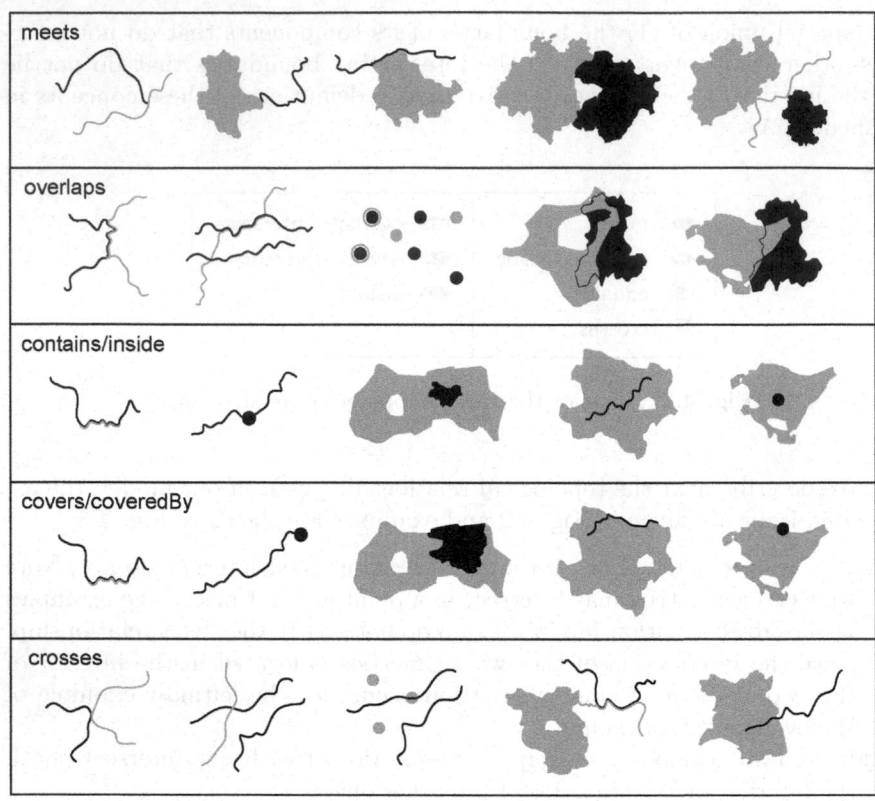

Fig. 4.3. Examples of the various topological relationships. The two objects in the relationship are drawn in black and in gray, respectively

4.1.5 Conceptual Models for Spatial Data

Various conceptual models have been proposed in the literature for representing spatial and spatiotemporal data. These models usually extend existing conceptual models, for example, the ER model (e.g., [142, 227, 300]) or UML (e.g., [17, 20, 246]). However, these conceptual models vary significantly in many aspects, and none of them has yet been widely adopted by the user or research communities. An analysis of the expressiveness of various spatial and spatiotemporal models is given in [233, 309].

4.1.6 Implementation Models for Spatial Data

Object- and field-based data models are used to represent spatial data at an abstract level. These abstractions of space are represented at the implementation level by the raster and vector data models.

The **raster data model** is structured as an array of cells, where each cell represents the value of an attribute for a real-world location [315]. A cell is addressed by its position in the array (row and column number). Usually cells represent square areas of the ground, but other shapes can also be used. The raster data model can be used to represent various types of spatial objects: for example, a point may be represented by a single cell, a line by a sequence of neighboring cells, and a surface by a collection of contiguous cells. However, storage of spatial data in the raster model can be extremely inefficient for a large uniform area with no special characteristics. For example, representing a city in the raster model will include a set of cells that covers the city's interior, all cells having its name as an attribute value.

In the **vector data model**, objects are constructed using points and lines as primitives [256]. A point is represented by a pair of coordinates, whereas more complex linear and surface objects use structures (lists, sets, or arrays) based on the point representation. The vector data representation is inherently more efficient in its use of computer storage than the raster one, because only points of interest need to be stored. However, it is not adequate for representing phenomena for which clear boundaries do not necessarily exist, such as temperature.

4.1.7 Models for Storing Collections of Spatial Objects

Several different models can be used to store a collection of spatial objects using the vector representation: spaghetti, network, and topological models [256]. By considering collections, we focus not only on the individual objects but also on the relationships between them.

In the **spaghetti** model, the geometry of any spatial object in the collection is described independently of other objects. No topology is stored in such a model, and it must be computed "on the fly" if required for analysis purposes. This structure introduces redundancy into the representation. For example, if two surfaces representing states share a common boundary line, this line will be defined twice, once for each surface. Therefore, a risk of inconsistency is incurred if, for example, the same boundary has slightly different coordinates, since the two definitions of the boundary come from different sources of information. On the other hand, this approach is simple, since it facilitates insertion of new objects, i.e., users do not need to know a priori the topological relationships that exist between different spatial objects. In spite of the redundant representation, in many applications it requires a smaller volume of data compared with the network and topological models, to which we refer next.

The **network** and **topological** models describe geometries representing the topology that exists between lines and between surfaces, respectively. These models define a set of specific geometry types (such as nodes and arcs) and use them to represent a collection of spatially related objects. Using the previous example, to define a common boundary between two surfaces, first the definition of the arc representing the boundary is introduced, and then the definition of each surface refers to this arc. Queries about topological relationships are very efficient using this representation. For example, to retrieve states that have a common boundary, it suffices to search for surfaces referencing a common arc. However, as already said, the network and topological models are more complex and they have less flexibility for introducing new spatial objects [74].

4.1.8 Architecture of Spatial Systems

Regardless of the structure chosen for storing spatial objects or collections of them, i.e., raster or vector, or spaghetti or topological, two different types of computer architectures may be used: dual and integrated [305]. The former is based on separate management systems for spatial and nonspatial data, whereas the latter extends DBMSs with spatial data types (for example point, line, and surface) and functions (for example overlap, distance, and area).

A dual architecture is used by traditional geographic information systems (GISs). This requires heterogeneous data models to represent spatial and nonspatial data. Spatial data is usually represented using proprietary data structures, which implies difficulties in modeling, use, and integration, i.e., it leads to increasing complexity of system management.

On the other hand, spatially extended DBMSs provide support for storing, retrieving, querying, and updating spatial objects while preserving other DBMS functionalities, such as recovery techniques and optimization [256]. These spatially extended DBMSs allow users to define an attribute of a table as being of spatial data type, to retrieve **topological relationships** between spatial objects using **spatial operators** and **spatial functions**, to speed up spatial queries using **spatial indexes**, and so on. They also facilitate spatial enabling of applications, since spatial data is stored together with nonspatial data.

Several commercial DBMSs support the management of spatial data, for example Oracle Spatial [151, 216] and IBM DB2 Spatial Extender [110]. This support is based on existing standards [304]. For example, the Open Geospatial Consortium (OGC) defined a standard SQL schema that supports storing, retrieving, querying, and updating of spatial data (called feature collections in OGC's terminology) via the SQL Call-Level Interface (SQL/CLI) [210]. In particular, this standard defines a hierarchy of geometry types with associated methods. Further, the SQL:2003 standard has been extended by the ISO/IEC 13249 SQL/MM standard for managing multimedia and application-specific packages. This extension includes several parts, one of which [128] defines spatial user-defined types, routines, and schemas for generic spatial-data handling. It addresses the need to store, manage, and retrieve information on the basis of aspects of spatial data such as geometry, location, and topology.

4.2 Spatial Extension of the MultiDim Model

In this section, we present the spatial extension of the MultiDim model. The formal definition of the model can be found in Appendix A. To describe our model we use the schema shown in Fig. 4.4, which can be used for the analysis of highway maintenance costs. Highways are divided into highway sections, which in turn are divided into highway segments. For each segment, information about the number of cars and the cost of repairs during various periods of time is available. Since the maintenance of highway segments is the responsibility of the counties through which the highway passes, the analysis should consider the administrative division of the territory into counties and states. The analysis should also help to reveal how different types of road coating influence the maintenance costs.

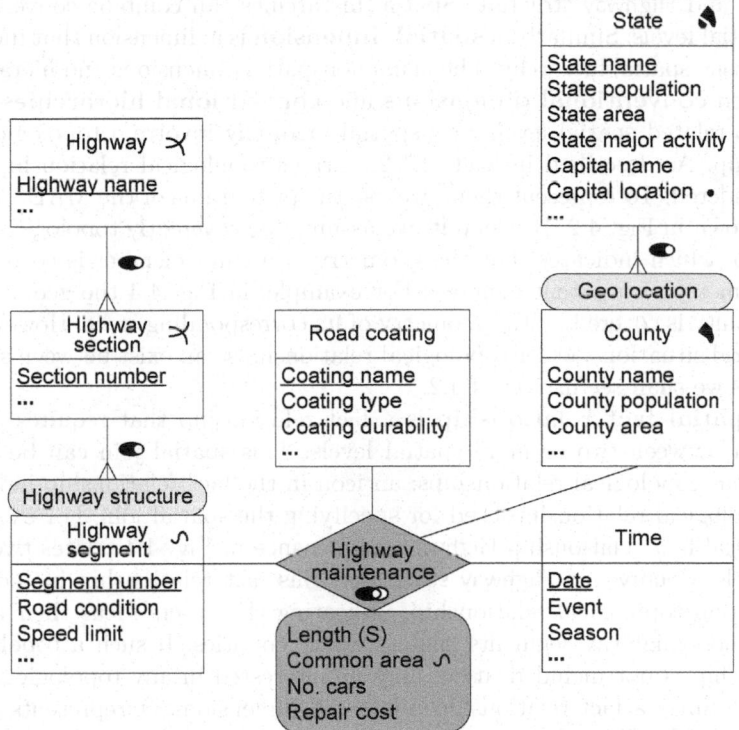

Fig. 4.4. A multidimensional schema with spatial elements

Since the spatially extended MultiDim model can contain both spatial and nonspatial elements, the definitions of schemas, levels, hierarchies, cardinalities, and fact relationships with measures remain the same as those presented in Sect. 3.1.

In addition, we define a **spatial level** as a level for which the application needs to store spatial characteristics. This is captured by its **geometry**, which is represented using one of the spatial data types described in Sect. 4.1.2. In Fig. 4.4, we have five spatial levels: County, State, Highway segment, Highway section, and Highway. The usual nonspatial levels are called **conventional levels**. For instance, Road coating and Time in Fig. 4.4 are conventional levels.

A **spatial attribute** is an attribute that has a spatial data type as its domain. We adopt an orthogonal approach, where a level may be spatial independently of the fact that it has spatial attributes. This achieves maximal expressive power when, for example, a level such as State may be spatial or not depending on application requirements, and may have spatial attributes such as Capital location (see Fig. 4.4).

We define a **spatial hierarchy** as a hierarchy that includes at least one spatial level. For example, in Fig. 4.4 we have two spatial hierarchies: Geo location and Highway structure. Spatial hierarchies can combine conventional and spatial levels. Similarly, a **spatial dimension** is a dimension that includes at least one spatial hierarchy. The usual nonspatial dimensions and hierarchies are called **conventional dimensions** and **conventional hierarchies**.

Two related spatial levels in a spatial hierarchy involve a topological relationship. As described in Sect. 4.1.2, various topological relationships can be considered. To represent them, we use the pictograms of the MADS model [227] shown in Fig. 4.2. By default, we assume the coveredBy topological relationship, which indicates that the geometry of a child member is covered by the geometry of a parent member. For example, in Fig. 4.4 the geometry of each county is covered by the geometry of its corresponding state. However, in real-world situations, other topological relationships can exist between spatial levels as we shall see in Sect. 4.4.2.

A **spatial fact relationship** is a fact relationship that requires a spatial join between two or more spatial levels. This spatial join can be based on various topological relationships: an icon in the fact relationship indicates the topological relationship used for specifying the spatial join. For example, the spatial fact relationship Highway maintenance in Fig. 4.4 relates two spatial levels: County and Highway structure. This fact relationship includes an intersection topological relationship, indicating that users focus their analysis on those highway segments that intersect counties. If such a topological relationship is not included, users may be interested in any topological relationship. Since a fact relationship relates all dimensions, it represents a join of the usual kind for the conventional dimensions (i.e., for the Road coating and Time dimensions in Fig. 4.4) and an intersection topological relationship for the spatial dimensions (i.e., for the Highway structure and County dimensions in Fig. 4.4). Note that, without considering any measure, the schema in Fig. 4.4 can answer queries indicating, for example, whether all highway segments pass through some counties or whether some highway segments belong to more than one county.

As described in Sect. 3.1, fact relationships, whether spatial or not, may contain measures that represent data that is meaningful for leaf members that are aggregated when a hierarchy is traversed. Measures can be **conventional** or **spatial**: the former are numerical data, while the latter are represented by a geometry. Note that conventional measures can be calculated using **spatial operators**, such as distance and area. To indicate that a conventional measure is calculated using spatial operators, we use a symbol "(S)" next to the measure's name. For example, Fig. 4.4 contains two measures. Length is a conventional measure obtained by use of spatial operators, representing the length of the part of a highway segment that belongs to a county, and Common area represents the geometry of the common part.

Measures require the specification of the function used for aggregation along the hierarchies. By default we assume sum for numerical measures and spatial union for spatial measures. For example, in Fig. 4.4, when users roll up from the County to the State level, for each state the measures Length, No. cars, and Repair cost of the corresponding counties will be summed, while the Common area measure will be a LineSet resulting from the spatial union of the lines representing highway segments for the corresponding counties.

In the following sections, we refer in more detail to the various spatial elements of the MultiDim model. We first refer to them at the conceptual level and then consider the mapping to the logical level, discussing also implementation considerations.

4.3 Spatial Levels

A spatial level is represented in the MultiDim model using the icon of its associated spatial type to the right of the level name. For the example in Figs. 4.5a,b, the SurfaceSet icon is used to represent the geometry of State members, since states are formed by several counties, some of which may be islands.

A level may be spatial independently of the fact that it has spatial attributes. For example, depending on application requirements, a level such as State may be spatial (Figs. 4.5a,b) or not (Fig. 4.5c) and may have spatial attributes such as Capital location (Figs. 4.5b,c).

4.4 Spatial Hierarchies

4.4.1 Hierarchy Classification

A spatial hierarchy is composed of several related levels, of which at least one is spatial. If two related levels in a hierarchy are spatial, a pictogram indicating the **topological relationship** between them should be placed on the link between the levels. If this symbol is omitted, the coveredBy topological

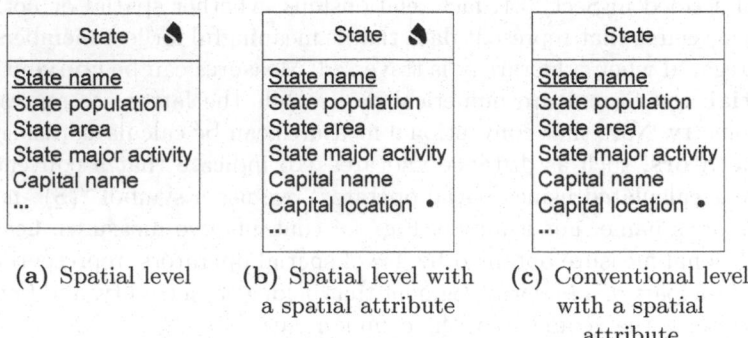

(a) Spatial level   (b) Spatial level with   (c) Conventional level
                        a spatial attribute   with a spatial
                                                    attribute

Fig. 4.5. Examples of levels with spatial characteristics

relationship is assumed by default. For example, in Fig. 4.6 the surfaces representing the geometries of the County members are covered by the geometry of the State members, which is represented by a surface set.

We define next the various types of spatial hierarchies. Although their definitions are based on those given for conventional hierarchies in Sect. 3.2, we briefly present them in this section to facilitate reading.

Simple Spatial Hierarchies

Simple spatial hierarchies are those hierarchies where, if all its component parent-child relationships are one-to-many, the relationship between their members can be represented as a tree. Further, these hierarchies use only one criterion for analysis. Simple spatial hierarchies can be further categorized into balanced, unbalanced, and generalized spatial hierarchies.

Balanced spatial hierarchies have, at the schema level, only one path where all levels are mandatory (Fig. 4.6a). At the instance level (Fig. 4.6b), the members form a tree where all the branches have the same length. As implied by the cardinalities, all parent members must have at least one child member, and a child member belongs to exactly one parent member. Note that different spatial data types are associated with the levels of the hierarchy: point for Store, surface for County, and surface set for State.

Unbalanced spatial hierarchies have only one path at the schema level (Fig. 4.7a) but, as implied by the cardinalities, some lower levels of the hierarchy are not mandatory. Thus, at the instance level, the members represent an unbalanced tree, i.e., the branches of the tree have different lengths (Fig. 4.7b). The example of Fig. 4.7 represents an unbalanced spatial hierarchy for division of a forest consisting of little cells, cells, segments, and regions. Since some parts of the forest are located in mountain areas and are difficult to access, detailed representations of all areas are not available for analysis purposes and some members of the hierarchy are leaves at the segment or cell level.

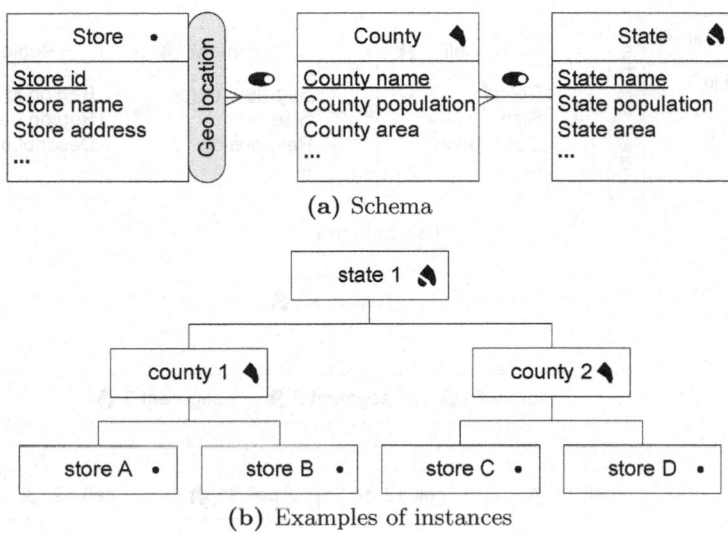

(a) Schema

(b) Examples of instances

Fig. 4.6. A balanced spatial hierarchy

Generalized spatial hierarchies contain multiple exclusive paths sharing some levels (Fig. 4.8a). All these paths represent one hierarchy and account for the same analysis criterion. At the instance level, each member of the hierarchy belongs to only one path (Fig. 4.8b). The symbol ⊗ indicates that for every member, the paths are exclusive. In the example, it is supposed that road segments can belong to either city roads or to highways, where the management of city roads is the responsibility of districts while that of highways is privatized. Note that the geometry associated with the Company level (a surface) represents the spatial extent that a company is responsible for maintenance of.

Some other examples of generalized hierarchies can be found in the data model of the U.S. Census administrative boundaries [302]. One of these represents a spatial hierarchy containing a county level. However, in Maryland, Missouri, Nevada, and Virginia the county level is replaced by independent cities or places, and in American Samoa, counties are replaced by districts and islands.

Generalized spatial hierarchies include a special case commonly referred to as **noncovering hierarchies**. In these hierarchies some paths skip one or several levels.

Nonstrict Spatial Hierarchies

Until now we have assumed that the parent-child relationships have one-to-many cardinalities, i.e., a child member is related to at most one parent member and a parent member may be related to several child members. However,

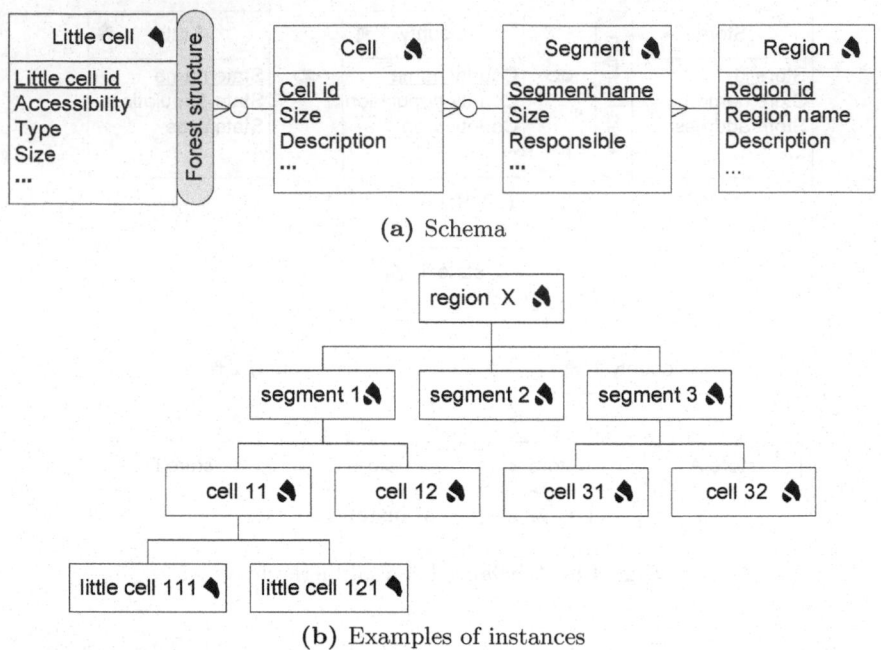

(a) Schema

(b) Examples of instances

Fig. 4.7. An unbalanced spatial hierarchy

many-to-many relationships are very common in real-world spatial applications. For example, a mobile phone network cell may belong to several ZIP areas [133], and several tribal subdivisions in the U.S. Census hierarchy belong both to an American Indian reservation and to the Alaska Native Areas [302].

We call a spatial hierarchy **nonstrict** if it has at least one many-to-many relationship (Fig. 4.9a); it is called **strict** if all cardinalities are one-to-many. The members of a nonstrict hierarchy form a graph (Fig. 4.9b). The fact that a hierarchy is strict or not is orthogonal to its type. Thus, the various kinds of spatial hierarchies presented earlier can be either strict or nonstrict.

Figure 4.9 shows a balanced nonstrict spatial hierarchy. The many-to-many cardinality represents the fact that a lake may belong to more than one city. This hierarchy may be used, for example, when one is interested in controlling the level of pollution in a lake caused by neighbor cities, to indicate the acidity or the percentage of sodium or dissolved carbon dioxide. Most nonstrict hierarchies arise when a partial containment relationship exists [133], for example when only part of a lake belongs to a city or when only part of a highway belongs to a state. In real situations it is difficult to find nonstrict hierarchies with a full containment relationship, i.e., when a spatial member of a lower level belongs wholly to more than one spatial member of a higher level.

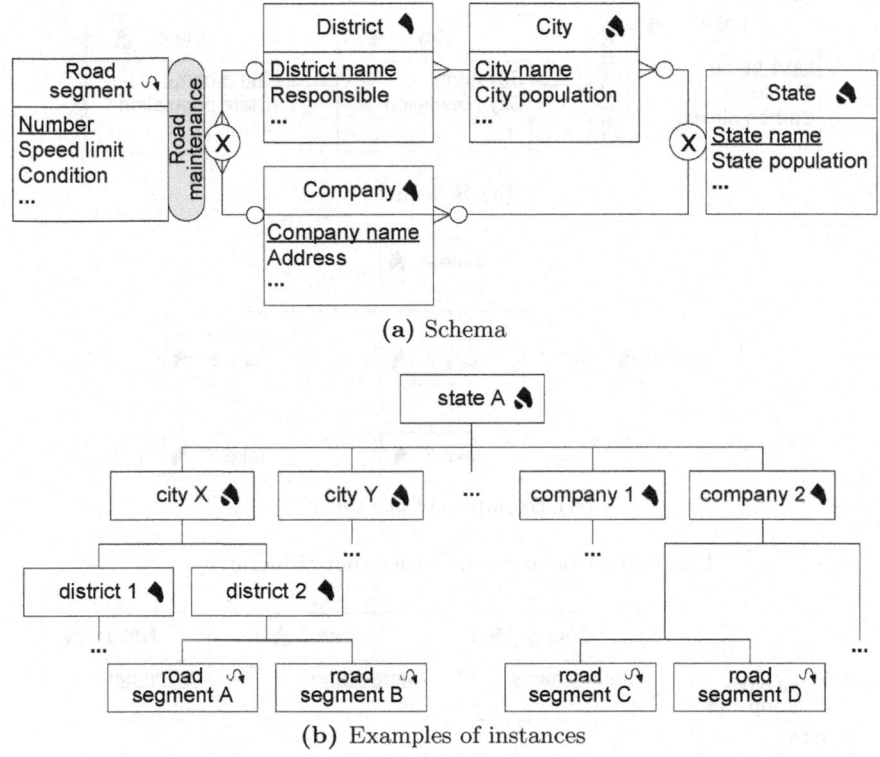

(a) Schema

(b) Examples of instances

Fig. 4.8. A generalized spatial hierarchy

As explained in Sect. 3.2.2, nonstrict spatial hierarchies may include a **distributing factor**, denoted by the symbol ⊕. This indicates how measures should be distributed between different parent members when a many-to-many relationship is reached during a roll-up operation.

Alternative Spatial Hierarchies

Alternative spatial hierarchies have in them several nonexclusive simple spatial hierarchies sharing some levels. However, all these hierarchies account for the same analysis criterion. At the instance level, these hierarchies form a graph, since a child member can be associated with more than one parent member belonging to different levels. In a set of alternative spatial hierarchies, it is not semantically correct to simultaneously traverse different component hierarchies. The user must choose one of the alternative aggregation paths for analysis.

The example given in Fig. 4.10 represents part of the set of hierarchies used by the U.S. Census Bureau [302]. The hierarchy for American Indian and

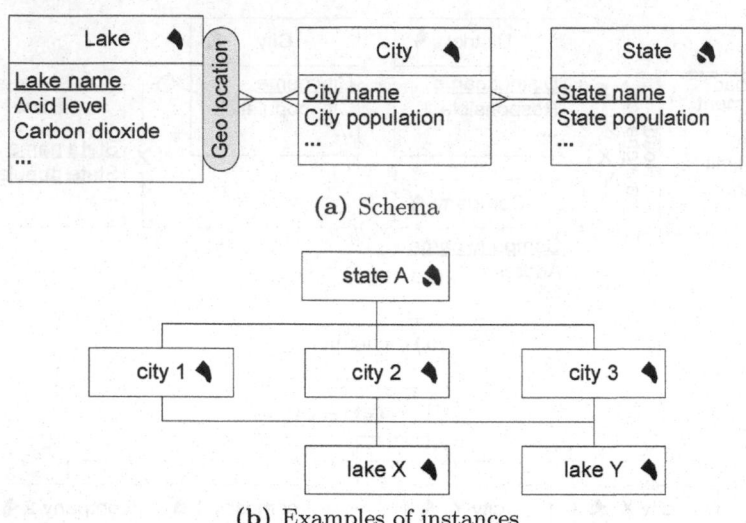

(a) Schema

(b) Examples of instances

Fig. 4.9. A balanced nonstrict spatial hierarchy

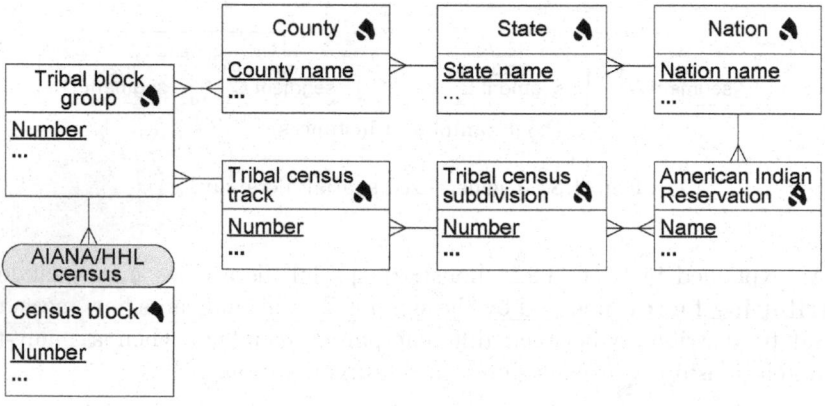

Fig. 4.10. A set of alternative spatial hierarchies formed by two nonstrict
balanced hierarchies

Alaska Native Areas and the Hawaii Home Land (AIANA/HHL census) uses a
particular subdivision of the territory (the lower path in the figure). However,
the usual hierarchy composed of the County, State, and Nation levels[1] (the
upper path in the figure) provides another subdivision of the same territory.
This path can be used to obtain statistics about American Indian territories by

---

[1] To simplify the example, we ignore the fact that some states are not divided into
counties.

county and state. It is obvious that both hierarchies cannot be simultaneously used during analysis.

Parallel Spatial Hierarchies

Parallel spatial hierarchies arise when a dimension has associated with it several spatial hierarchies accounting for different analysis criteria. Such hierarchies can be independent or dependent. In a set of **parallel independent spatial hierarchies**, the different hierarchies do not share levels, i.e., they represent nonoverlapping sets of hierarchies. The example given in Fig. 4.11 includes two independent hierarchies, Geo location and Organizational structure. The former represents the administrative division of the territory into cities, states, and regions, while the latter represents the division according to various sales structures.

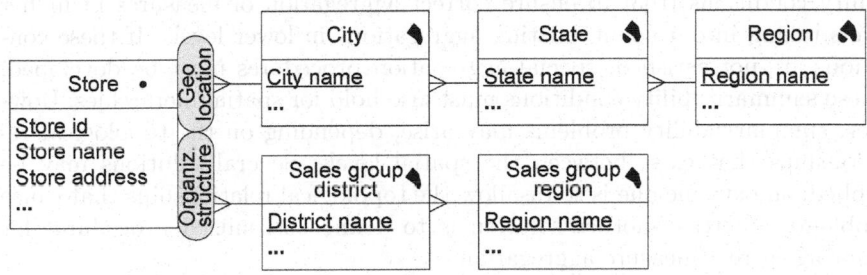

Fig. 4.11. A set of parallel independent spatial hierarchies associated with one dimension

In contrast, **parallel dependent spatial hierarchies**, have in them different hierarchies sharing some levels. The example in Fig. 4.12 represents an insurance company that provides hospitalization services for clients. The Client dimension contains two spatial balanced hierarchies. One of them, Hospitalization, represents the hospitalization structure; it is composed of the Client, Hospitalization area, City, Hospitalization region, and State levels. The other hierarchy, Residence, represents the geographic division of the client's address; it includes the Client, Municipality, City, and State levels. The two spatial hierarchies share the City and State levels.

4.4.2 Topological Relationships Between Spatial Levels

The spatial levels forming a hierarchy can be topologically related. Since hierarchies are used for aggregating measure values when levels are traversed, it is therefore necessary to determine what kinds of topological relationships should be allowed between spatial levels.

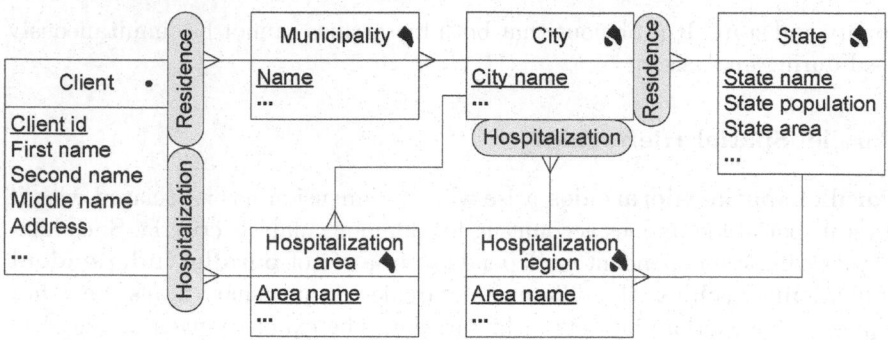

Fig. 4.12. A set of parallel dependent spatial hierarchies

As explained in Sect. 2.6.2, nonspatial hierarchies must satisfy summarizability conditions [163] to ensure correct aggregation of measures in higher levels taking into account existing aggregations in lower levels. If these conditions are not satisfied, special aggregation procedures must be developed. These summarizability conditions must also hold for spatial hierarchies. However, summarizability problems may arise, depending on the topological relationships that exist between the spatial levels. Several solutions may be applied: an extreme one is to disallow the topological relationships that cause problems, whereas another solution is to define customized procedures for ensuring correct measure aggregation.

We give next a classification of topological relationships according to the procedures required for establishing measure aggregation. Our classification, shown in Fig. 4.13, is based on the intersection between the spatial union of the geometries of child members, denoted by $SU(C_{geom})$, and the geometry of their associated parent member, denoted by P_{geom}.

The disjoint topological relationship is not allowed between spatial-hierarchy levels, since during a roll-up operation the next hierarchy level cannot be reached. Thus, a nonempty intersection between $SU(C_{geom})$ and P_{geom} is required.

Various topological relationships may exist if the intersection of P_{geom} and $SU(C_{geom})$ is not empty. If $SU(C_{geom})$ coveredBy P_{geom}, the spatial union of the geometries of child members (and consequently the geometry of each child member) is covered by the geometry of their parent. In this case, the aggregation of measures from a child to a parent level can be done safely using the traditional approach. The situation is similar if $SU(C_{geom})$ inside P_{geom} or if $SU(C_{geom})$ equals P_{geom}. Note, however, that users must be aware of the semantics of the coveredBy and inside topological relationships. For example, suppose that we have a hierarchy composed of the spatial levels City and County and a measure Population. The aggregation of all cities' populations does not necessary give the population of the county, since counties may include other administrative entities.

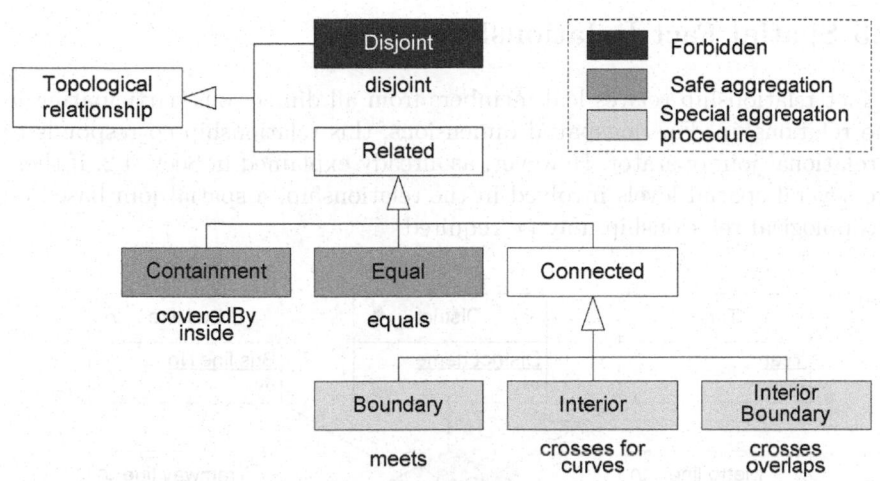

Fig. 4.13. Classification of topological relationships for the purpose of aggregation procedures

The situation is different if the geometries of the parent and child members are related by a topological relationship distinct from coveredBy, inside, or equals. As can be seen in Fig. 4.13, various topological relationships belong to this category, for example, meets and crosses. As in [247], we distinguish three possibilities, depending on whether a topological relationship exists between the boundaries, the interiors, or both the boundaries and the interiors of the geometries of the parent and child members. For example, this distinction may be important in the case of Fig. 4.9 for determining how to perform measure aggregation if a lake meets one city and overlaps another.

For example, if $SU(C_{geom})$ intersects P_{geom} but the topological relationships coveredBy and inside do not hold between them, the geometries of some (or all) child members are not completely covered by the geometry of their parent member. Therefore, the topological relationship between the geometries of individual child members and their parent members must be analyzed in order to determine which measure values can be considered in their entirety for aggregation and which must be partitioned. For example, if in the hierarchy in Fig. 4.6 the spatial union of the points representing stores is not covered by the geometries of their counties, every individual store must be analyzed to determine how measures (for example, taxes to be paid) should be distributed between counties. Therefore, an appropriate procedure for measure aggregation according to the specific details of the application must be developed, such as the one proposed in [133] for the partial-containment topological relationship.

4.5 Spatial Fact Relationships

A fact relationship relates leaf members from all dimensions participating in the relationship. For nonspatial dimensions, this relationship corresponds to a relational join operator. However, as already explained in Sect. 4.2, if there are several spatial levels involved in the relationship, a spatial join based on a topological relationship may be required.

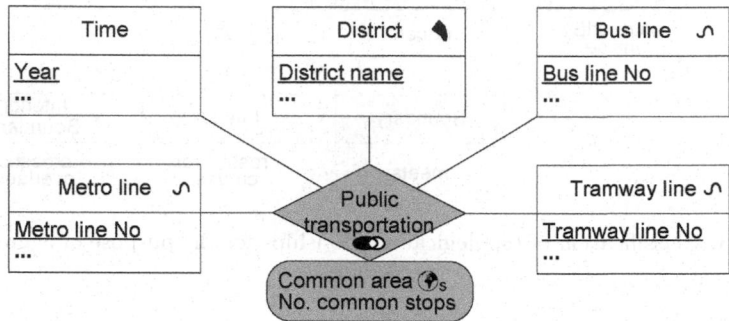

Fig. 4.14. Schema for analysis of transportation services in a city

Figures 4.4 and 4.14 show examples of spatial fact relationships. Since these relationships involve several spatial dimensions, a topological relationship is included to specify the spatial join operation required. In both examples, the intersection relationship is specified. As described previously, in Fig. 4.4 this indicates that users need to focus their analysis on those highway segments that intersect counties.

The schema in Fig. 4.14 could be used in a study aimed at improving the transportation network in a city and decreasing the number of overlapping routes of different kinds of transportation. The city is divided into districts through which bus, metro, and tramway lines pass.[2] In this schema, the topological relationship in the spatial fact relationship indicates that users are interested in analyzing different transportation lines that intersect. This fact relationship includes a numeric measure that stores the number of common stops, and a spatial measure Common area, to which we refer in the next section.

As illustrated by Fig. 4.14, spatial data warehouses include a feature not commonly found in spatial databases: n-ary topological relationships. In spatial databases, topological relationships are usually binary relationships. However, in spatial data warehouses, where the focus is on topological analysis, the topological relationships of interest can relate more than two spatial

---

[2] For simplicity, the hierarchies are not taken into account.

dimensions. There has not been extensive research on n-ary topological relationships. To the best of our knowledge, only [247] refers to them.

4.6 Spatiality and Measures

4.6.1 Spatial Measures

Figures 4.4 and 4.14 include spatial measures, i.e., measures represented by a geometry. In Fig. 4.4, the Common area measure represents the geometry (a line) of the part of a highway segment belonging to a county. In Fig. 4.14, the spatial measure represents the geometry resulting from the intersection of different transportation lines. The resulting spatial measure in Fig. 4.14 is represented by a SimpleGeo, since it may be a point or a line.

In current OLAP systems, numeric measures require the specification of an aggregation function that is used during the roll-up and drill-down operations. If a function is not specified, the sum is applied by default. Recall that several different types of aggregation functions exist: distributive, algebraic, and holistic [85] (see Sect. 2.6.2). For example, distributive functions, such as the sum, minimum, and count, reuse aggregates for a lower level of a hierarchy in order to calculate aggregates for a higher level. Algebraic functions, such as the average, variance, and standard deviation, need an additional manipulation in order to reuse values; for example, the average for a higher level can be calculated taking into account the values of the sum and count for a lower level. On the other hand, holistic functions, such as the median, most frequent, and rank require complete recalculation using data from the leaf level.

Similarly, spatial measures require the specification of a spatial aggregation function. Several different aggregation functions for spatial data have been defined [268]. For example, the spatial distributive functions include the convex hull, the spatial union, and the spatial intersection. The results of these functions are represented by simple or complex geometries. Some examples of spatial algebraic functions are center of n points and the center of gravity, and some examples of spatial holistic functions are the equipartition and the nearest-neighbor index [268]. In the MultiDim model, the spatial union is used by default for aggregating spatial measures.

As another example, in the schema in Fig. 4.15 the user is interested in analyzing the locations of road accidents taking into account the various insurance categories (full coverage, partial coverage, etc.) and the particular client data. The schema includes a spatial measure representing the locations of accidents. If the spatial union is used as the aggregation function, when a user rolls up to the Insurance category level, the accident locations corresponding to each category will be aggregated and represented as a set of points. Other **spatial operators** can also be used, such as the center of n points. As shown by the example, our model allows spatial measures independently of the fact

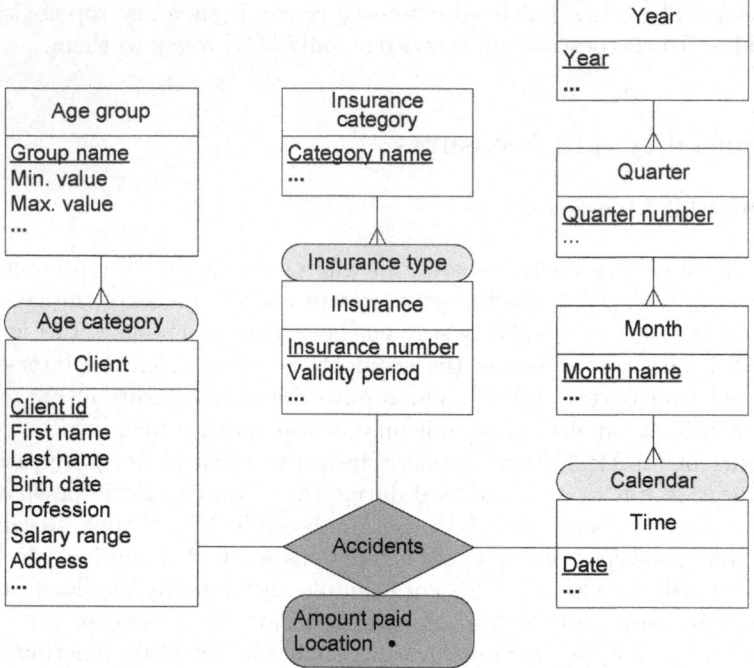

Fig. 4.15. A fact relationship with a spatial measure

that there are spatial dimensions. This not considered by existing models for spatial data warehouses, which only allow spatial measures when there are spatial dimensions.

Spatial measures allow richer analysis than do nonspatial measures; for example, the spatial union of the various accident locations is more informative than aggregating nonspatial measures from the fact relationship. Retrieving a map representation of the locations obtained can show regions of a city with high accident rate; in contrast, retrieving zip codes or location names still needs additional processing for analysis purposes.

An alternative schema for the analysis of road accidents is shown in Fig. 4.16; this schema does not include a spatial measure, and the focus of analysis has been changed to the amount of insurance payments made according to the various geographic locations. Furthermore, the spatial hierarchy allows spatial data manipulation using slice-and-dice, drill-down, and roll-up operations.

We compare now the two schemas in Figs. 4.15 and 4.16 with respect to various queries that can be addressed to them. Although these schemas are similar, different analyses can be performed when a location is handled as a spatial measure or as a spatial hierarchy. The designer of the application must determine which of these schemas better represents users' needs.

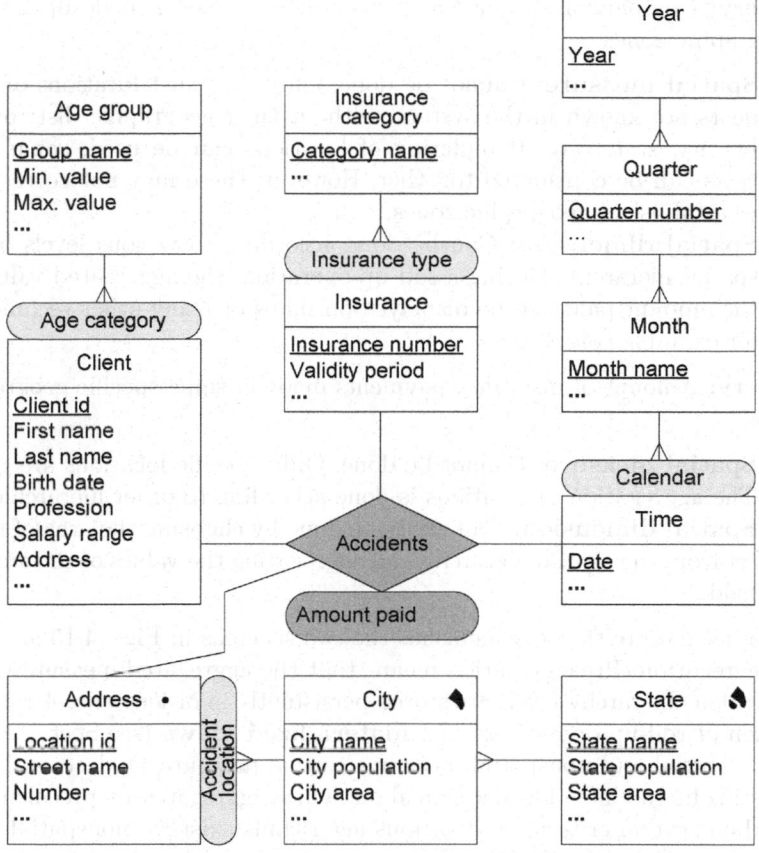

Fig. 4.16. A variant of the schema of Fig. 4.15

- **Query:** Visualize locations of accidents.

  - **Spatial measure:** Can be done by displaying the location measure on a map.
  - **Spatial dimension:** Can be done by showing on the map the points of the addresses participating in the fact relationship.

- **Query:** Aggregate locations of accidents (by using spatial union) according to some predicate involving time or insurance.

  - **Spatial measure:** Can be done when a roll-up operation in the time or insurance hierarchy is executed.
  - **Spatial dimension:** Cannot be done. The dimensions are independent, and traversing a hierarchy in one of them does not aggregate data in another hierarchy. The aggregation is done according to measures, for example the amount paid.

- **Query:** Comparison of amount of insurance payments made in different geographic zones.

  - **Spatial measure:** Cannot be done. Only the exact locations of accidents are known in the system, without their geographic distribution by city, state, etc. If buffering of locations can be used, some locations can be considered together. However, these may not correspond to established geographic zones.
  - **Spatial dimension:** Can be done according to various levels of the spatial hierarchy. During a roll-up operation, the aggregated values of the amount paid can be displayed on maps or using other techniques, for example coloring.

- **Query:** Amount of insurance payments made in some specific geographic area.

  - **Spatial measure:** Cannot be done. Only specific locations are given. The aggregation of locations is done according to other hierarchies.
  - **Spatial dimension:** Can easily be done by choosing the area of interest from the spatial hierarchy and aggregating the values of the amount paid.

Another feature that distinguishes the two schemas in Figs. 4.15 and 4.16 is preaggregation. Preaggregation means that the aggregated measure values of the various hierarchy levels are stored persistently in order to accelerate the execution of roll-up operations, i.e., **materialized views** (see Sect. 2.8) are created. Spatial and nonspatial measures can be preaggregated according to the existing hierarchies. The decision about which aggregates are precalculated is based on several criteria, and various algorithms exist for nonspatial (e.g., [255]) and spatial (e.g., [222, 284]) measures.

On the other hand, if instead of spatial measures we use spatial hierarchies (for example, the Accident location hierarchy in Fig. 4.16), the hierarchy levels representing different degrees of generalization will be included in the system a priori[3] without the necessity to incur additional calculations, as was done for measures.

4.6.2 Conventional Measures Resulting from Spatial Operations

Conventional measures may be obtained by applying **spatial operators**. The example in Fig. 4.4 includes the Length measure, which is a number representing the length of a highway segment that belongs to a county. The example in Fig. 4.17 includes a measure that indicates the minimum distance from a city to a highway segment. In both schemas (Figs. 4.4 and 4.17), the measures are conventional; nevertheless they require a spatial operator for their calculation. The symbol "(S)" is used to distinguish this kind of measures.

---

[3] To highlight the similarity between OLAP and spatial OLAP systems, we ignore the issue of aggregations created dynamically through window queries [222].

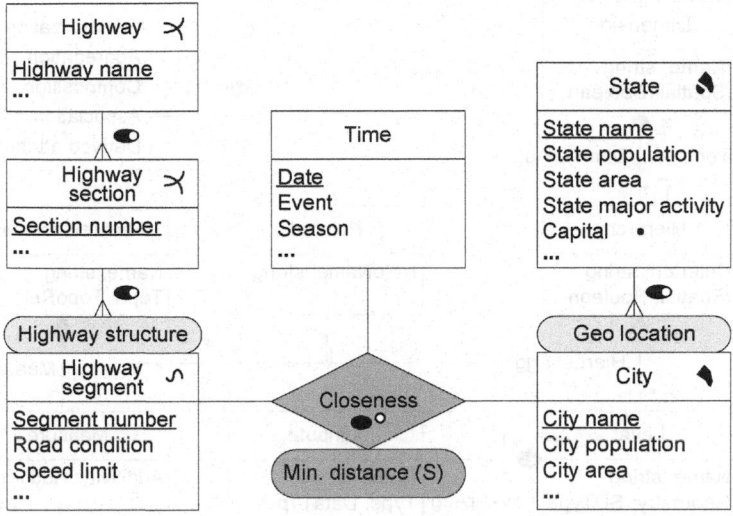

Fig. 4.17. A schema for analyzing the closeness of cities to highways

As is the case for traditional measures, conventional measures calculated from spatial operators may be distributive, algebraic, or holistic. For example, the measure **Length** in Fig. 4.4 is distributive, i.e., it can be aggregated taking into account the lengths calculated for a lower hierarchy level. However, if a measure represents the percentage of a highway segment that belongs to a county, that measure cannot be aggregated for the **State** level by reusing existing measures for the **County** level. On the other hand, the measure **Min. distance** in Fig. 4.17 is holistic: the minimum distance from a state to a highway section cannot be obtained from the minimum distances between cities and highway segments.

4.7 Metamodel of the Spatially Extended MultiDim Model

In this section, we present the metamodel of the spatially extended MultiDim model using UML notation, as shown in Fig. 4.18. This metamodel is fairly similar to that for the conventional version of the model, which was presented in Fig. 3.29.

A dimension is composed of either one level or one or more hierarchies, and each hierarchy belongs to only one dimension. A hierarchy contains two or more related levels that can be shared between different hierarchies. A criterion name identifies a hierarchy, and each level has a unique name. The leaf-level name is used to derive the name of a dimension: this is represented by the derived attribute **Name** in **Dimension**. A level may have a geometry,

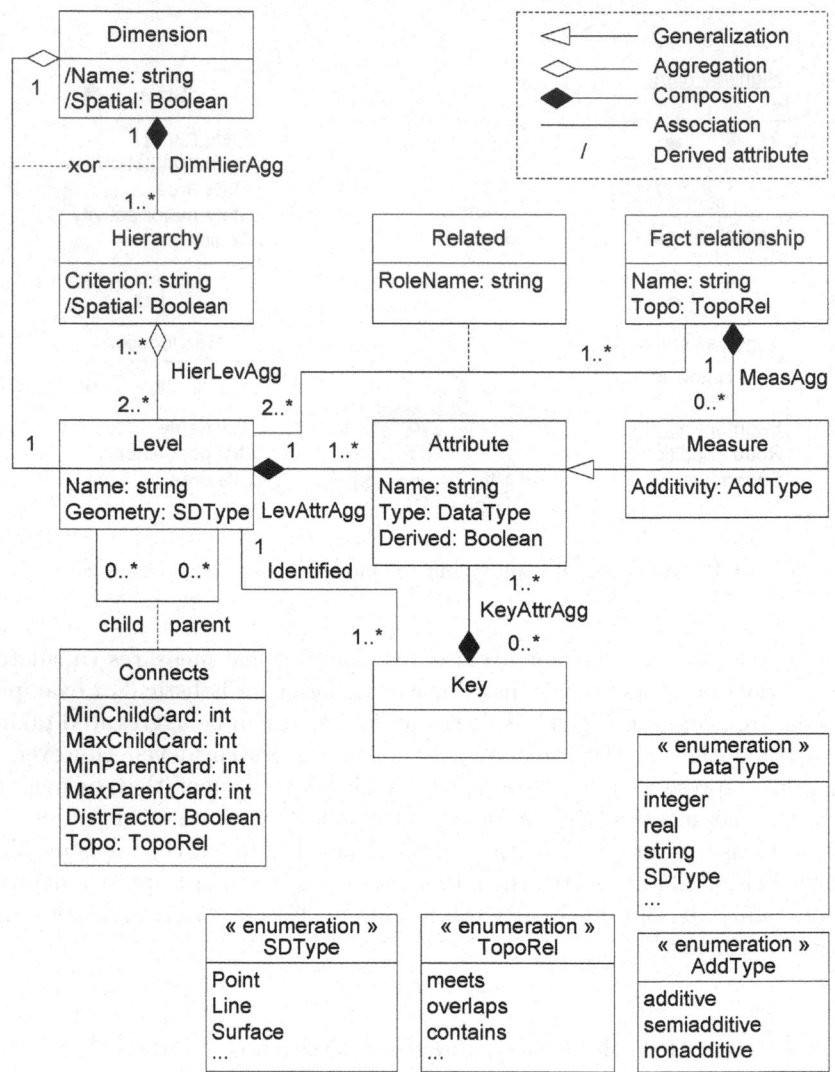

Fig. 4.18. Metamodel of the spatially extended MultiDim model

which must be a spatial data type (SDType). A level that has a geometry is called a spatial level. A dimension is spatial if it has at least one spatial hierarchy, and a hierarchy is spatial if it has at least one spatial level. This is represented by the derived attributes Spatial in Dimension and in Hierarchy.

Levels include attributes, some of which are key attributes used for aggregation purposes, while others are descriptive attributes. Attributes have a name and a type, which must be a DataType, i.e., integer, real, string, etc., or a spatial data type (SDType). An attribute that has a spatial data type as its type is called a spatial attribute; otherwise, it is called a conventional attribute. Further, an attribute may be derived. A constraint not shown in the diagram specifies that only conventional attributes can belong to a key.

The levels forming hierarchies are related through the Connects association class. This association class is characterized by the minimum and maximum cardinalities of the parent and child roles, an optional distributing factor, and an optional topological relationship. A constraint not shown shown in the diagram specifies that a parent-child relationship has a distributing factor only if the maximum cardinalities of the child and the parent are equal to "many". Another constraint specifies that a parent-child relationship has a value for a topological relationship only if both related levels are spatial.

A fact relationship represents an n-ary association between levels with $n > 1$. Since levels can play several different roles in this association, the role name is included in the Related association class. A fact relationship may contain attributes, which are commonly called measures, and these may be additive, semiadditive, or nonadditive (AddType). Further, a fact relationship may have a topological relationship (TopoRel). A constraint specifies that a fact relationship has a topological relationship only if it relates at least two spatial levels.

4.8 Rationale of the Logical-Level Representation

In this section, we present our rationale for transforming a multidimensional model with spatial support into a logical model using an object-relational representation. We also refer to the advantages of using an integrated architecture provided by a spatially extended DBMS (e.g., Oracle Spatial). Finally, we refer to the importance of preserving semantics during the transformation from a conceptual to a logical schema.

4.8.1 Using the Object-Relational Model

In general, logical-level implementations of spatial databases can be based on the relational, object-oriented, or object-relational approach [309]. The relational model has well-known limitations when it is used for representing complex, nonatomic data. For example, since spatial features are modeled using only conventional atomic attribute types, a boundary of a surface must be stored as a set of rows, each one containing the coordinates of two points representing a line segment that forms part of the boundary [69, 268]. Therefore, the relational model imposes on users the responsibility of knowing and maintaining the grouping of rows representing the same real-world object in all their interactions with the database [48].

The object-oriented approach can solve these problems by using complex objects to represent geometries. However, a pure object-oriented approach has not yet been integrated into current DBMSs. On the other hand, the object-relational model preserves the foundations of the relational model while extending its modeling power by organizing data using an object model [154]. The object-relational model allows attributes to have complex types, thus grouping related facts into a single row [48]; for example, the boundary of a surface can be represented in one row. In addition, object-relational features are also included in the SQL:2003 standard [192, 193] and in leading DBMSs, such as Oracle and IBM DB2.

As an example of object-relational DBMSs, we have used Oracle 10g [213] here. As already said in Sect. 2.11, Oracle allows users to define **object types** to represent real-world objects. For example, an object type describing an address can contain several attributes that make it up, such as street, number, city, and province. Object types can be used as a data type of a column in a relational table, as a data type of an attribute in other object types, thus allowing nested structures, or as a type for a table. The latter table is called an **object table**. It is similar to a conventional relational table, with the difference that the columns of an object table correspond to the attributes of the object type used for its declaration. Object tables include additionally an **object identifier**, also called a **surrogate**. Object identifiers can be based on primary keys or can be automatically generated by the DBMS. They can be used for establishing links between tables using a special **ref** type. A ref type is always associated with a specified object type. For example, in a table representing client data, the address column will refer to the object type "address".

4.8.2 Using Spatial Extensions of DBMSs

Current DBMSs, such as Oracle and IBM DB2, provide spatial extensions that allow the storage, retrieval, and manipulation of spatial data [304]. These systems provide raster and/or vector representations of spatial data as described in Sect. 4.1. We have chosen a vector representation, since our model uses an object-based approach for modeling spatial data. Furthermore, the vector representation is inherently more efficient in its use of computer storage, because only points of interest need to be stored.

Oracle Spatial 10g [151, 216] represents geometries using an object-relational model. The **basic geometric types** provided are points, line strings, and polygons. **Points** are elements composed of coordinates, often corresponding to longitude and latitude. **Line strings** are composed of two or more pairs of points that define line segments, which can be of straight or curved shape. **Polygons** are composed of connected line strings. Circles and rectangles are particular kinds of polygons. Compound line strings and polygons are made of a mix of line strings.

The model in Oracle Spatial is organized hierarchically in layers, composed of geometries, which in turn are composed of elements. **Elements** are the basic components of geometries. Elements can be of any of the basic geometric types. A **geometry** consists of either a single element or an ordered list of elements. Complex geometries are modeled using a list of elements; for example, a set of islands is modeled by a list of polygons, and a lake containing islands is also represented by a list of polygons, where one of these represents the exterior boundary and the other polygons represents the interior boundaries. The ordered list of elements in a geometry may be heterogeneous, i.e., it may be made of elements of different types. A **layer** is a set of elements sharing the same attributes, i.e., a layer is represented in one table.

Oracle Spatial includes a spaghetti model [216], and a network and a topological model [215], as described in Sect. 4.1. We have chosen the spaghetti model, since it is simpler and it represents in a better way the intrinsic semantics of multidimensional models. One reason for this is that using a topological model to represent hierarchies guarantees only "horizontal consistency" [74], i.e., maintaining topological relationships between members belonging to the same level. However, in data warehouses and OLAP systems, hierarchies are used for traversing from one level to another, and information about the topological relationships between spatial members belonging to the same level is rarely required.

Additionally, Oracle Spatial extends SQL with **spatial operators** and **spatial functions**. The difference between the two is that spatial operators use **spatial indexes**, whereas functions do not. For example, the operator sdo_relate tests whether two geometries satisfy a topological relationship, whereas the operator sdo_within_distance tests whether two geometries are at certain distance from each other. Some functions offer additional functionality with respect to the corresponding operators: for example, sdo_geom.relate, unlike sdo_relate, allows one to determine the name of the topological relationship that exists between two geometries. Oracle Spatial also includes spatial aggregation functions. For example, sdo_aggr_union and sdo_aggr_centroid return, respectively, the spatial union of the given geometries and their centroid.

4.8.3 Preserving Semantics

Conceptual models, including the MultiDim model, provide constructs for representing in a more direct way the semantics of the modeled reality. However, much of this semantics may be lost when a conceptual schema is translated into a logical schema, since only the concepts supported by the target DBMS can be used. To ensure semantic equivalence between the conceptual and logical schemas, integrity constraints must be introduced. The idea of having explicit integrity constraints for spatial data is not new in the spatial-database community (e.g., [30, 140, 324]).

Current DBMSs provide support for some declarative integrity constraints, such as keys, referential integrity, and check constraints. However, as we have seen in Sect. 2.3.1, in many cases this support is not sufficient and integrity constraints must be implemented using triggers. The SQL:2003 standard and major commercial DBMSs support triggers. As a result of using declarative integrity constraints and/or triggers, the semantics of an application domain is kept in the database as opposed to keeping it in the applications accessing the database. In this way, constraints are encoded once and are available to all applications accessing the database, thereby enforcing data quality and facilitating application management.

4.9 Object-Relational Representation of Spatial Data Warehouses

4.9.1 Spatial Levels

In the MultiDim model, a level corresponds to an entity type in the ER model. Therefore, using the mapping rules given in Sect. 2.3.2, a level is represented by a table in the object-relational model. However, since in our model the spatial support is represented implicitly (i.e., using pictograms), spatial levels require an additional attribute for storing the geometry of their members. Thus, the rules for mapping levels and conventional attributes are as follows.

Rule 1: A conventional level is represented in the object-relational model as a typed table that contains all attributes of the level, and an additional attribute for the surrogate.

Rule 2: A spatial level is represented in the object-relational model as a conventional level using Rule 1, and includes an additional attribute of a spatial data type to represent the geometry of its members.

Rule 3: A conventional attribute is represented in the object-relational model as a monovalued attribute of one of the traditional data types, such as integer, string, and date.

Figure 4.19a shows the object-relational representation with some members of the State level shown in Fig. 4.5a. The definition of a table representing this level in Oracle Spatial is as follows:

```
create type StateType as object (
      Geometry mdsys.sdo_geometry,
      Name varchar2(25),
      Population number(10),
      Area number,
      MajorActivity varchar2(50),
      Capital name varchar2(25) );
create table State of StateType (
      constraint statePK primary key (Name) )
      object identifier is system generated;
```

As shown above, we use an object table in order to have surrogates for the State level. The declaration of an object table requires first the definition of

State				
<u>Sid</u>	Name	Geometry	Population	...
1	San José	⬠	134543	...
2	Cartago	◁	76547	...
3	Heredia	▽	187111	...

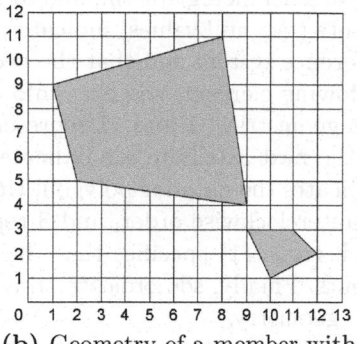

(a) Examples of members

(b) Geometry of a member with an island

Fig. 4.19. Object-relational representation of a spatial level

its associated type. The clause object identifier is system generated indicates that the surrogate attribute is automatically generated by the system.

Oracle Spatial provides a unique data type mdsys.sdo_geometry for representing the spatiality of objects. It is a complex type composed of the following elements:

- sdo_gtype defines the geometry type, i.e., point, line string, etc.
- sdo_srid identifies a spatial reference system. If it is null, the usual Cartesian system is used.
- sdo_point defines a point by its coordinates. If the geometry type is different from the type point, this attribute is ignored.
- sdo_elem_info_array is an array of numbers that allows one to interpret the next attribute.
- sdo_ordinates_array contains an array of coordinates describing the geometry.

In the type definition above, the attribute Geometry is used for storing the geometries of states. Since the type mdsys.sdo_geometry is generic, i.e., it can be used to represent any kind of geometry (such as a line or a surface), the specific geometries of states are defined and instantiated during insert operations. For example, the insertion of a state member composed of two polygons as shown in Fig. 4.19b is defined as follows:

```
insert into State values (
        mdsys.sdo_geometry (2007, null, null,
        mdsys.sdo_elem_info_array(1,1003,1,11,1003,1),
        mdsys.sdo_ordinate_array (2,5,9,4,8,11,1,9,2,5,10,1,12,2,11,3,9,3,10,1)),
        'San José', 134543, 45, 'Tourism', 'San José');
```

The first element of the mdsys.sdo_geometry declaration above (2007) defines several components: 2 indicates the number of dimensions, 0 refers to a linear referencing system, and 07 represents a multipolygon. The next two elements (two null values) indicate that the Cartesian system is used as a spatial reference system and that the geometry is different from the point type. The following element, sdo_elem_info_array, contains two triple values for each of the geometries: 1 and 11 represent the starting positions for the coordinates of the next attribute sdo_ordinate_array, 1003 indicates the type of element (1 indicates the exterior polygon, the coordinates of which should be specified in counterclockwise order, and 3 represents a polygon), and the last number of each triple (1) specifies that the polygon is represented by straight-line segments. Finally, sdo_ordinate_array contains an array of coordinates describing the geometry.

As shown above, when the specific spatial types defined in the conceptual schema, such as SurfaceSet for the State level in Fig. 4.5a, are transformed into Oracle Spatial, their semantics is lost. This would allow one to insert a value which is not of the spatial data type specified in the conceptual schema (for example, a line). Therefore, a trigger must be added to enforce the condition that the geometries of a state are of type multipolygon (type 7 in Oracle Spatial):

```
create trigger ValidGeometryState
    before insert or update on State
    for each row
begin
    if :new.Geometry.get_gtype() <> 7 then
        raise_application_error(-2003,'Invalid Geometry for State');
    end if;
end;
```

4.9.2 Spatial Attributes

Both conventional and spatial levels may contain spatial attributes. Therefore, Rules 1 or 2 given in Sect. 4.9.1 should be applied to the level depending on whether it is conventional or spatial, and Rule 3 should be applied to the conventional attributes. In addition, the mapping of a spatial attribute requires the following rule:

Rule 4: A spatial attribute can be represented in the object-relational model in two different ways:

> **Rule 4a:** as an attribute of a spatial data type to store its geometry, or
> **Rule 4b:** as part of the geometry of the level, if the attribute belongs to a spatial level.

For example, using Rule 4a, the attribute Capital location in Figs. 4.5b,c will be represented by a spatial attribute of type Point. Alternatively, when using Rule 4b, the geometry of the State level in Fig. 4.5b will include the geometry of the attribute Capital location, thus forming a heterogeneous spatial data type composed of a surface set and a point.

The solution to be chosen depends on users' requirements and implementation considerations. For example, the first solution that includes a separate spatial attribute allows topological constraints to be represented explicitly: for example, one can ensure that the geometry of a state contains the geometry of its capital. Enforcing such a constraint in the second solution requires specific functions to extract the geometry of the capital from the overall geometry of the level. On the other hand, the second solution ensures that the geometry of a state always includes a capital. This constraint can be enforced in the first solution by specifying a not null constraint on the spatial attribute.

The definitions of the State level and of the spatial attribute Capital location (Fig. 4.5b) in Oracle Spatial will be slightly different from those given in Sect. 4.9.1. If Rule 4a is used, the declaration of the StateType object will include an additional attribute Capital location of type mdsys.sdo_geometry. Alternatively, if Rule 4b is used, in the insertion described in Sect. 4.9.1 the first parameter of mdsys.sdo_geometry will be equal to 2004, which represents a collection, and the coordinates of the geometry of the level and of the point representing the capital location must be encoded in the mdsys.sdo_elem_info_array and mdsys.sdo_ordinate_array attributes.

4.9.3 Spatial Hierarchies

We proposed in Sect. 3.6 a mapping of various kinds of hierarchies that are composed of parent-child relationships between nonspatial (i.e., conventional) levels. In this section, we refer to the mapping of relationships between nonspatial and spatial levels and between two spatial levels. For the latter, we consider the topological relationships that exist between the spatial levels.

Mapping of Relationships Between Levels

A relationship between levels in a hierarchy corresponds to a binary relationship in the ER model. The rules described in Sects. 3.5.2 and 3.6 can thus be applied for mapping relationships between levels.

Rule 5: A one-to-many parent-child relationship is represented in the object-relational model by extending the type corresponding to the child level with an attribute that references the parent level.

Rule 6: A many-to-many parent-child relationship is represented in the object-relational model in two different ways:

> **Rule 6a:** The relationship may be represented in a type that corresponds to a bridge table (see Sect. 3.6.2). This type contains the attributes referencing the parent and child levels, and contains an additional attribute for the distributing factor, if any.
>
> **Rule 6b:** The relationship may be represented in the type corresponding to the child level. If the relationship does not have a distributing factor, the type corresponding to the child level is extended with a multiset

attribute that references the parent level. In there is a distributing factor, the type corresponding to the child level is extended with a complex multiset attribute, composed of the reference to the parent level and the distributing factor.

Note that these mapping rules are used independently of the fact that the related levels are spatial or not.

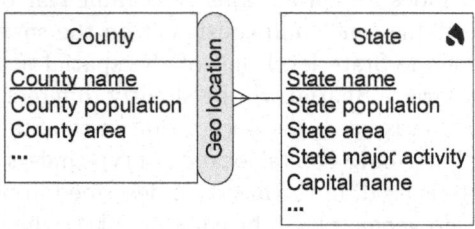

Fig. 4.20. A relationship between nonspatial and spatial levels

For example, the mapping of the binary one-to-many relationship between the County and State levels in Fig. 4.20 adds to the table corresponding to the child level (County) an attribute referencing the parent level (State) as follows:

```
create type CountyType as object (
     Name varchar2(25),
     Population number(10),
     Area number,
     StateRef ref StateType );
create table County of CountyType (
     StateRef not null,
     constraint CountyPK primary key (Name),
     constraint CountyFK foreign key (StateRef) references State );
```

In the declarations above, not allowing the attribute StateRef to have null values and enforcing referential integrity constraints ensure that every county member will be related to a valid state member.

However, when data is inserted into the County table, the surrogates of the corresponding state members must be known. To facilitate this operation, we first create a view that allows the user to introduce a state name instead of a state surrogate:

```
create view CountyView(Name, Population, Area, StateName) as
     select C.Name, C.Population, C.Area, C.StateRef.Name
     from County C;
```

Since Oracle does not allow one to update views defined on two tables, an instead of trigger is needed for one to be able to insert data into the County table using the view CountyView above, as in the following statement:

```
insert into CountyView values ('County1', 100000, 3000, 'StateA');
```

This kind of triggers, which can only be used for views, performs specified actions instead of the operation that activates the trigger. The following trigger first checks if the state name exists in the State table and then inserts the reference into the County table or raises an error message:

```
create trigger CountyViewInsert
        instead of insert on CountyView
        for each row
declare
        NumRows number(5);
begin
        select count(*) into NumRows
        from State S
        where :new.StateName = S.Name;
        if NumRows = 1 then
            insert into County
                select :new.Name, :new.Population, :new.Area, ref(S)
                from State S
                where S.name = :new.StateName;
        else
            raise_application_error(-2000, 'Invalid state name: ' || :new.StateName);
        end if;
end;
```

Similar triggers can be created for the update and delete operations to facilitate operations and to ensure data integrity.

Representing Topological Relationships Between Spatial Levels

Topological relationships between spatial levels in a hierarchy should be considered in order to prevent the inclusion of incorrect data and to indicate what kinds of aggregation procedures should be developed. Two solutions can be envisaged: (1) constrain the geometry of child members during insert operations, or (2) verify the topological relationships that exist between the spatial union of the geometries of the child members and the geometry of their associated parent member after the insertion of all child members. We describe these two solutions next. For this purpose, we suppose that the tables representing the parent and child spatial levels and the relationships between them (for example, for the County and State levels in Fig. 4.21) have been created using Rules 1–6 above. Then, we create a view CountySpatialView similar to the previous view CountyView for facilitating data insertions:

```
create view CountySpatialView(Geometry, Name, Population, Area, StateName) as
        select C.Geometry, C.Name, C.Population, C.Area, C.StateRef.Name
        from County C;
```

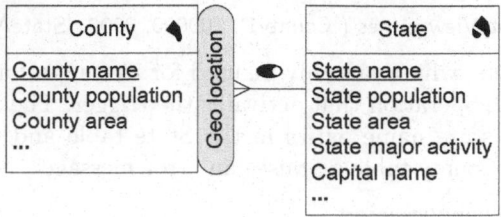

Fig. 4.21. A topological relationship between two spatial levels

The first solution, which constrains the geometry of child members during insert operations, requires us to extend the previous instead of trigger (CountyViewInsert) to verify the topological relationship between the geometries of county and state members:

```
create trigger CountySpatialViewInsert
      instead of insert on CountySpatialView
      for each row
declare
      StateGeometry State.Geometry%Type;
begin
      select S.Geometry into StateGeometry
      from State S
      where S.Name = :new.StateName;
      if SQL%found then
            if sdo_geom.relate(StateGeometry, 'anyinteract',
                  :new.Geometry, 0.005) = 'TRUE'
            then
                  insert into County
                        select :new.Geometry, :new.Name, :new.Population,
                        :new.Area, ref(S)
                        from State S
                        where S.Name = :new.StateName;
            else
                  raise_application_error(-2002, 'Invalid topological relationship:
                        The county and the state are disjoint');
            end if;
      else
            raise raise_application_error(-2000, 'Invalid state name: ' ||
                  :new.StateName);
      end if;
end;
```

This trigger raises an error if either the state name is invalid or the geometries of a county member and its related state member are disjoint. Otherwise, it inserts the new data into the County table. In the example, to check the topological relationship we use the sdo_geom.relate function with an 'anyinteract' mask, which accepts any topological relationship other than disjoint between parent and child members. However, a specific topological relationship can be used instead of anyinteract, for example covers as in the following:

```
if sdo_geom.relate(StateGeometry, 'covers', :new.Geometry, 0.005) = 'covers'
then ...
```

Here, the sdo_geom.relate function returns covers if the second object is entirely inside the first object and their boundaries touch at one or more points. The value 0.005 indicates the tolerance, which reflects the distance that two points can be apart and still be considered the same.

In the second solution, we allow child members to be inserted without activating an instead of trigger. When all child members have been inserted, verification of the topological relationship between the spatial union of the geometries of the child members and the geometry of their associated parent member must be performed. We now give an example of this verification. First, we define a function that returns 1 if the geometry of a given State member is equal to the spatial union of the geometries of its County members:

```
create function ChildrenEqualParent (StateName State.Name%Type)
    return Number is
    StName State.Name%Type;
begin
    select S1.Name into StName
    from State S1, (
        select C.StateRef.Name as Name,
        sdo_aggr_union(sdoaggrtype(C.Geometry, 0.005)) as Geometry
        from County C
        where C.StateRef.Name = StateName
        group by C.StateRef.Name) SU
    where S1.Name = StateName and
    sdo_geom.relate(S1.Geometry, 'equal', SU.Geometry, 0.005)= 'equal';
    if SQL%found then return 1;
    else return 0;
    end if;
end;
```

Here, we use the sdo_aggr_union function, which returns the spatial union of the given spatial objects, in our example the members of County. This function is similar to the traditional (nonspatial) aggregation functions: it uses a group by clause to define groups of values over which the aggregation function is applied. The inner select statement creates a temporary table SU with two attributes StName and Geometry. The latter is the spatial union of the counties grouped by a state name. Then, this table is used in the where clause of the outer select statement to test the equals topological relationship.

We show next an example of the use of the function ChildrenEqualParent to display a message stating whether the spatial union of the geometries of child members is equal to the geometry of their parent member. If this is not the case, we display the topological relationship that exists between every state and its associated county.

```
declare
     StateName State.Name%type;
     TopoRelName varchar2(20);
     CountyName CountySpa.Name%type;
     cursor RetrieveState is
          select S.Name
          from State S;
     cursor RetrieveCounty (StateName2 State.Name%type) is
          select C.Name,
          sdo_geom.relate(C.StateRef.Geometry, 'determine', C.Geometry, 0.005)
          from CountySpa C
          where C.StateRef.Name=StateName2;
begin
     open RetrieveState;
     loop
          fetch RetrieveState into StateName;
          exit when RetrieveState%notfound;
          if (ChildrenWithinParent (StateName) = 1) then
               dbms_output.put_line('The geometry of ' || StateName || ' is equal to
               the spatial union of the geometries of its counties');
          else
               dbms_output.put_line('The geometry of ' || StateName || ' is not equal
                    to the spatial union of the geometries of its counties');
               open RetrieveCounty(StateName);
               loop
                    fetch RetrieveCounty into CountyName, TopoRelName;
                    exit when RetrieveCounty%notfound;
                    dbms_output.put_line(StateName || TopoRelName || CountyName);
               end loop;
               close RetrieveCounty;
          end if;
     end loop;
     close RetrieveState;
end;
```

The **else** branch uses the cursor RetrieveCounty to display the topological relationship between the individual child members and their parent member. This topological relationship is obtained using the function sdo_geom.relate with the parameter 'determine'. For every state member, the state name, the county name, and the topological relationship existing between them are shown.

4.9.4 Spatial Fact Relationships

A fact relationship in the MultiDim model corresponds to an n-ary relationship in the ER model. Thus, the traditional mapping to the object-relational model as described in Sect. 2.3.2 (Rule 3) gives the following rule:

Rule 7: A fact relationship is represented in the object-relational model as a typed table that contains reference attributes pointing to the participating levels, and includes all measures as attributes.

Consider for example the spatial fact relationship given in Fig. 4.4, ignoring for the moment the measures Length and Common area. Suppose that tables corresponding to the levels and the hierarchies have already been defined using Rules 1–6 above (Sects. 4.9.1–4.9.3). In Oracle Spatial, we can define a table for the fact relationship as follows:

```
create type HighwayMaintenanceType as object (
    HighwaySegmentRef ref HighwaySegmentType,
    RoadCoatingRef ref RoadCoatingType,
    CountyRef ref CountyType,
    TimeRef ref TimeType,
    NoCars integer,
    RepairCost real );
create table HighwayMaintenance of HighwayMaintenanceType (
    constraint HighwayMaintenancePK primary key
        (HighwaySegmentRef, RoadCoatingRef, CountyRef, TimeRef),
    /* foreign key and not null constraints */ )
    object identifier is system generated;
```

The topological relationship included in a spatial fact relationship (e.g., intersection in Fig. 4.4) imposes a constraint on the related level members: their geometries must satisfy the topological relationship indicated. The following statement allows one to select the surrogates of members of the HighwaySegment and County tables whose geometries are not disjoint:

```
select ref(HS), ref(C)
from HighwaySegment HS, County C, table(sdo_join('HighwaySegment',
    'Geometry', 'County', 'Geometry', 'mask=anyinteract')) j
where j.rowid1 = HS.rowid and j.rowid2 = C.rowid;
```

In Oracle Spatial, sdo_join is not an operator but a table function. This function is recommended when full table joins are required, i.e., each of the geometries in one table is compared with each of the geometries in the other table. The function sdo_join returns a table that contains pairs of row identifiers (rowid1,rowid2) from participating tables that satisfy the topological relationship specified in the mask. In the example, we have specified anyinteract, which accepts any topological relationship other than disjoint; however, other topological relationships can also be used, for example covers.

However, it is more complex to enforce a topological relationship when a spatial fact relationship involves more than two spatial dimensions, as in the example of Fig. 4.14. Since Oracle Spatial only supports spatial joins between two spatial attributes, this operation must be performed in the application. For example, the following statement retrieves the identifiers of three spatial objects that intersect:

```
select ref(BL), MT.MetroLineRef, MT.TramwayLineRef
from BusLine BL,
    (select ref(ML) as MetroLineRef, ref(TL) as TramwayLineRef,
    sdo_geom.sdo_intersection(ML.Geometry, TL.Geometry, 0.005) as Geometry
    from MetroLine ML, TramwayLine TL) MT
where sdo_anyinteract(BL.Geometry, MT.Geometry)= 'TRUE';
```

The inner select statement creates a temporary table MT with three attributes MetroLineRef, TramwayLineRef, and Geometry. The first two attributes are surrogates of metro and tramway line members and the last one is the geometry of their intersection (if they are not disjoint). The table MT is used in the where clause of the outer select statement to test whether there exists a topological relationship distinct from disjoint between the resulting geometry of the table MT and the geometry of the bus line members.

4.9.5 Measures

Since measures are attributes of a fact relationship, they can be mapped using the rules for mapping attributes of a level.

Rule 8: A measure is mapped in the object-relational model using either Rule 3 (Sect. 4.9.1) if it is a numeric measure (which may be calculated using spatial operators), or Rule 4a (Sect. 4.9.2) if it is a spatial measure represented by a spatial data type.

For example, the Common area measures in Figs. 4.4 and 4.14 are represented by a line and a heterogeneous collection of geometries, respectively. The values for these spatial measures can be either derived from the geometries of related level members or given explicitly by providing their geometries.

The derivation of the Common area measure in Fig. 4.4 is performed as follows:

```
select sdo_geom.sdo_intersection(HM.HighwaySegmentRef.Geometry,
    HM.CountyRef.Geometry, 0.005),
    from HighwayMaintenance HM;
```

Additionally, if users require the name of the topological relationship between a highway segment and a county, the select clause may include the following column:

```
sdo_geom.relate(HM.HighwaySegmentRef.Geometry, 'determine',
    HM.CountyRef.Geometry, 0.005).
```

The derivation is more complex for spatial measures obtained from three or more spatial attributes, as in the case of the Common area measure in Fig. 4.14. In this case, a strategy similar to that described for the last example in Sect. 4.9.4 should be applied. For example, the intersection of the geometries of the members of Bus line, Metro line, and Tramway line for the example in Fig. 4.14 can be obtained as follows:

```
select sdo_geom.sdo_intersection(BL.Geometry, MT.Geometry, 0.005)
from BusLine BL,
     (Select ref(ML) as MetroLineRef, ref(TL) as TramwayLineRef,
      sdo_geom.sdo_intersection(ML.Geometry, TL.Geometry, 0.005) as Geometry
      from MetroLine ML, TramwayLine TL) MT
where sdo_anyinteract(BL.Geometry, MT.Geometry)= 'TRUE';
```

On the other hand, in the absence of spatial dimensions, spatial measures can be introduced by providing their geometries. For example, for the Location spatial measure in Fig. 4.15, the coordinates can be given by the user or brought in from a source system. They can be inserted in a similar way to that described for the insertion of the geometry of the State level in Fig. 4.19b.

Another aspect of spatial measures is their aggregation during roll-up operations. We now give a simplified example of aggregation for the schema in Fig. 4.15. We suppose that the fact relationship is represented by the Accidents table, which contains references to the Client and AgeGroup tables in the attributes ClientRef and AgeGroupRef:

```
select A.ClientRef.AgeGroupRef.GroupName, sum(A.AmountPaid),
       sdo_aggr_union(sdoaggrtype(A.Location, 0.005))
from Accidents A
group by A.ClientRef.AgeGroupRef.GroupName;
```

The above query performs a roll-up operation in the Client dimension to obtain for each age group the sum of the amounts paid and the spatial union of the locations where accidents have occurred. If a roll-up operation is to be applied to more than one dimension, for example Client and Time, the group by rollup operator can be used as follows:

```
select A.ClientRef.AgeGroupRef.GroupName, A.TimeRef.MonthRef.Name,
       sum(A.AmountPaid), sdo_aggr_union(sdoaggrtype(A.Location, 0.005))
from Accidents A
group by rollup (A.ClientRef.AgeGroupRef.GroupName,
       A.TimeRef.MonthRef.Name);
```

This query aggregates measures with respect to age groups and months. For example, if there are three age groups G1, G2, and G3, and two months May and June, the query gives aggregated measures for all six combinations of age group and month, as well as subtotals for the three age groups and the grand total.

On the other hand, the group by cube operator performs **cross-tabulations**, where the subtotals for all combinations of members in a group are calculated. In the previous example, in addition to the aggregations obtained by use of the group by rollup operator, subtotals for the two months are also given.

Note that various spatial aggregation functions can be used. For example, a user can require the center of n points instead of the spatial union for the above queries as follows:

sdo_aggr_centroid(sdoaggrtype(A.Location, 0.005)).

The example in Fig. 4.4 shows a conventional measure calculated from spatial operators, namely the Length measure. This measure may be calculated from the spatial dimensions in the schema. Alternatively, if the schema does not include spatial dimensions, the measure must be calculated on the basis of spatial data in the source systems. In Oracle Spatial, the Length measure can be obtained as follows:

sdo_geom.sdo_length(sdo_geom.sdo_intersection(H.Geometry,
    C.Geometry,0.005), 0.005)

where H and C indicate the tables HighwaySegment and County, respectively.

4.10 Summary of the Mapping Rules

In this section, we recapitulate all of the mapping rules given in the previous sections for transforming the spatially extended MultiDim model into the object-relational model.

Rule 1: A conventional level is represented in the object-relational model as a typed table that contains all attributes of the level, and an additional attribute for the surrogate.

Rule 2: A spatial level is represented in the object-relational model as a conventional level using Rule 1, and includes an additional attribute of a spatial data type to represent the geometry of its members.

Rule 3: A conventional attribute is represented in the object-relational model as a monovalued attribute of one of the traditional data types, such as integer, string, and date.

Rule 4: A spatial attribute can be represented in the object-relational model in two different ways:

> **Rule 4a:** as an attribute of a spatial data type to store its geometry, or
> **Rule 4b:** as part of the geometry of the level, if the attribute belongs to a spatial level.

Rule 5: A one-to-many parent-child relationship is represented in the object-relational model by extending the type corresponding to the child level with an attribute referencing the parent level.

Rule 6: A many-to-many parent-child relationship is represented in the object-relational model in two different ways:

> **Rule 6a:** The relationship may be represented in a type that corresponds to a bridge table. This type contains the attributes referencing the parent and child levels, and contains an additional attribute for the distributing factor, if any.

Rule 6b: The relationship may be represented in the type corresponding to the child level. If the relationship does not have a distributing factor, the type corresponding to the child level is extended with a multiset attribute referencing the parent level. In there is a distributing factor, the type corresponding to the child level is extended with a complex multiset attribute, composed of the reference to the parent level and the distributing factor.

Rule 7: A fact relationship is represented in the object-relational model as a typed table that contains reference attributes pointing to the participating levels, and includes all measures as attributes.

Rule 8: A measure is mapped in the object-relational model using either Rule 3 if it is a numeric measure (which may be calculated using spatial operators), or Rule 4a if it is a spatial measure represented by a spatial data type.

4.11 Related Work

Spatial databases have been investigated over the last few decades (e.g., [64, 93]). Various aspects have been considered, such as conceptual and logical modeling, specification of topological constraints, query languages, spatial index structures, and efficient storage management. Rigaux et al. [256] have referred in more detail to these and other aspects of spatial-database research. Further, Viqueira et al. [309] have presented an extensive evaluation of spatial data models considering spatial data types, data structures used for their implementation, and spatial operations for GIS-centric and DBMS-centric architectures.

Several publications have proposed conceptual models for spatial databases (e.g., [226]) and for data warehouses (e.g., [298]) on the basis of either the ER model or UML. However, a multidimensional model has seldom been used for spatial data modeling. Moreover, although organizations such as ESRI have recognized the necessity for conceptual modeling by introducing templates for spatial data models for various application domains [70], these models often refer to particular aspects of the logical-level design and are too complex to be understood by decision-making users.

To the best of our knowledge, very few conceptual multidimensional models with spatial support have been proposed. Pedersen and Tryfona [232] extended the work of Pedersen et al. [231] by the inclusion of spatial measures. They focused on the problem of aggregation in the presence of various topological relationships between spatial measures. On the other hand, Jensen et al. [133] extended the model proposed by Pedersen and Tryfona [232] to allow one to include spatial objects in hierarchies with partial containment relationships, i.e., where only part of a spatial object belongs to a higher hierarchy level (for example, only part of a street might belong to a higher hierarchy level represented by a ZIP area). They also classified spatial hierarchies on the basis of the work of Pedersen et al. [231]. However, in their proposal it is not clear whether the levels are spatial or conventional. They focused mostly on imprecision in aggregation paths and on transformation of hierarchies with partial containment relationships to simple hierarchies.

Other publications have considered the integration between GISs and data warehouse or OLAP environments. Pourabbas [240] and Ferri et al. [72] referred to common key elements of spatial and multidimensional databases: time and space. They formally defined a geographic data model including "contains" and "full-contains" relationships between hierarchy levels. On the basis of these relationships, integration between GISs and data warehouse or OLAP environments can be achieved by a mapping between the hierarchical structures of the two environments. Moreover, on the basis of a "full-contains" function, these authors were able to provide data lacking in one environment using data from another environment.

The concept of a mapping between hierarchies was also exploited by Kouba et al. [152]. To ensure consistent navigation in a hierarchy between OLAP systems and GISs, these authors proposed a dynamic correspondence through classes, instances, and action levels defined in a metadata repository.

Several authors have defined elements of spatial data warehouses, i.e., spatial measures and dimensions. For example, Stefanovic et al. [284] proposed three types of spatial dimensions based on the spatial references of the hierarchy members: nonspatial (the usual conventional hierarchy), spatial-to-nonspatial (a level has a spatial representation that rolls up to a nonspatial representation), and fully spatial (all hierarchy levels are spatial). We consider that nonspatial-to-spatial references should also be allowed, since a nonspatial level (for example, an address represented as an alphanumeric data type) can roll up to a spatial level. Further, we have extended the classification of spatial dimensions by allowing a dimension to be spatial even when the dimension has only one spatial level.

Regarding measures, Stefanovic et al. [284] distinguished numerical and spatial measures; the latter represent the collection of pointers to spatial objects. Rivest et al. [257] extended the definition of spatial measures and included measures that were represented as spatial objects or calculated using spatial metric or topological operators. However, in their approach, the inclusion of spatial measures represented by a geometry requires the presence of spatial dimensions. In contrast, in our model a spatial measure can be included in a schema with only conventional dimensions (e.g., Fig. 4.15).

On the other hand, Fidalgo et al. [73] excluded spatial measures from spatial data warehouses. Instead, they created spatial dimensions that contain spatial objects previously represented as measures. They extended a star schema by including two new dimensions for managing spatial hierarchies: geographic and hybrid dimensions. Both of these dimensions were subsequently divided into more specialized structures. However, this proposal has several drawbacks. For example, it does not allow sharing of spatial objects represented by a point. This is very restrictive for some kinds of applications, for example when a city is represented by a point that is shared between Store and Client dimensions. Further, we consider that spatial measures should be allowed in spatial data warehouses and we have shown in Sect. 4.6 several spatial data warehouse scenarios that justify their inclusion.

The extensions proposed by Stefanovic et al. [284], Rivest et al. [257], and Fidalgo et al. [73] are based mainly on the star schema. This logical representation lacks expressiveness, as it does not allow one to distinguish between spatial and nonspatial data or to include topological constraints such as those proposed in our model. Further, none of the current work considers spatial fact relationships as proposed in this book.

Van Oosterom et al. [304] have evaluated different DBMSs that include spatial extensions, i.e., Oracle, IBM Informix, and Ingres, with respect to their functionality and performance. The functionality was compared with the Simple Feature Specification (SFS) for SQL defined by the Open GIS Consortium (OGC) [211] and the performance was evaluated by inserting and querying high volumes of spatial data. These authors concluded that, currently, spatial DBMSs are sufficient for the storage, retrieval, and simple analysis of spatial data. The analysis also indicated that Ingres provided the least richness with respect to functionality, that Informix was the only system compliant with the OpenGIS specifications, and that Oracle did not always have the best performance, but had the richest functionality.

Multidimensional modeling concepts applied to spatial data have been used in various spatial OLAP prototypes, such as in [99, 257, 268]. Rivest et al. [257] defined features that spatial OLAP tools should include in order to explore their potential. These authors categorized them into various groups, such as data visualization, data exploration, and data structures. On the basis of these features, they developed a spatial OLAP tool, currently commercialized under the name JMap [144, 258]. Shekhar and Chawla [268] developed a map cube operator, extending the concepts of a data cube and aggregation to spatial data. Further, on basis of the classification used for nonspatial data, these authors presented a classification and examples of various types of spatial measures, for example spatial distributive, algebraic, and holistic functions. Han et al. [99] developed GeoMiner, a spatial data-mining system for online spatial data analysis. This system includes spatial as well as tabular and histogram representations of data.

The extension of OLAP systems to include spatial data is also a concern for commercial software companies. Companies traditionally involved in business intelligence and companies dedicated to spatial data manipulation are combining their efforts to manage spatial and nonspatial data [138]. For example, Business Objects, which specializes in OLAP tools, provides an interface between ESRI's ArcView GIS and its Business Query tool, thus allowing not only multidimensional reporting and analysis, but also geographic analysis. Another company, Dimensional Insight, Inc., allows MapInfo files to be integrated into its analytical tool. SAP has implemented a mapping technology from ESRI in its Business Information Warehouse, enabling interaction through a map-based interface. Finally, Sybase Industry Warehouse Solutions has developed warehouse geocoded data models for industries such as utilities and telecommunications [138].

Other publications in the field of spatial data warehouses relate to new index structures for improving performance (e.g., [197, 223, 252]), materialization of aggregated measures to manage high volumes of spatial data (e.g., [222, 284]), the extension of spatial query languages to query spatial multidimensional data (e.g., [241]), and implementation issues for spatial OLAP (e.g., [99, 257, 268]).

4.12 Summary

In this chapter, we have presented various elements that should be included in a spatial multidimensional model, such as spatial levels, spatial hierarchies, spatial fact relationships, and spatial measures.

First, we referred to the conceptual representation of spatial levels. We presented various cases where these levels may or may not have spatial attributes. We also referred to the case of conventional levels that have spatial attributes. For spatial hierarchies, we showed that the classification of the various kinds of hierarchies proposed in Sect. 3.2 is also applicable to spatial levels. For hierarchies that have related spatial levels, we emphasized that the summarizability problem may also occur, depending on the topological relationship that exists between hierarchy levels. We classified these relationships according to the complexity of developing procedures for measure aggregation.

Next, we presented the concepts of spatial fact relationships and spatial measures. When fact relationships relate more than one spatial dimension, we proposed an extension of the usual join operation for nonspatial fact relationships by inclusion of the topological relationship defining the spatial join to be performed. We also referred to spatial measures that are represented by a geometry, as well as conventional measures that result from a calculation using spatial or topological operators. We discussed the necessity of having spatial aggregation functions defined for measures when hierarchies are included in the schema. Moreover, we relaxed the requirement to have a spatial dimension to represent a spatial measure.

Finally, we presented the mapping to the object-relational model of spatial levels, spatial hierarchies, and spatial fact relationships with spatial measures. Further, we gave examples of integrity constraints that are needed to ensure semantic equivalence between the conceptual and logical schemas; these constraints were implemented using mainly triggers. Using Oracle Spatial 10g, we discussed implementation considerations in relation to representing the semantics of a spatial multidimensional model. We also showed examples of various spatial functions, including spatial aggregations functions, that are useful for spatial-data-warehouse applications.

In proposing a spatial extension of the MultiDim model, we have provided a concise and organized data representation for spatial data warehouse applications [18] that will facilitate the delivery of data for spatial OLAP systems. Further, since our model is platform-independent, it allows a communication bridge to be established between users and designers. It reduces the difficulties of modeling spatial applications, since decision-making users do not usually possess the expertise required by the software currently used for managing spatial data. Additionally, this conceptual model and the classification of topological relationships between hierarchy levels according to the procedure required for measure aggregation will help implementers of spatial OLAP tools to have a common vision of spatial data in a multidimensional model and to develop correct and efficient solutions for manipulation of spatial data.

The described mappings to the object-relational model, along with the examples using a commercial system, namely Oracle Spatial 10g, show the applicability of the proposed solutions to real-world situations and the feasibility of implementing spatial data warehouses in current DBMSs. Further, integrated architectures, where spatial and conventional data is defined within the same DBMS, facilitate system management and simplify data definition and manipulation. However, even though the mapping to the logical level is based on well-known rules, it does not completely represent the semantics expressed in the conceptual level. Therefore, additional programming effort is required to ensure equivalence between the conceptual and logical schemas.

Since the MultiDim model is independent of any implementation, another system could have been used in place of Oracle Spatial. However, the implementation details would have been different, as different systems provide different object-relational features and spatial extensions. The proposed mapping may also vary according to the expected usage patterns, for example use with data-mining algorithms.

5

Temporal Data Warehouses

Current data warehouse and OLAP models include a time dimension that, like other dimensions, is used for grouping purposes (using the roll-up operation) or in a predicate role (using the slice-and-dice operation). The time dimension also indicates the time frame for measures (for example, in order to know how many units of a product were sold in March 2007). However, the time dimension cannot be used to keep track of changes in other dimensions, for example, when a product changes its ingredients or its packaging. Consequently, the "nonvolatile" and "time-varying" features included in the definition of a data warehouse (Sect. 2.5) apply only to measures, and this situation leaves to applications the responsibility of representing changes in dimensions. Kimball et al. [146] proposed several solutions for this problem in the context of relational databases, the slowly changing dimensions. Nevertheless, these solutions are not satisfactory, since they either do not preserve the entire history of the data or are difficult to implement. Further, they do not take account of all research that has been done in the field of temporal databases.

Temporal databases are databases that provide structures and mechanisms for representing and managing information that varies over time. Much research has been done in the field of temporal databases over the last few decades (e.g., [54, 71, 273, 274]). Therefore, combination of the research achievements in temporal databases and data warehouses has led to the new field of temporal data warehouses. Temporal data warehouses raise many issues, including consistent aggregation in the presence of time-varying data, temporal queries, storage methods, and temporal view materialization. Nevertheless, very little attention has been given by the research community to conceptual and logical modeling for temporal data warehouses, or to the analysis of what temporal support should be included in temporal data warehouses.

In this chapter, we propose a temporal extension of the MultiDim model. The chapter starts by giving in Sect. 5.1 a general overview of slowly changing dimensions, which is the mostly used approach in the data warehouse community for managing changes in dimension data. Section 5.2 briefly introduces some concepts related to temporal databases. In Sect. 5.3, we give a general overview of the proposed temporal extension. Then, we refer to the inclusion of temporal support in the various elements of the model, i.e., levels, hierarchies, fact relationships, and measures. In Sect. 5.4, we discuss temporal support for attributes and for a level as a whole, and in Sect. 5.5 we refer to temporal hierarchies and present different cases considering whether it is important to store temporal changes to levels, to the links between them, or to both levels and links. Temporal fact relationships are discussed in Sect. 5.6, and temporal measures in Sect. 5.7. Section 5.8 examines the issue of different temporal granularities: for example, the source data may be introduced on a daily basis but the data in the data warehouse may be aggregated by month. The temporal support in the MultiDim model is summarized in Sect. 5.9, where the metamodel is presented.

This chapter also includes a mapping of our temporal multidimensional model into the classical (i.e., nontemporal) entity-relationship and object-relational models. After describing in Sect. 5.10 the rationale for mapping to these two models, Sect. 5.11 presents the logical representation of temporality types, temporal levels, temporal hierarchies, temporal fact relationships, and temporal measures. This section refers also to the inadequacy of relational databases for representing temporal data. After summarizing the mapping rules in Sect. 5.12, we present in Sect. 5.13 various implementation considerations that should be taken into account in performing aggregation of measures. Finally, Sect. 5.14 surveys work related to temporal data warehouses and temporal databases, and Sect. 5.15 concludes the chapter.

5.1 Slowly Changing Dimensions

The problem of managing changes to dimension data represented as a star or snowflake schema was addressed by Kimball et al. [146]. These authors defined **slowly changing dimensions** as dimensions where attributes do not change over time very frequently and users need to store these changes in a data warehouse. They proposed three solutions, called type 1, type 2, and type 3, to which we refer next.

In the **type 1** or overwrite model, when an attribute changes, the new value overwrites the previous value. This is the current situation in traditional data warehouses, where attributes always reflect the most recent value without maintaining the history of changes. Suppose that in Fig. 5.1a the first product changes its size from 375 to 500. In this case, when the record of the product is updated with the new value, 500, we lose track of the previous value, 375. Note that this solution may give incorrect analysis results, since measures may be affected by changes to dimension data and these cannot be observed.

In the **type 2** or conserving-history model, every time an attribute changes, a new record is inserted containing the new value of the attribute that has changed and the same values as before for the remaining attributes.

Product					
Surr. key	Prod. number	Name	Size	Description	...
101	QB876	Muesli	375	...	...
102	QD555	Olive Oil	750	...	...

(a) Type 1

Product					
Surr. key	Prod. number	Name	Size	Description	...
101	QB876	Muesli	375	...	...
102	QD555	Olive Oil	750	...	...
103	QB876	Muesli	500	...	...

(b) Type 2

Product							
Surr. key	Prod. number	Name	Current size	Original size	Effective date	Description	...
101	QB876	Muesli	500	375	8/11/2006	...	...
102	QD555	Olive Oil	750	750	7/10/2005	...	...

(c) Type 3

Fig. 5.1. Three different implementation types of slowly changing dimensions

The new record will have a surrogate which is different from that of the previous record. This is shown in Fig. 5.1b, where at time t_1 the first product, with surrogate 101, changes its size from 375 to 500. In this case, a new record with a new surrogate, 103, is inserted; this record includes the same values for all attributes except the size attribute.

Since the fact table refers to dimension members through their surrogates, it will contain links to the record with surrogate 101 before time t_1 and links to the record with surrogate 103 after time t_1. Using this solution, users can pose queries that span time periods in which attributes have different values. For example, a query about sales for product QB876 will add all sales of that product regardless of whether they have a surrogate 101 or 103. Note that this query can be formulated because every dimension member includes one or several key attributes (beside the surrogate) that uniquely identifies it, such as the product number in our example. On the other hand, if an analysis is performed according to the sizes of products, the product will be considered in two groups corresponding to the values 375 and 500.

Even though this solution allows changes to dimension data to be stored, however, it has several drawbacks. First, for each member that stores the history of attribute changes, the data staging area must include information about the value of the currently valid surrogate key. This is required for assigning a correct foreign key in the fact table for measures loaded from an operational database. Second, since the same member participates in a fact table with as many surrogates as there are changes in its attributes, counting different members over specific time periods cannot be done in the usual way, i.e., by counting the appearance of a particular member in a fact table. In our example, if in order to know how many products were sold during a time period that includes t_1 we count the number of different surrogates in the fact table, the product QB876 will be counted twice. Another drawback of the type 2 solution is that an analysis of how changes in attribute values affect measures requires significant programming effort, because there is no direct relationship between the different surrogates assigned to the same member. Finally, since a new record is inserted every time an attribute value changes, the dimension can grow considerably, decreasing the performance during join operations with the fact table.

Finally, the **type 3** or limited-history model includes two columns for each attribute for which changes must be stored; these columns contain the current value and the previous value of the attribute. An additional column storing the time at which the change occurred may also be used. This is shown in Fig. 5.1c. Note that this solution will not store all the history of an attribute if it is changed more than once. For example, if the product's size later changes to 750, the value 500 will be lost. In addition, the history of changes is limited to the number of additional columns. According to Kimball et al. [145], this solution is used when users require to see measures as if changes to dimension data have never occurred. This would be the case, for example, if a sales organization were to change its structure and users required an analysis of current sales in terms of the previous organizational structure.

As can be seen, the proposed solutions are not satisfactory, since they either do not preserve the history of the data (type 1), keep only part of this history (type 3), or are difficult to implement (type 2). Further, as was explained above, types 2 and 3 are used for specific kinds of applications; they do not provide a general solution that allows users to keep together a member and all changes to its attributes. Further, these solutions refer to implementation details applied to relational databases; they do not provide an abstract view for representing users' requirements that indicates for which data elements the data warehouse will keep track of the evolution and for which elements only the most recent value will be kept. It is worth noting that leading data warehouse platforms, such as Oracle, IBM, and SAP, provide support for slowly changing dimensions.

5.2 Temporal Databases: General Concepts

Temporal databases are databases that allow one to represent and manage information that varies over time. In addition to storing current data, temporal databases allow previous or future data to be stored, as well as the times when the changes in this data occurred or will occur. Thus, temporal databases enable users to know the evolution of information as required for solving complex problems in many application domains, for example environmental, land management, financial, and healthcare applications.

Time can be represented in a database in various ways. The vast majority of approaches in the field of temporal databases (e.g., [71, 86, 135, 273]), as well as this book, assume a discrete model of time where the instants in the time line are isomorphic (i.e., structurally identical) to the natural numbers. The time line is then represented by a sequence of nondecomposable, consecutive time intervals of identical duration, called **chronons**, which correspond to the smallest time unit that the system is able to represent. Depending on application requirements, consecutive chronons can be grouped into larger units called **granules**, such as seconds, minutes, or days. The **granularity** represents the time unit used for specifying the duration of a granule.

5.2.1 Temporality Types

They are several ways of interpreting the time frame associated with the facts contained in a temporal database. These interpretations are captured by several temporality types.[1]

Valid time (VT) specifies the period of time in which a fact is true in the modeled reality; for example, it allows one to capture when a specific salary was paid to an employee. The valid time is usually supplied by the user.

Transaction time (TT) indicates the period of time in which a fact is current in the database and may be retrieved. The transaction time of a fact begins at the time when it is inserted or updated, and ends when the fact is deleted or updated. The transaction time is generated by the database system.

Valid time and transaction time can be combined to define **bitemporal time** (BT). This indicates both when a fact is true in reality and when it is current in the database.

In many applications it is necessary to capture the time during which an object exists. This is represented by the **lifespan** (LS) or **existence time**; for example, it can be used to represent the duration of a project. The lifespan of an object o may be seen as the valid time of the related fact, "o exists." Lifespan also applies to relationships, for example, it may be used to capture the time during which an employee has worked for a project. In addition to a lifespan, an object or relationship can also record a transaction time, indicating the time when it is current in the database.

---

[1] These are usually called *temporal dimensions*; however, we use here the term *dimension* in the multidimensional context.

5.2.2 Temporal Data Types

Temporal databases allow one to associate a temporal extent with a real-world phenomenon. These temporal extents may represent either events or states. **Events** correspond to phenomena occurring at a particular instant, for example the time when a car accident occurred (Fig. 5.2a). **States** represent phenomena that last over time, such as the duration of a project (Fig. 5.2b).

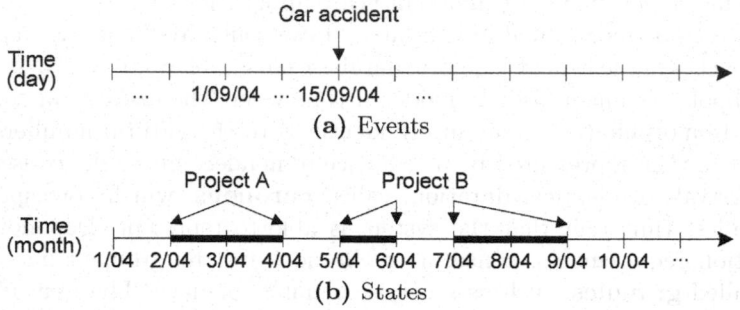

Fig. 5.2. Representing the temporal characteristics of real-world phenomena as events or as states

Like the case for spatial databases, temporal data types are needed to specify the temporal extent of real-world phenomena. In this book, we use the temporal data types defined by the spatiotemporal conceptual model MADS [227]. These temporal data types are organized in a hierarchy, as shown in Fig. 5.3.

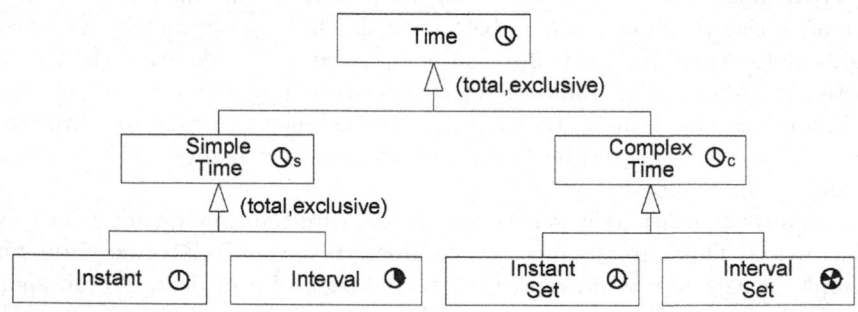

Fig. 5.3. Temporal data types

An Instant denotes a single point in time according to a specific granularity. It has no duration. An instant can have a special value **now** [49], which is used for indicating the current time. Instants are used to represent events, for example the time at which a car accident occurred as in Fig. 5.2a.

An Interval denotes a set of successive instants enclosed between two instants. Intervals are used to represent states, for example the duration of project A in Fig. 5.2b.

SimpleTime is a generic type that generalizes the Instant and Interval data types. It is an abstract type and thus, when a value associated to SimpleTime is created, the specific subtype (either Instant or Interval) must be specified. A SimpleTime can be used, for example, to represent the time (with a granularity of one day) at which an event such as a conference occurs, where one-day conferences are represented by an Instant and other conferences, spanning two or more days, are represented by an Interval.

An InstantSet is a set of instants, which could be used to represent, for example, the instants at which car accidents have occurred in a particular location.

An IntervalSet is a set of simple intervals (sometimes called a temporal element) and may be used, for example, to represent discontinuous durations, such as the duration of project B in Fig. 5.2b.

A ComplexTime denotes any heterogeneous set of temporal values that may include instants and intervals.

Finally, Time is the most generic temporal data type, meaning "this element has a temporal extent" without any commitment to a specific temporal data type. It is an abstract type that can be used, for example, to represent the lifespan of projects, where it may be either an Interval or an IntervalSet, as in Fig. 5.2b.

We allow empty temporal values, i.e., values that represent an empty set of instants. This is needed, in particular, to express the fact that the intersection of two temporal values may be an empty set of instants.

5.2.3 Synchronization Relationships

Synchronization relationships[2] specify how two temporal extents relate to each other. They are essential in temporal applications, since they allow one to determine, for example, whether two events occur simultaneously or whether one precedes the other.

The synchronization relationships for temporal data correspond to the topological relationships for spatial data, which were described in Sect. 4.1.4. They are defined in a similar way, on the basis of the concepts of the boundary, interior, and exterior. They generalize Allen's temporal predicates [9] for intervals. Intuitively, the **exterior** of a temporal value is composed of all the instants of the underlying time frame that do not belong to the temporal value. On the other hand, the **interior** of a temporal value is composed of all

---

[2] These are usually called *temporal relationships*, but here we use this term for relationships that have temporal support.

its instants that do not belong to the boundary. The **boundary** is defined for the different temporal data types as follows. An instant has an empty boundary. The boundary of an interval consists of its start and end instants. The boundary of a ComplexTime value is (recursively) defined by the union of the boundaries of its components that do not intersect with other components. We give precise definitions of these concepts in Appendix A.

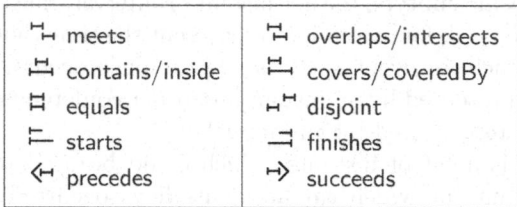

Fig. 5.4. Icons for the various synchronization relationships

We describe commonly used synchronization relationships below; their icons are shown in Fig. 5.4.

meets: Two temporal values meet if they intersect in an instant but their interiors do not. Note that two temporal values may intersect in an instant but do not meet.

overlaps: Two temporal values overlap if their interiors intersect and their intersection is not equal to either of them.

contains/inside: contains and inside are symmetric predicates: a contains b if and only if b inside a. A temporal value contains another one if the interior of the former contains all instants of the latter.

covers/coveredBy: covers and coveredBy are symmetric predicates: a covers b if and only if b coveredBy a. A temporal value covers another one if the former includes all instants of the latter. This means that the former contains the latter, as defined above, but without the restriction that the boundaries of the temporal extents do not intersect. As a particular case, the two temporal values may be equal.

disjoint/intersects: disjoint and intersects are inverse temporal predicates: when one applies, the other does not. Two temporal values are disjoint if they do not share any instant.

equals: Two temporal values are equal if every instant of the first value belongs also to the second and conversely.

starts/finishes: A temporal value starts another if the first instants of the two values are equal. Similarly, a temporal value finishes another if the last instants of the two values are equal.

precedes/succeeds: A temporal value precedes another if the last instant of the former is before the first instant of the latter. Similarly, a temporal value succeeds another if the first instant of the former is later than the last instant of the latter.

As in the spatial case, some of the above temporal predicates do not apply for particular temporal data types. We refer to [227] for a detailed description of all these restrictions.

5.2.4 Conceptual and Logical Models for Temporal Databases

Various conceptual models have been proposed to represent time-varying information in databases. These temporal models usually extend existing conceptual models, for example the ER model (e.g., [86, 325]) or UML (e.g., [38]). The proposed extensions either change the semantics of the existing constructs or introduce new constructs into the model [87]. However, these conceptual models vary significantly in many aspects, and, in addition, none of them has been widely adopted by the user or research communities. An analysis of the expressiveness of various temporal extensions to the ER model has been given in [88].

Once a conceptual schema for a temporal database has been created, it must be translated into a logical schema for implementation in a DBMS. However, current DBMSs, and SQL in particular, provide little support for dealing with time-varying data: They provide only standard data types for encoding dates or timestamps. Querying and updating time-varying data using standard SQL is a challenging task. This was the reason that motivated the research performed in recent decades in the field of temporal databases. A temporal extension of SQL-92 called TSQL2 [273] was proposed to the international standardization committees [275, 276], leading to a dedicated chapter of the SQL:1999 standard called "Part 7, SQL/Temporal". However, such an extension has not yet passed the standardization process [52, 277]. An alternative approach to coping with temporal data in relational databases was proposed in [54]. The consequence of this state of affairs is that, today, database practitioners are left with standard relational databases and SQL for storing and manipulating time-varying data. An account of how to deal with these issues is presented in [274], discussing in what situations and under what assumptions the various solutions can be applied.

There is still not a well-accepted procedure for mapping temporal conceptual models into the relational model. This issue has been studied in, for example, [88, 89]. Another approach to logical-level design for temporal databases is to use temporal normal forms (e.g., [135, 310]). However, achieving logical-level design of temporal databases using temporal normal forms is a difficult process. This can be easily understood, since the process of normalizing standard (i.e., nontemporal) databases is already complex, and this was one of the reasons for proposing the ER conceptual model for database design.

In any case, the relational representation of temporal data produces a significant number of tables. Consequently, the semantics of the modeled reality is dispersed over many tables, which causes performance problems owing to the multiple join operations required for reassembling this information. In addition, the resulting relational schema has to be supplemented with many integrity constraints that encode the underlying semantics of time-varying data. The object-relational model partially solves the first of these problems, since related temporal data can be grouped into a single table. Nevertheless, as with the relational case, many integrity constraints must be added to the object-relational schema. In addition, managing time-varying data in the object-relational model needs to be further studied, in particular with respect to performance issues.

5.3 Temporal Extension of the MultiDim Model

In this section, we briefly present the temporal extension of the MultiDim model. The formal definition of the model can be found in Appendix A, its metamodel is given in Sect. 5.9, and the graphical notation is described in Sect. B.5.

5.3.1 Temporality Types

The MultiDim model provides support for **valid time** (VT), **transaction time** (TT), and **lifespan** (LS). However, these temporality types are not introduced by users (in the case of valid time and lifespan) or generated by the DBMS (in the case of transaction time) as is done in temporal databases; instead, they are brought from the source systems, provided they exist. Table 5.1, which is based on [3, 131], describes several different types of source systems and lists the temporal support that can be obtained from them when one is building a temporal data warehouse. The temporality types in parentheses indicate their possible existence in the source system.

Supporting these temporality types is important in data warehouses for several reasons. First, valid time and lifespan allow users to analyze measures taking changes in dimension data into account. Second, these temporality types help implementers to develop procedures for correct measure aggregation during roll-up operations [25, 63, 195, 318]. Finally, transaction time is important for traceability applications, for example for fraud detection, when the changes to the data in the operational databases and the time when they occurred are required for the process of investigation.

In addition, since data in data warehouses is neither modified nor deleted,[3] we proposed the **loading time** (LT), which indicates the time since when the data is current in a data warehouse. This time can differ from the transaction time of the source systems owing to the delay between the time when a change

---

[3] We ignore modifications due to errors during data loading, and deletion for the purpose of purging data in a data warehouse.

Table 5.1. Temporal support provided by various types of source systems

Type of source system	Description	Temporal support in sources
Snapshot	Data is obtained by dumping the source system. Changes are found by comparing current data with previous snapshots. The time when a snapshot is taken determines neither the transaction time nor the valid time. Valid time and/or lifespan may be included in user-defined attributes.	(VT), (LS)
Queryable	The source system offers a query interface. Detection of changes is done by periodic polling of data from the system and by comparing it with previous versions. Valid time and/or lifespan may also be provided. Queryable systems differ from snapshot systems only in providing direct access to data.	(VT), (LS)
Logged	All data modifications are recorded by the system in log files. Periodic polling is required for discovering the data changes that have occurred. Transaction times can be retrieved from log files. As in the previous cases, valid time and/or lifespan may also be included in the system.	TT, (VT), (LS)
Callback and internal actions	The source system provides triggers, active capabilities, or a programming environment that can be used to automatically detect changes of interest and notify them to interested parties, for example, to a temporal data warehouse. The data changes and the time when they have occurred are detected without any delay.	TT, (VT), (LS)
Replicated	The detection of changes is done by analyzing the messages sent by the replication system. This may happen manually, periodically, or using specific criteria. Depending on the features of the change monitor, systems of this kind can be considered as snapshot or callback systems.	(TT), (VT), (LS)
Bitemporal	The source systems are temporal databases including valid time, and/or lifespan, as well as transaction time.	TT, VT, LS

occurs in a source system and the time when this change is integrated into a temporal data warehouse. The loading time can help users to know the time since when an item of data has been available in a data warehouse for analysis purposes.

5.3.2 Overview of the Model

Even though most real-world phenomena vary over time, storing their evolution in time may be not necessary for an application. Therefore, the choice of the data elements for which the data warehouse will keep track of their evolution in time depends on both the application requirements and the availability of this information in the source systems. The MultiDim model allows users to determine which temporal data they need by including in the schema the symbols of the corresponding temporality types. For example, in the schema in Fig. 5.5, users are not interested in keeping track of changes to clients' data; thus, this dimension does not include any temporal support. On the other hand, changes in the values of measures and changes in data related to products and stores are important for analysis purposes, as indicated by the various temporality types included in the schema.

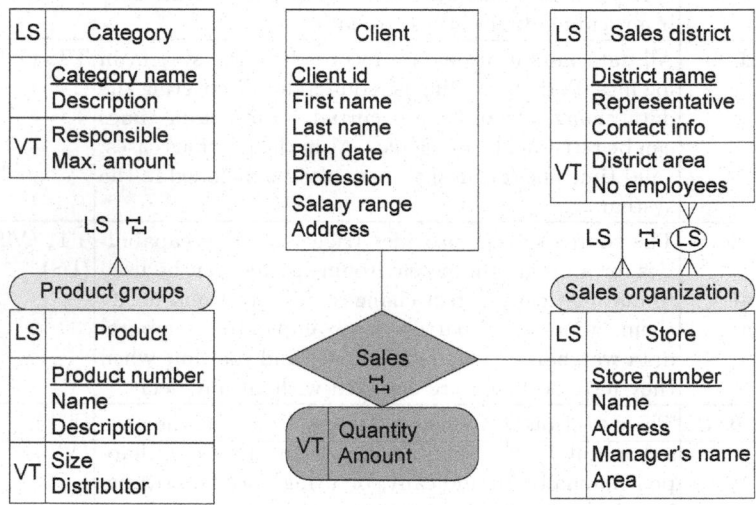

Fig. 5.5. A conceptual schema for a temporal data warehouse

As can be seen in the schema in Fig. 5.5, the MultiDim model allows both temporal and nontemporal attributes, levels, parent-child relationships, hierarchies, and dimensions. The definitions of the nontemporal elements of the model remain the same as those presented in Sect. 3.1.

In addition, we define a **temporal level** as a level for which the application needs to store the time frame associated with its members. The schema in Fig. 5.5 includes four temporal levels (as shown by the LS symbol next to the level name): Product, Category, Store, and Sales district. This allows users to track changes in a member as a whole, for example inserting or deleting a product or a category. The usual nontemporal levels are called **conventional levels**. For instance, Client in Fig. 5.5 is a conventional level.

A **temporal attribute** is an attribute that keeps track of the changes in its values and the time when they occur. For instance, the valid time support for Size and Distributor in the Product level indicates that the history of changes in these attributes will be kept. As in the spatial case, in our model we adopt an orthogonal approach where a level may be temporal independently of the fact that it has temporal attributes.

A **temporal parent-child relationship** keeps track of the time frame associated with the links relating a child and a parent member. For example, in Fig. 5.5 the symbol LS in the relationship linking Product and Category indicates that the evolution in time of the assignments of products to categories will be stored.

Recall that **cardinalities** in parent-child relationships constrain the minimum and the maximum number of members in one level that can be related to a member in another level. Temporal support for parent-child relationships leads to two interpretations of cardinalities. The **instant cardinality** is valid at every time instant, whereas the **lifespan cardinality** is valid over the entire member's lifespan. The instant cardinality is represented using the symbol for the temporality type (for example, LS in Fig. 5.5) next to the line for the parent-child relationship, whereas the lifespan cardinality includes the symbol LS surrounded by an ellipse. When the two cardinalities are the same, this is represented using only one cardinality symbol. In Fig. 5.5, the instant cardinality between Store and Sales district levels is one-to-many, while the lifespan cardinality is many-to-many. These cardinalities indicate, in particular, that a store belongs to only one sales district at any time instant, but may belong to many sales districts over its lifespan, i.e., its assignment to sales districts may change. On the other hand, the instant and lifespan cardinalities between the Product and Category levels are both one-to-many. They indicate, in particular, that products may belong to only one category over their lifespan (and consequently at every instant).

We define a **temporal hierarchy** as a hierarchy that has at least one temporal level or one temporal parent-child relationship. Thus, temporal hierarchies can combine temporal and nontemporal levels. Similarly, a **temporal dimension** is a dimension that has at least one temporal hierarchy. The usual nontemporal dimensions and hierarchies are called **conventional dimensions** and **conventional hierarchies**.

Two related temporal levels in a hierarchy define a synchronization relationship. As described in Sect. 5.2.3, several different synchronization relationships can be considered. To represent them, we use the pictograms shown in Fig. 5.4. By default we assume the overlaps synchronization relationship, which indicates that the lifespan of a child member overlaps the lifespan of a parent member. For example, in Fig. 5.5 the lifespan of each product overlaps the lifespan of its corresponding category, i.e., each valid product belongs to a valid category. However, in real-world situations, other synchronization relationships may exist between temporal levels, as we shall see in Sect. 5.5.

A **temporal fact relationship** is a fact relationship that requires a temporal join between two or more temporal levels. This temporal join can be based on various synchronization relationships: an icon in the fact relationship indicates the synchronization relationship used for specifying the join. For example, the temporal fact relationship Sales in Fig. 5.5 relates two temporal levels: Product and Store. The overlaps synchronization icon in the relationship indicates that users focus their analysis on those products whose lifespan overlaps the lifespan of their related store. If a synchronization icon is not included in a fact relationship, there is no particular synchronization constraint in the instances of that relationship. In our example, this could allow users to analyze whether the exclusion of some products from stores affects sales.

In a temporal multidimensional model, it is important to provide similar temporal support for different elements of the model, i.e., levels, hierarchies, and measures. We would like to avoid mixing two different approaches such that dimensions include explicit temporal support (as described above) while measures require the presence of the traditional time dimension to keep track of changes. Therefore, since measures can be considered as attributes of fact relationships, we provide temporal support for measures in the same way as is done for attributes of levels. For example, in the example in Fig. 5.5, valid time support is used to keep track of changes in measure values for Quantity and Amount.

An important question is, thus, whether it is necessary to have a time dimension in the schema when temporality types are included for measures. If all attributes of the time dimension can be obtained by applying time manipulation functions, such as the corresponding week, month, or quarter, this dimension is not required anymore. However, in some temporal-data-warehouse applications this calculation can be very time-consuming, or the time dimension may contain data that cannot be derived, for example events such as promotional seasons. Thus, the time dimension can be included in a schema depending on users' requirements and the capabilities provided by the underlying DBMS.

Table 5.2 summarizes the temporality types of the MultiDim model. For example, lifespan support is allowed for levels and parent-child relationships, while valid time support is allowed for attributes and measures. Note that transaction time support for measures may be allowed or not. As we will see in Sect. 5.8.3, this depends on whether the measures are aggregated or not before they are loaded into the data warehouse.

In the following sections, we refer in more detail to the temporal elements of the MultiDim model described above. For levels and hierarchies, we give examples using valid time; nevertheless, the results may be straightforwardly generalized to the other temporality types. We first show the conceptual representation of the temporal elements and then provide their mapping into the ER and object-relational models.

Table 5.2. Temporality types of the MultiDim model

Temporality types	Levels	Attributes	Measures	Parent-child relationships
LS				
VT				
TT			/	
LT				

5.4 Temporal Support for Levels

Changes in a level can occur either for a member as a whole (for example, inserting or deleting a product in the catalog of a company) or for attribute values (for example, changing the size of a product). Representing these changes in a temporal data warehouse is important for analysis purposes, for example to discover how the exclusion of some products or changes to the size of a product influence sales. As shown in Fig. 5.6, in the MultiDim model, a level may have temporal support independently of the fact that it has temporal attributes.

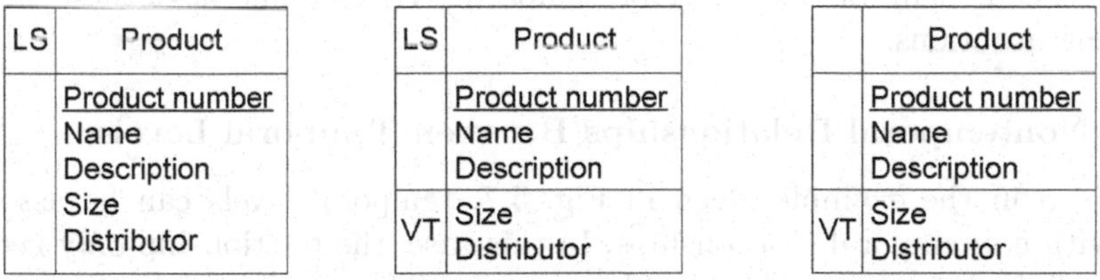

LS	Product
	Product number
	Name
	Description
	Size
	Distributor

(a) Temporal level

LS	Product
	Product number
	Name
	Description
VT	Size
	Distributor

(b) Temporal level with temporal attributes

	Product
	Product number
	Name
	Description
VT	Size
	Distributor

(c) Nontemporal level with temporal attributes

Fig. 5.6. Types of temporal support for a level

Temporal support for a level allows a time frame to be associated with its members. This is represented by including the symbol for the temporality type next to the level name, as shown in Fig. 5.6a. Various temporality types are possible for levels. Lifespan support is used to store the time of existence of the members in the modeled reality. On the other hand, transaction time and loading time indicate when members are current in a source system and in a temporal data warehouse, respectively. These three temporality types can be combined.

On the other hand, temporal support for attributes allows one to store changes in their values and the times when these changes occurred. This is represented by including the symbol for the corresponding temporality type next to the attribute name. For attributes we allow valid time, transaction time, loading time, or any combination of these. Figure 5.6b shows an example of a Product level that includes temporal attributes Size and Distributor. We group temporal attributes, firstly, to ensure that the two kinds of attributes (temporal and nontemporal) can be clearly distinguished, and, secondly, to reduce the number of symbols.

Many existing temporal models impose constraints on temporal attributes and the lifespan of their corresponding entity types. A typical constraint is that the valid time of the attribute values must be covered by the lifespan of their entity. As was done in [227], in our model we do not impose such constraints a priori. In this way, various situations can be modeled, for example when a product does not belong to a store's catalog but it is on sales on a trial basis. For this product, the valid time of its temporal attributes may not be within the product's lifespan. On the other hand, temporal integrity constraints may be explicitly defined, if required, using a calculus that includes Allen's operators [9].

5.5 Temporal Hierarchies

The MultiDim model allows us to represent hierarchies that contain several related levels. Given two related levels in a hierarchy, the levels, the relationship between them, or both may be temporal. We examine next these three different situations.

5.5.1 Nontemporal Relationships Between Temporal Levels

As shown in the example given in Fig. 5.7, temporal levels can be associated with nontemporal relationships. In this case, the relationship only keeps *current* links between products and categories. Therefore, in order to ensure correctness of the roll-up and drill-down operations, we assume a constraint stating that a product may be related with a category provided that the two are currently valid. As a consequence, the lifespans of a child member and its associated parent member must overlap. To indicate this we include the icon of the synchronization relationship in the temporal link as shown in the figure. If there is no synchronization constraint specified, by default we assume the overlaps synchronization relationship between related temporal levels. Nevertheless, other synchronization relationships between temporal levels may be specified. For example, if for administrative reasons a store must be related to at most one sales district throughout its lifespan, the synchronization relationship to be enforced is that the lifespan of a store is covered by the lifespan of its associated sales district.

A nontemporal relationship between temporal levels may induce incorrect analysis scenarios if the relationship changes. For example, consider the situation depicted in Fig. 5.8a, where at time t_1 product P is assigned to category

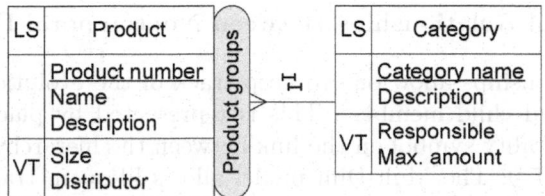

Fig. 5.7. A nontemporal relationship between temporal levels

C, and later on, at time t_2, category C ceases to exist. In order to have meaningful roll-up operations, product P must be assigned to another category C1 at the instant t_2. However, in a nontemporal relationship only the last modification is kept. Therefore, if the relationship P–C is replaced by P–C1, two incorrect aggregation scenarios may occur: either (1) measures cannot be aggregated before t_2 if category C1 did not exist before that instant (Fig. 5.8b), or (2) there will be an incorrect assignment of product P to category C1 before the instant t_2 if C1 existed before that instant. Therefore, users and designers must be aware of the consequences of allowing changes in nontemporal parent-child relationships between temporal levels. Note, however, that this problem also arises in conventional (i.e., nontemporal) data warehouses.

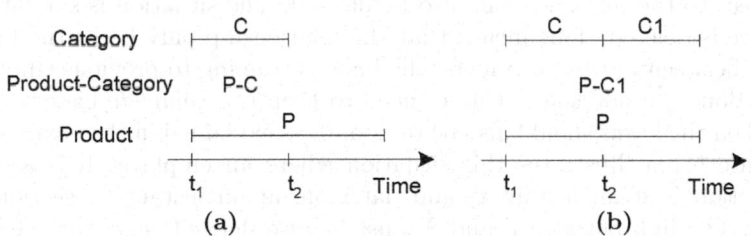

Fig. 5.8. An example of an incorrect analysis scenario when a nontemporal relationship between temporal levels changes

In our model, we use time-invariant identifiers for members, which, in particular, allow us to store a relationship between members independently of the evolution of its attributes. However, key attributes are typically used to display relationships between members for the roll-up and drill-down operations. In the example of Fig. 5.7, Category name is the key of the Category level. Suppose now that valid time support is added to the Category name attribute and that at time t_2, category C is renamed as category C1. In this case, two names for the category will exist: C before t_2 and C1 after t_2. Therefore, in order to adequately display the values of key attributes for different periods of time, special aggregation procedures must be developed for the roll-up operation.

5.5.2 Temporal Relationships Between Nontemporal Levels

Temporal relationships allow one to keep track of the evolution of links between parent and child members. This is represented by placing the corresponding temporality symbol on the link between the hierarchy levels, as can
be seen in Fig. 5.9. The MultiDim model allows lifespan, transaction time,
loading time, or any combination of these for representing temporal relationships between levels.

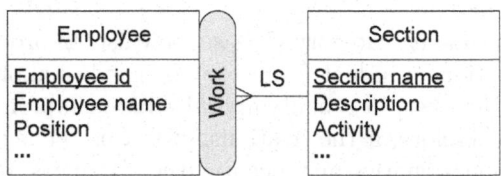

Fig. 5.9. A temporal relationship between nontemporal levels

Consider the example in Fig. 5.9 where a temporal relationship links nontemporal levels. In order to avoid dangling references, upon deletion of a
member of the related levels all its links must also be deleted. In Fig. 5.9 this
means that if a section S is deleted, then all the history of assignments of
employees to the section S will also be deleted. The situation is similar if an
employee is deleted. This means that the relationship only keeps the history
of links between *current* members; the links pertaining to previous employees
and sections are not kept. If users need to keep the *complete* history of the
links, then the levels should also be temporal, as explained in the next section.

Figure 5.10a illustrates this situation where an employee E is assigned
to a section S at an instant t_1 and, later on, at an instant t_2, section S is
deleted. The link between E and S must be also deleted, since, as the levels
are nontemporal, there is no more information about the existence of section
S. To ensure meaningful roll-up operations, employee E must be assigned to
another section at the instant t_2 (Fig. 5.10b). As can be seen in the figure,
part of the history of the assignment of employees to sections is lost.

Note also that when levels include valid time support for key attributes,
special aggregation procedures may be required if these key attributes change
their values, as was explained in Sect. 5.5.1.

5.5.3 Temporal Relationships Between Temporal Levels

Temporal relationships that link temporal levels help to avoid incorrect analysis scenarios and partial history loss of the kind described in Sects. 5.5.1 and
5.5.2. Consider the example in Fig. 5.11, which models a sales company that
is in a state of active development: changes to sales districts may occur to

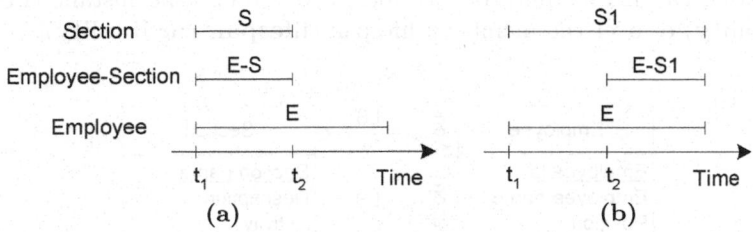

Fig. 5.10. Links kept by a temporal relationship between nontemporal levels: (a)
before and (b) after deleting a section

improve the organizational structure. The application needs to store the lifespans of districts in order to analyze how organizational changes affect sales.
Similarly, new stores may be created or existing ones may be closed; thus the
lifespans of stores are also stored. Finally, the application needs to keep track
of the evolution of assignments of stores to sales districts.

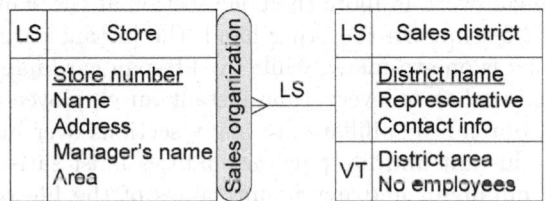

Fig. 5.11. A temporal relationship between temporal levels

In order to ensure correctness of roll-up and drill-down operations, we
impose a constraint on temporal relationships between temporal levels stating
that the lifespan of a relationship instance must be covered by the intersection
of the lifespans of the participating objects. In the example of Fig. 5.11 this
means that a store S and a sales district D may be related by a relationship
R provided that the store and the sales district exist throughout the lifespan
of the relationship R linking them.

As was the case in Sect. 5.5.1, when levels include valid time support
for key attributes, special aggregation procedures may be required if the key
attributes change their values.

5.5.4 Instant and Lifespan Cardinalities

Cardinalities in a nontemporal model indicate the number of members in
one level that can be related to members in another level. In our temporal

model, this cardinality may be considered for every time instant (**instant cardinality**) or over the members' lifespan (**lifespan cardinality**).

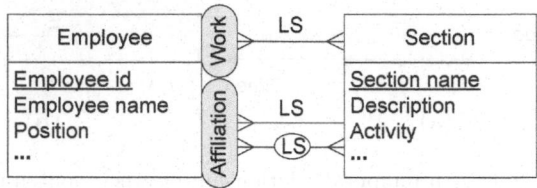

Fig. 5.12. Instant and lifespan cardinalities between hierarchy levels

In the MultiDim model, we assume by default that the instant cardinality is equal to the lifespan cardinality. However, if these cardinalities are different, the lifespan cardinality is represented by an additional line with the symbol LS surrounded by an ellipse, as shown in Fig. 5.12. In this example, the instant and lifespan cardinalities for the Work hierarchy are many-to-many, indicating that an employee can work in more than one section at the same time instant and over his/her lifespan. On the other hand, the instant cardinality for the Affiliation hierarchy is one-to-many, while the lifespan cardinality is many-to-many; this indicates that at every time instant an employee is affiliated to only one section, but can be affiliated to many sections over his/her lifespan.

Note that the instant and lifespan cardinalities must satisfy a constraint stating that the minimum and maximum values of the lifespan cardinality are greater than or equal to the minimum and maximum values, respectively, of the instant cardinalities. For example, if the instant cardinality between employees and sections is many-to-many, then the lifespan cardinality will also be many-to-many.

Consider now the cardinalities in the case of a temporal relationship between temporal levels, as in Fig. 5.11. In this case, the instant cardinalities must be interpreted with respect to the lifespan of the participating members, i.e., a child member must be related to a parent member *at each instant of its lifespan*; nevertheless, this parent member may change. Similarly, a parent member must have at least one child member assigned to it *at each instant of its lifespan*. This means that

- the lifespan of a child member (e.g., a store S) is covered by the union of the lifespans of all its links to a parent member (e.g., a sales district D); and
- the lifespan of a parent member is covered by the union of the lifespans of all its links to a child member.

Note that combining the two constraints above with the constraint stated in Sect. 5.5.3 that constrains the lifespan of a relationship instance to be covered by the intersection of the lifespans of the participating objects, then the lifespan of a child member is equal to the union of the lifespans of all its links to a parent member and vice versa.

5.6 Temporal Fact Relationships

An instance of a fact relationship relates leaf members from all its participating dimensions. If some of these members are temporal, they have an associated lifespan. In order to ensure correct aggregation, in our model we impose the constraint that the valid time of measures must be covered by the intersection of the lifespans of the related temporal members. This constraint is similar to the one given in Sect. 5.5.3 that constrains the lifespan of an instance of a parent-child relationship to be covered by the intersection of the lifespans of the participating objects.

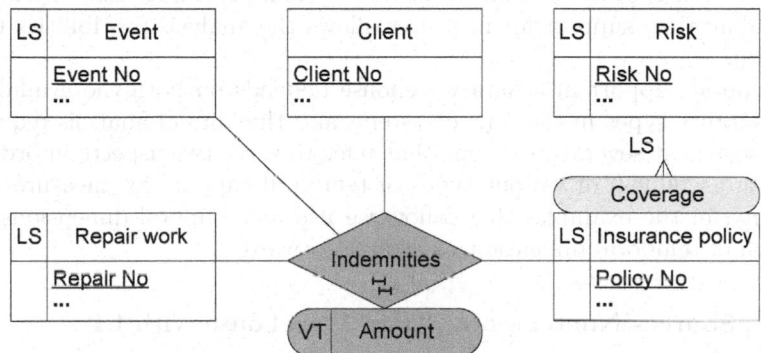

Fig. 5.13. Schema for analysis of indemnities paid by an insurance company

Consider the example in Fig. 5.13, which could be used by an insurance company to analyze the indemnities paid under insurance policies covering various risks. The measure in the fact relationship determines the amount of indemnities. In this example, the constraint stated above means that for an instance of the fact relationship, the valid time of the Amount measure must be covered by the lifespans of the related Repair work and Insurance policy members.

When there are two or more temporal levels that participate in a fact relationship, then a temporal join based on different synchronization relationships may be required. In Fig. 5.13, the fact relationship includes an intersection synchronization relationship that specifies the temporal join operation required. In this schema, the synchronization relationship in the fact relationship indicates that in order to relate an insurance policy, an event, and a repair work, the lifespans of the three members must have a nonempty intersection.

Nevertheless, in some situations this synchronization constraint does not correspond to reality. Although in Fig. 5.13 the events considered must occur within the lifespan of their related insurance policy, it might be the case that a client had decided to terminate an insurance policy some time after the event has occurred; therefore the repair work that determines the amount of indemnities paid would be done after the end of the policy. In order to be able to include such cases in the fact relationship in Fig. 5.13, the synchronization constraint must be removed.

5.7 Temporal Measures

5.7.1 Temporal Support for Measures

Current multidimensional models provide only valid time support for measures. Nevertheless, as we shall see in this section, providing transaction time and loading time support for measures allows the analysis possibilities to be expanded.

Temporal support in a data warehouse depends on both the availability of temporality types in the source systems and the kind of analysis required. We present next several situations that refer to these two aspects in order to show the usefulness of various types of temporal support for measures. For simplicity, in the examples that follow we use nontemporal dimensions; the inclusion of temporal dimensions is straightforward.

Case 1. Sources Nontemporal, Data Warehouse with LT

In real-world situations, the sources may be nontemporal or temporal support may be implemented in an ad hoc manner that can be both inefficient and difficult to obtain. Further, even though the sources may have temporal support, their integration into the data warehouse may be too costly; for example, this may be the case when checking the time consistency between different source systems. If decision-making users require a history of how the source data has evolved, measure values can be timestamped with the loading time, indicating the time when the data was loaded into the warehouse.

In the example in Fig. 5.14, users require a history of measures related to the inventory of products in relation to different suppliers and warehouses. The abbreviation "LT" indicates that measure values will be timestamped when they are loaded into the temporal data warehouse.

Case 2. Sources and Data Warehouse with VT

In some situations the source systems can provide valid time and this is required in the temporal data warehouse. Such valid time can represent events or states. Figure 5.15a gives an example of an event model used for analysis

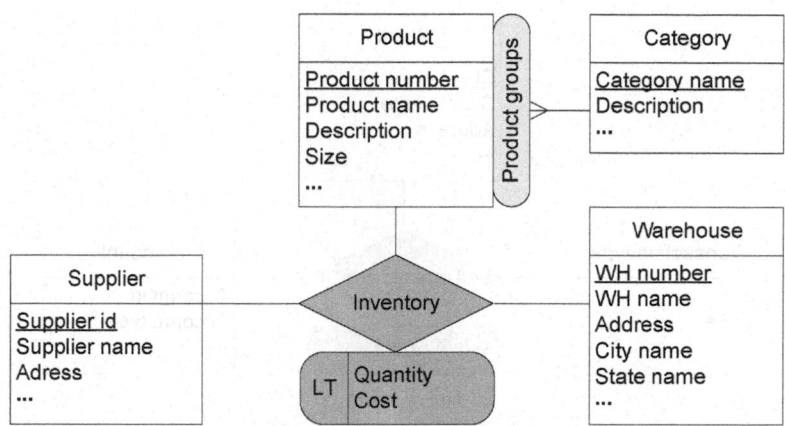

Fig. 5.14. Inclusion of loading time for measures

of banking transactions, and Fig. 5.15b shows a state model for analysis of employees' salaries. Note that the difference between events and states is not explicit in the graphical notation of our model, but can be specified in its textual representation, which is shown in Appendix A.

Various types of queries can be formulated for these schemas. For example, in Fig. 5.15a we can analyze clients' behavior related to the times between operations, the maximum or minimum withdrawal, the total amount involved in withdrawal operations, the total number of transactions during lunch hours, the frequency with which clients use ATMs, etc. This model also can be used to analyze clients' sequential behavior to avoid, for example, cancellation of an account at the bank, or to promote new services. On the other hand, the model in Fig. 5.15b could be used to analyze the evolution of the salaries paid to employees according to various criteria, for example changes in professional skills or participation in various projects.

Note that the event and state models are complementary, in the sense that from the events we can construct the states that span two events, and, likewise, from the states we can construct the events that represent changes from one state to another [23]. For example, the schema in Fig. 5.15a can be transformed into a state model that has a measure indicating the balance of an account between two events.

Case 3. Sources with TT, Data Warehouse with VT

In this case, the users require to know the time when data is valid in reality but the source systems can provide only the time when the data was available in the system, i.e., the transaction time. Thus, it is necessary to analyze whether the transaction time can be used to approximate the valid time. For example, if a measure represents the balance of an account, the valid time for

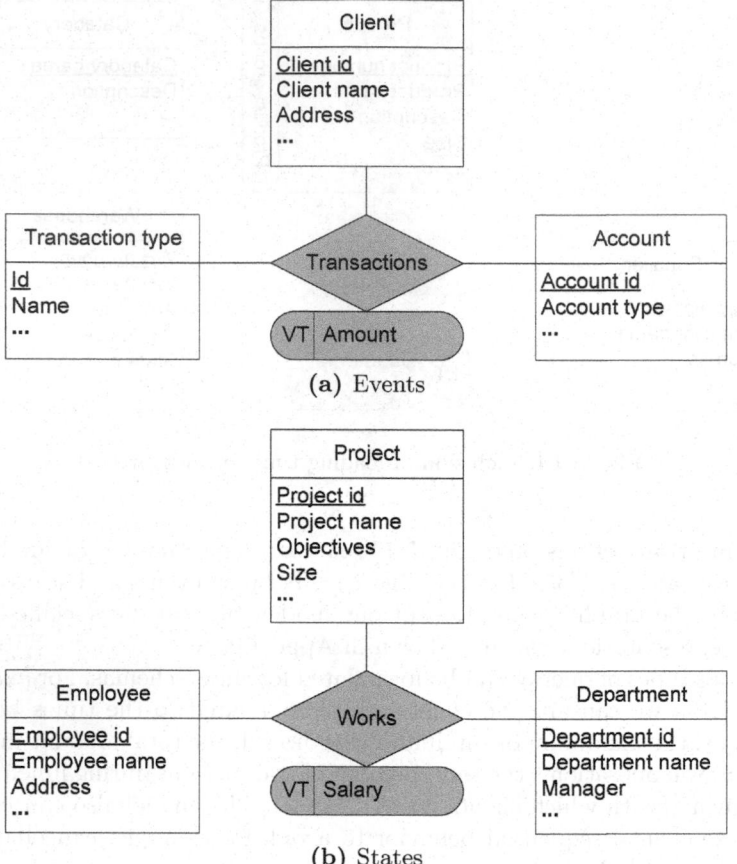

Fig. 5.15. Inclusion of valid time for measures

this measure can be calculated from the transaction times of two consecutive operations.

Nevertheless, in the general case the transaction time cannot be used for calculating the valid time, since data can be inserted into a source system (generating a transaction time) when it is no longer or not yet valid in the modeled reality, for example when one is recording an employee's previous or future salary. Since in many applications the valid time can be determined only by the user, it is incorrect to assume that if valid time is not available, the data should be considered valid when it is current in a source system [187]. The use of transaction time to approximate valid time must be a careful decision, and users must be aware of the imprecision that this may introduce.

Case 4. Sources with VT, Data Warehouse with VT and LT

The most common practice is to include valid time in a temporal data warehouse if it is available in the source systems. However, including in addition the loading time for measures can give information about the time since when data has been available for the purpose of a decision-making process.

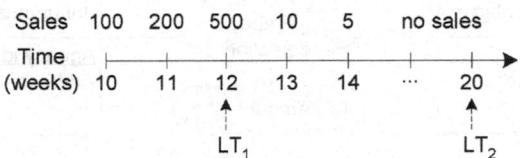

Fig. 5.16. Usefulness of including both valid time and loading time

The inclusion of loading time can help one to better understand decisions made in the past and to adjust loading frequencies. For the example shown in Fig. 5.16, suppose that at the loading time LT_1 it was decided to increase the inventory of a product on the basis of an increasing trend in its sales during weeks 10, 11, and 12. However, a sudden decrease in sales was revealed in the next data warehouse load at a time LT_2 eight weeks later. Thus, an additional analysis can be performed to understand the causes of these changes in sales behavior. Further, a decision about more frequent loads might be taken.

Case 5. Sources with TT, Data Warehouse with TT (and optionally LT and VT)

When a data warehouse is used for traceability applications, for example for fraud detection, the changes to the data and the time when they occurred should be available. This is possible if the source systems record the transaction time, since in this case past states of the database are kept.

The example given in Fig. 5.17 could be used by an insurance company that wishes to perform an analysis focused on the amount of insurance payments made. If there is a suspicion of an internal fraud that is modifying the insurance payments made to clients, it is necessary to obtain detailed information indicating when changes in measure values have occurred, i.e., transaction times. Note that including in addition the loading time would give information about the time since when data has been available to the investigation process. Further, the inclusion of valid time would allow one to know when payments were received by clients. In many real systems, a combination of both transaction time and valid time is included.

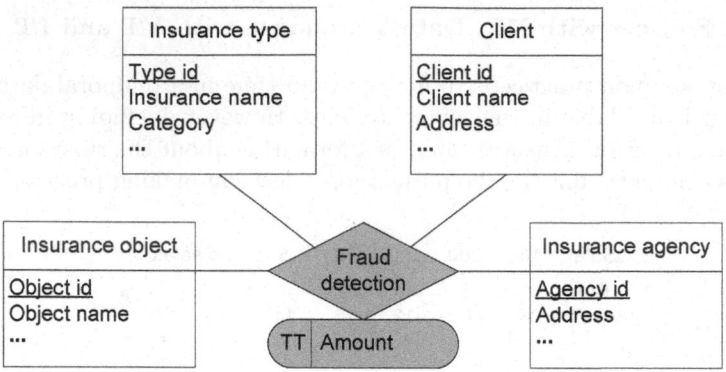

Fig. 5.17. A temporal data warehouse schema for an insurance company

Case 6. Sources with BT, Data Warehouse with BT and LT

Data in temporal data warehouses should provide a timely consistent representation of information [32]. Since some delay may occur between the time when an item of data is valid in reality, when it is known in a source system, and when it is stored in the data warehouse, it is sometimes necessary to include valid time, transaction time, and loading time.

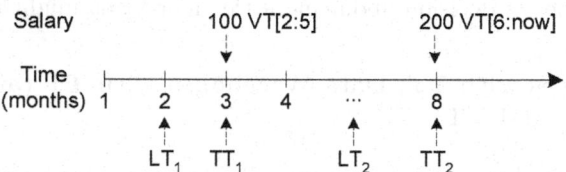

Fig. 5.18. Usefulness of valid time, transaction time, and loading time

Figure 5.18 shows an example inspired by [32] of the usefulness of these three temporality types. In this example, a salary of 100 with a valid time from month 2 to month 5 was stored at month 3 (TT_1) in a source system. Afterwards, at month 8 (TT_2), a new salary was inserted, with a value 200 and a valid time from month 6 until now. As shown in the figure, data was loaded into the temporal data warehouse at times LT_1 and LT_2. Different values of the salary can be retrieved depending on which instant of time the user wants to analyze; for example, the salary at month 1 is unknown, but at month 7 the value 100 is retrieved, since this is the last value available in the temporal data warehouse even though a new salary has already been stored in the source system. For more details, readers can refer to [32], where conditions are specified to ensure timely correct states for analytical processing.

5.7.2 Measure Aggregation for Temporal Relationships

Temporal relationships may require special aggregation procedures that adequately distribute measures between members of the related levels. Consider the example in Fig. 5.19, which is based on the schema in Fig. 5.11. In the example, a store S changes its assignment from sales district SD1 to SD2. The figure illustrates different cases of how the validity of the measures may be related to the lifespan of the link. In Fig. 5.19a the validity of the measures is equal to the lifespans of the links; therefore the aggregation can be developed as for conventional data warehouses. On the other hand, Figs. 5.19b,c illustrate different ways in which the validity of the measures may overlap the lifespans of the links. As can be seen, in Fig. 5.19b the measure 20 must be distributed between the two districts. This situation is similar in Fig. 5.19c for the measure 30. In the figures, we have distributed the measures between the two districts according to the percentage of the interval of the measure that is included in the interval of the link between stores and sales districts. However, other solutions can be applied depending on the particular semantics of the application.

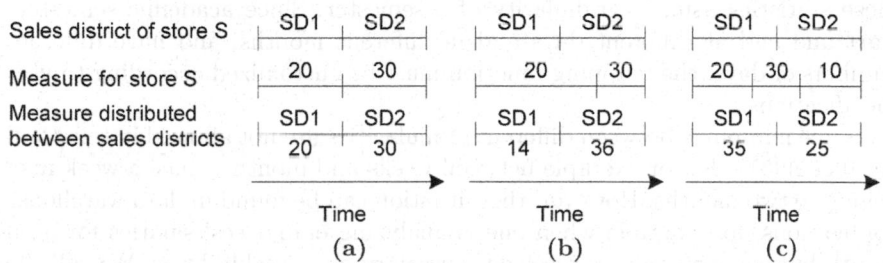

Fig. 5.19. An example of distribution of measures in the case of temporal relationships

As we shall see in the following section, the issue of measure aggregation in the presence of temporal relationships is somehow related to the problem of managing different temporal granularities.

5.8 Managing Different Temporal Granularities

Temporal data warehouses must usually deal with different temporal granularities. For example, the temporal granularities of measures and of dimensions may differ. Also, the temporal granularity of measures may be finer in the source systems than in the temporal data warehouse. In this section, we examine how to address this issue.

5.8.1 Conversion Between Granularities

Two types of mappings can be distinguished for converting between different temporal granularities: regular and irregular [59].

In a **regular** mapping, some conversion constant exists, i.e., one granule is a partitioning of another granule, and so if one granule is represented by an integer, this granule can be converted to another one by a simple multiply or divide strategy. Typical examples are conversion between minutes and hours and between days and weeks.

In an **irregular** mapping, granules cannot be converted by a simple multiply or divide, for example, when one is converting between months and days, since not all months are composed of the same number of days. Other examples include granularities that include gaps [21], for example business weeks that contain five days separated by a two-day gap. Thus, the mapping between different time granules must be specified explicitly and this requires customized functions with a detailed specification to obtain the desired conversion [59]. Such mapping functions may be quite complicated. For example, at a university the statistics for the use of a computer laboratory may be captured on a daily basis; however, a temporal data warehouse may include these statistics using a granularity of a semester. Since academic semesters start and end at different days and in different months, and have different numbers of days, the mapping function must be customized according to user specifications.

Some mappings between different granularities are not allowed in temporal databases [51, 59], for example between weeks and months, since a week may belong to two months. However, this situation can be found in data warehouse applications, for example when one is analyzing employees' salaries for each month but some employees receive their salary on a weekly basis. We call the mapping of such granularities **forced**. This requires special handling during measure aggregation, to which we refer next.

5.8.2 Different Granularities in Measures and Dimensions

In a temporal data warehouse, the temporal granularities of measures and dimensions may be different. Suppose that in the example shown in Fig. 5.11 the temporal granularities for measures and for dimension data are both equal to a month. This indicates that even if a store is assigned to a new sales district in the middle of a month, the timestamps are adjusted to the established granularity, i.e., one month. In this case, the aggregation of measures can be done safely, since the granularities are the same. A similar situation occurs when there is a regular mapping between the temporal granularities of the dimension data and of the measures, i.e., when one granule is a partitioning of another granule. For example, this is the case when measures are included on a monthly basis while changes to the association of stores with sales districts are included on a quarterly basis. We suppose that the temporal granularity for measures is finer than for dimensions. This is a realistic assumption, since the dimension data changes less frequently than the measures.

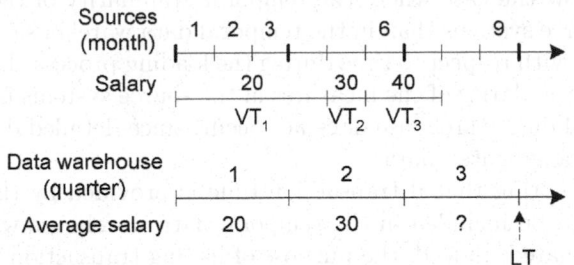

Fig. 5.20. Different temporal granularities in dimensions and measures

However, the situation is different when a forced mapping is required for converting between the temporal granularities of dimensions and of measures, i.e., when the temporal granularity of the measures is not a multiple of that of the dimensions. This situation is depicted in Fig. 5.20, where the measure is registered on a weekly basis but changes to dimension data are included on a monthly basis. Various kinds of changes can be considered, such as changing the assignment of a store from one sales district SD to another one SD1, or splitting a sales district SD into SD1 and SD2. Since some weeks may cross boundaries between months, it is not clear how to deal with the measure equal to 8 in the figure, i.e., whether include it in the aggregation for the sales district SD or for the sales district SD1, or distribute it between the two sales districts.

Various solutions can be applied in this case. An extreme one is to forbid this kind of situation by imposing a regular mapping between the temporal granularities of all multidimensional elements. Another solution is to specify **coercion functions** or **semantic assumptions** [21, 196, 310], which specify how to calculate aggregated values attached to multiple temporal granularities. In our example, we could indicate that if at least 70% of the time period for a specific measure value is included in the lifespan of a sales district SD, the measure should be considered in its totality for that district. An alternative solution could be to calculate the percentage of the week contained in each month and use this value to distribute measures between the two sales districts, as explained in Sect. 5.7.2.

All these solutions require a complex programming effort to manage multiple temporal granularities. This problem can be very complex and hard to solve if the multidimensional schema comprises several dimensions with multiple temporal granularities among them and different time granularities between the dimensions and the measures.

5.8.3 Different Granularities in the Source Systems and in the Data Warehouse

We consider now the case where the temporal granularity of the measures is finer in the source systems than in the temporal data warehouse, i.e., measures are aggregated with respect to time during the loading process. The case where the temporal granularity of the measures in the source systems is coarser than in the temporal data warehouse does not occur, since detailed data cannot be obtained from aggregated data.

It is worth noting that if transaction time is provided by the source systems, it will not be included in the temporal data warehouse when the measures are aggregated. Indeed, the purpose of having transaction time is to analyze changes that have occurred in individual items of data, and transaction time for aggregated data will not give useful information for decision-making users. Therefore, in this section we shall consider only measure aggregation with respect to valid time.

Note also that the loading frequency of a temporal data warehouse may be different from the temporal granularity used for measures: for example, data may be stored on a monthly basis while loading is performed every quarter. We assume that data can be kept in the source systems before it is loaded into a temporal data warehouse.

When measures must be aggregated prior to their loading into the data warehouse, this aggregation must take the type of measure into account. As already said in Sect. 2.6.2, three types of measures can be distinguished [146, 163]. **Additive measures**, for example monthly income, can be summarized along all dimensions, and in particular the time dimension. For example, if the time granularity in a temporal data warehouse is a quarter, three monthly incomes will be added together before being loaded into the temporal data warehouse. **Semiadditive measures**, for example inventory quantities, cannot be summarized along the time dimension, although they can be summarized along other dimensions. Therefore, it is necessary to determine what kind of function can be applied for aggregating them, for example, the average. Finally, **nonadditive measures**, for example item price, cannot be summarized along any dimension.

In some cases, the procedures for measure aggregation can be complex. A simple example is given in Fig. 5.21, where the sources have a granularity of a month and the data warehouse has a granularity of a quarter. This example includes three different cases: (1) periods of time with a constant salary that overlaps several quarters (where the salary is 20 or 40), (2) a quarter with different salaries (quarter 2), and (3) a quarter when the salary is not paid for several months (quarter 3). Suppose that a user requires as measure the average salary per quarter. For the first quarter, the average value is easily calculated. For the second quarter, a simple average does not work, and thus a weighted mean value may be given instead. However, for the third quarter, the user must indicate how the value is specified. In the example, we have

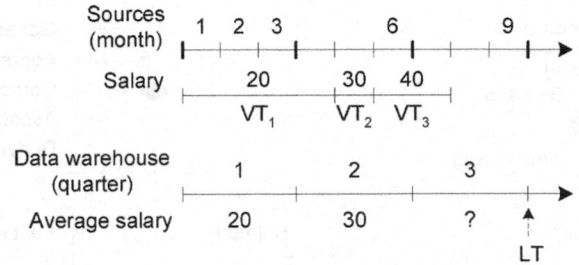

Fig. 5.21. Example of a coercion function for salary

opted to give an undefined value. Nevertheless, if instead of using the average salary we were to use the sum (the total salary earned during a quarter), the measure value for quarter 3 could be defined.

Real-world situations can be more complicated, demanding the specification of coercion functions or semantic assumptions to calculate values attached to multiple temporal granularities. In the example in Fig. 5.21, we have used a user-defined coercion function stating that if a temporal-data-warehouse granule is not totally covered by the valid time of one or several salaries, the average salary is undefined. It should be noted that coercion functions are always required for forced mappings of granularities, since a finer time granule can map to more than one coarser time granule; for example, a week may span two months. In this case, the measure values of the finer granule must be distributed. For example, suppose that a salary is paid on a weekly basis and that this measure is stored in a temporal data warehouse at a granularity of a month. If a week belongs to two months, the user may specify that the percentage of the salary that is assigned to a month is obtained from the percentage of the week contained in the month, for example 2 days out of 7, as explained in Sect. 5.7.2.

Coercion functions and semantic assumptions have been studied in the temporal-database community for several years. The proposed solutions can be used in the context of temporal data warehouses in order to develop aggregation procedures.

5.9 Metamodel of the Temporally Extended MultiDim Model

This section presents the metamodel of the temporally extended MultiDim model using UML notation, as shown in Fig. 5.22. This metamodel is based on that presented in Fig. 3.29 for conventional data warehouses.

As shown in the figure, a dimension is composed of either one level or one or more hierarchies, while each hierarchy belongs to only one dimension. A hierarchy contains two or more related levels that can be shared between

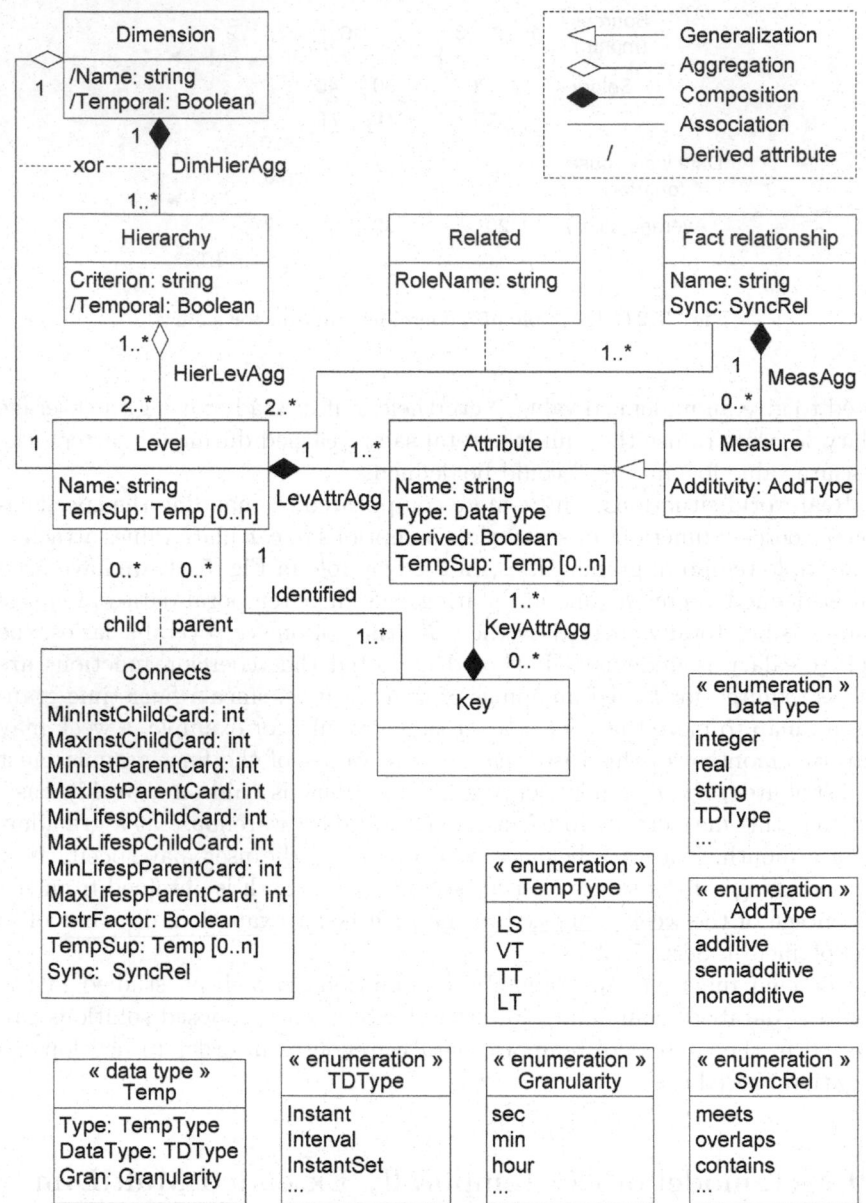

Fig. 5.22. Metamodel of the temporally extended MultiDim model

different hierarchies. A criterion name identifies a hierarchy, and each level has a unique name. The leaf-level name is used to derive a dimension's name: this is represented by the derived attribute Name in Dimension. Levels include attributes, some of which are key attributes used for aggregation purposes, while others are descriptive attributes. Attributes have a name and a type, which must be a data type, i.e., integer, real, string, etc. Further, an attribute may be derived.

Temporal support for levels and attributes is captured by the TempSup multivalued attribute of type Temp. The latter is a composite attribute with three components. The first one (Type) is of an enumerated type that represents the various kinds of temporal support, namely lifespan, valid time, transaction time, and loading time. The second component (DataType) is of an enumerated type containing various temporal data types defined in the model, i.e., instant, interval, set of instants, etc. Finally, the third component (Gran) is of an enumerated type containing the various granularities supported, i.e., seconds, minutes, hours, etc.

Parent-child relationships relating two levels are represented by the Connects association class. These relationships may be temporal, independently of whether the levels have temporal support or not. This is indicated by the attribute TempSup of the Connects association class. Additionally, parent-child relationships are characterized by instant and lifespan cardinalities. Both kinds of cardinalities include the minimum and maximum values expressed in the child and parent roles. The Connects association class includes also an optional distributing factor and an optional synchronization relationship. A constraint not shown in the diagram specifies that a parent-child relationship has a distributing factor only if the maximum cardinalities of the child and the parent are equal to "many". Another constraint specifies that a parent-child relationship has a value for the synchronization relationship only if both related levels are temporal.

A fact relationship represents an n-ary association between leaf levels with $n > 1$. Since leaf levels can play various roles in this association, the role name is included in the Related association class. A fact relationship may contain measures. These measures are temporal and may be additive, semiadditive, or nonadditive (AddType). Further, a fact relationship may have a synchronization relationship (SyncRel). A constraint specifies that a fact relationship has a synchronization relationship only if it relates at least two temporal levels.

A dimension is temporal if it has at least one temporal hierarchy. A hierarchy is temporal if it has at least one level with temporal support or one temporal parent-child relationship. This is represented by the derived attribute Temporal in Dimension and in Hierarchy.

5.10 Rationale of the Logical-Level Representation

As described in Chap. 3, the MultiDim model can be implemented by mapping its specifications into those of an operational data model, such as the relational or the object-relational model. In this chapter, we use a two-phase approach where a temporal MultiDim schema is transformed first into a classical entity-relationship (ER) schema and then into an object-relational (OR) schema.

We chose the ER model, since it is a well-known and widely used conceptual model. As a consequence, the ER representation of the MultiDim constructs allows a better understanding of their semantics. Further, the transformation of the ER model into operational data models is well understood (see, e.g., [66]) and this translation can be done by the usual CASE tools. Therefore, we have proposed a mapping in a second step that translates the intermediate ER schema into an object-relational schema.

We used the object-relational model to implement our temporal conceptual model because it allows attributes to have complex types. The object-relational model inherently groups related facts into a single row [48], thus allowing one to keep together changes to data and the times when they occurred. These facilities are not provided by the relational model, which imposes on users the responsibility of knowing and maintaining the groupings of tuples representing the same real-world fact in all their interactions with the database.

We chose a mapping approach instead of using normalization for several reasons. First, there are no well-accepted normal forms for temporal databases, even though some formal approaches exist (e.g., [135, 310, 312]). Further, the purpose of normalization is to avoid the problems of redundancy, inconsistency, and update anomalies. However, the usual practice in data warehouses is to denormalize relations to improve performance and to avoid the costly process of joining tables in the presence of high volumes of data. This denormalization can be done safely because the data in temporal data warehouses is integrated from operational databases, which are usually normalized, and thus there is no danger of incurring the problems mentioned above. Finally, a normalization approach may result in a number of artificial relations that do not correspond to real-world entities, making the system more complex for the purposes of designing, implementing, and querying.

5.11 Logical Representation of Temporal Data Warehouses

5.11.1 Temporality Types

Temporal support in the MultiDim model is represented in an implicit manner, using pictograms. Therefore, when the MultiDim model is mapped into the ER model, additional attributes are needed to keep this temporal support. Further, the mapping depends on whether the temporal elements represent events or states. The former are represented in the ER model by an instant or a set of instants, while the latter are represented by an interval or a set of intervals.

As already said, the MultiDim model provides several temporality types: valid time, transaction time, lifespan, and loading time. Valid time and lifespan are used for representing both events and states. Representing valid time by a set of instants or a set of intervals allows one to specify that an attribute has the same value in discontinuous time spans, for example when an employee works in the same section during different periods of time. Similarly, representing a lifespan by a set of intervals allows one to consider discontinuous lifespans, for example when a professor leaves on sabbatical during some period of time. To represent transaction time, the usual practice is to use an interval or a set of intervals. Finally, loading time is represented by an instant, since this temporality type indicates the time instant at which data was loaded into a temporal data warehouse.

The rule for mapping temporality types from the MultiDim model to the ER model is as follows:

Rule 1: A temporality type is represented in the ER model in one of the following ways:

> **Rule 1a:** An instant is represented as a monovalued attribute.
>
> **Rule 1b:** A set of instants is represented as a multivalued attribute.
>
> **Rule 1c:** An interval is represented as a composite attribute that has two attributes indicating the beginning and the end of the interval.
>
> **Rule 1d:** A set of intervals is represented as a multivalued composite attribute composed of two attributes indicating the beginning and the end of the interval.

We have used the SQL:2003 standard to specify the mapping to the object-relational model. As already said in Sect. 2.3.2, SQL:2003 represents collections using the array and multiset types. The **array type** allows variable-sized vectors of values of the same type to be stored in a column, while the **multiset type** allows unordered collections of values to be stored. Unlike arrays, multisets have no declared maximum cardinality. SQL:2003 also supports **structured user-defined types**, which are analogous to class declarations in object languages. Structured types may have attributes, which can be of any SQL type, including other structured types at any nesting. Structured types can be used as the domain of a column of a table, the domain of an attribute of another type, or the domain of a table.

Therefore, in the mapping to the object-relational model, (1) a multivalued attribute in the ER model is represented as a multiset (or array) attribute, and (2) a composite attribute in the ER model is represented as an attribute of a structured type. In this way, an instant, a set of instants, an interval, and a set of intervals can be represented in SQL:2003 as follows:

```
create type InstantType as date;
create type InstantSetType as InstantType multiset;
create type IntervalType as (FromDate date, ToDate date);
create type IntervalSetType as IntervalType multiset;
```

In the declaration of IntervalType above, FromDate and ToDate represent the beginning and the end instant of the interval. Note that in the above declarations we have used the date data type, which provides a granularity of a day. Alternatively, the timestamp data type could be used, which provides a granularity of a second or a fraction of a second. However, since SQL does not provide support for other granularities (such as a month or a year) these must be encoded in standard types (such as integers) and must be explicitly managed by the application.

As an example of a commercial object-relational DBMS, we have used Oracle 10g [213]. Oracle includes constructs that allow collections to be represented. A **varying array** stores an ordered set of elements in a single row, and a **table type** allows one to have unordered sets and to create nested tables, i.e., a table within a table. The former corresponds to the array type of SQL:2003 and the latter to the multiset type. Further, Oracle provides **object types** that are similar to structured user-defined types in SQL:2003. Thus, the declarations of the temporality types can be expressed in Oracle as follows:

```
create type InstantType as object (Instant date);
create type InstantSetType as table of InstantType;
create type IntervalType as object (FromDate date, ToDate date);
create type IntervalSetType as varray(10) of IntervalType;
```

In the above definition we used a varray for defining set of intervals. This avoids having two-level nesting of tables when defining temporal attributes.

5.11.2 Levels with Temporal Support

The transformation of levels and their attributes into the ER model is done according to the following rules:

Rule 2: A nontemporal level is represented in the ER model as an entity type.

Rule 3: A temporal level is represented in the ER model as an entity type with an additional attribute for each of its associated temporality types; the latter are mapped according to Rule 1 given in Sect. 5.11.1.

Rule 4: A nontemporal attribute is represented in the ER model as a mono-valued attribute.

Rule 5: A temporal attribute is represented in the ER model as a multivalued composite attribute that includes an attribute for the value and an additional attribute for each of its associated temporality types; the latter are mapped according to Rule 1.

For instance, the mapping of the Product level in Fig. 5.6c to the ER model is shown in Fig. 5.23a. For simplicity, we have not represented in the figure the Distributor attribute, which can be mapped in a way similar to the Size attribute. Note that in the figure the validity of attribute values is represented by an interval, which is a typical practice for dimension data in temporal data warehouses [23, 63]. If the valid time of the attribute Size is represented by a set of intervals, this results in the schema given in Fig. 5.23c.

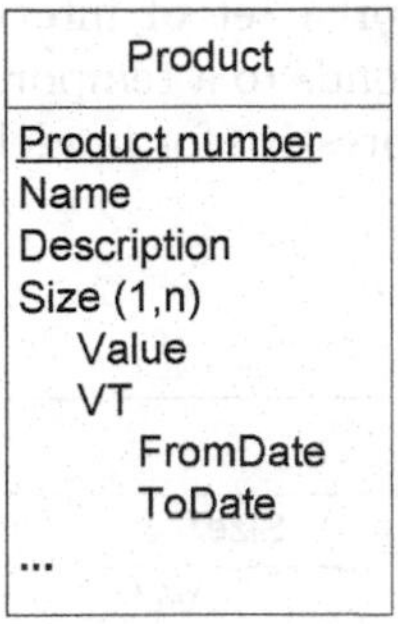

Product
Product number
Name
Description
Size (1,n)
Value
VT
FromDate
ToDate
...

Product

PId	Product number		Size*			
			Value	VT		...
		...		FromDate	ToDate	
1	QB876	...	10	05/2002	08/2002	...
			20	09/2002	07/2003	
2	QD555	...	18	05/2002	now	...

(a) ER schema (b) Object-relational representation

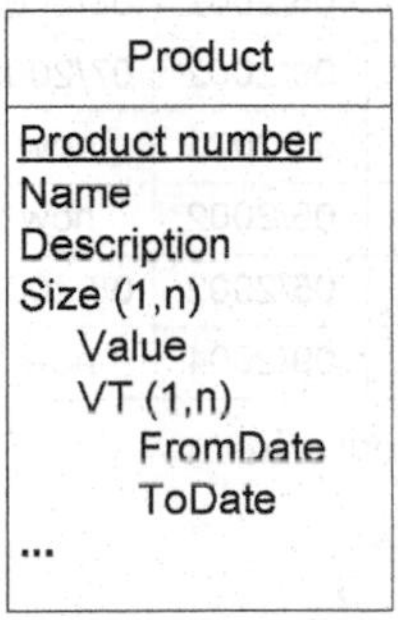

Product
Product number
Name
Description
Size (1,n)
Value
VT (1,n)
FromDate
ToDate
...

Product

PId	Product number		Size*			
			Value	VT*		...
		...		FromDate	ToDate	
1	QB876	...	10	05/2002	08/2002	...
				09/2002	07/2003	
			20	08/2003	now	
2	QD555	...	18	05/2002	now	...

(c) ER schema (d) Object-relational representation

Fig. 5.23. Mapping levels with temporal attributes

As can be seen by comparing the Product levels in Fig. 5.6c and Figs. 5.23a,c, the MultiDim model provides a better conceptual representation of time-varying attributes than does the ER model. It contains fewer elements, it allows one to clearly distinguish which data changes should be kept, and it leaves technical aspects such as multivalued or composite attributes outside the users' concern.

Applying the traditional mapping to the relational model (e.g., [66]) to the Product level in Fig. 5.23a gives three tables: one table for all monovalued attributes and one table for each multivalued attribute. All tables include the key attribute of the products. This relational representation is not very intuitive, since the attributes of a level are stored as separate tables. It also has well-known performance problems due to the join operations required, especially if the levels form hierarchies.

The object-relational representation allows these drawbacks to be overcome, keeping a level and its temporal attributes together in a single table. Figures 5.23b,d show object-relational schemas using a tabular representation. In the figure, we use the symbol * to denote collections. Two possible representations are included, depending on whether the valid time of the attribute Size is represented by an interval (Fig. 5.23b) or a set of intervals (Fig. 5.23c). The object-relational representation corresponds to a temporally grouped data model [48], which is considered as more expressive for modeling complex data [312].

(a) ER schema

Product
LS (1,n)
FromDate
ToDate
<u>Product number</u>
Name
Description
Size (1,n)
Value
VT (1,n)
FromDate
ToDate
...

(b) Object-relational representation

Product							
<u>Pld</u>	LS*		Product number	...	Size*		
	FromDate	ToDate			Value	VT*	
						FromDate	ToDate
1	05/2002	now	QB876	...	10	05/2002	08/2002
					20	09/2002	07/2003
					15	07/2003	now
2	05/2002	now	QD555	...	18	05/2002	now
3	05/2002	08/2003	QE666	...	25	05/2002	08/2003
	09/2004	now			18	09/2004	now

Fig. 5.24. Mapping a temporal level

For example, Fig. 5.24a shows the ER representation of the temporal level in Fig. 5.6b. Recall that in Sect. 5.11.1 we described the mapping of the lifespan temporality type to the ER model. In the example used there we represented the lifespan by a set of intervals, where FromDate and ToDate indicate, respectively, the beginning and the end instants of the intervals. Figure 5.24b shows the object-relational representation, where the surrogate attribute, the lifespan, and the attributes of all levels are kept together in the same table.

There are two ways to define the components of the temporal attribute Size in SQL:2003, corresponding, respectively, to Figs. 5.23b and 5.23d:

```
create type SizeValueType as (Value real, VT IntervalType);
create type SizeValueType as (Value real, VT IntervalSetType);
```

Since Size is a multivalued attribute, we represent it in SQL:2003 as a collection type using either an array or a multiset, i.e., using one of the following:

```
create type SizeType as SizeValueType multiset;
create type SizeType as SizeValueType array(7);
```

A level is represented in the object-relational model as a table containing all its attributes and an additional attribute for its key. Recall from Sect. 2.3.2 that two kinds of tables can be defined in SQL:2003. **Relational tables** are the usual kind of tables, although the domains for attributes are all predefined or user-defined types. **Typed tables** are tables that use structured types for their definition. In addition, typed tables contain a **self-referencing column** containing a value that uniquely identifies each row. Such a column may be the primary key of the table, it could be derived from one or more attributes, or it could be a column whose values are automatically generated by the DBMS, i.e., surrogates.

Since surrogates are important in data warehouses, we have used a typed table to represent the Product level.[4] Recall that the declaration of a typed table requires a previous definition of a type for the elements of the table:

```
create type ProductType as (
    ProductNumber integer,
    Name character varying(25),
    Description character varying(255),
    Size SizeCollType )
    ref is system generated;
create table Product of ProductType (
    constraint ProductPK primary key (ProductNumber),
    ref is PId system generated );
```

The clause ref is PId system generated indicates that PId is a surrogate attribute automatically generated by the system.

The definitions in Oracle differ slightly from those in SQL:2003. Given the specification of temporality types in Sect. 5.11.1, we define the type SizeType required for the Size attribute as follows:

```
create type SizeValueType as object (Value number, VT IntervalSetType);
create type SizeType as table of SizeValueType;
```

The Product table is defined in Oracle using an object table as follows:

```
create type ProductType as object (
    ProductNumber number(10),
    Name varchar2(25),
    Description varchar2(255),
    Size SizeType );
create table Product of ProductType (
    constraint ProductPK primary key (ProductNumber) )
    nested table Size store as SizeNT
    object identifier is system generated;
```

As said in Sect. 5.11.1, two different types of collections can be used in Oracle, i.e., varying arrays and table types. However, it is important to consider the differences at the physical level between these two options. Varying

---

[4] For simplicity, in the examples here we omit the full specification of constraints and additional clauses required by the SQL:2003 standard.

arrays are, in general, stored "inline" in a row[5] and cannot be indexed. On the other hand, rows in a nested table can have identifiers, can be indexed, and are not necessarily brought into memory when the main table is accessed (if the field defined as a nested table is not specified in the query). Nested tables require one to specify their physical location. For example, as shown in the above declaration, the physical location of the nested table Size must be explicitly defined when the Product table is created. Therefore, the choice between nested tables and varying arrays must be made according to the specific details of the application, for example which data is accessed and which operations are required, as well as taking performance issues into account.

5.11.3 Parent-Child Relationships

This section presents the mapping of parent-child relationships. We distinguish two cases, depending on whether the relationships are temporal or not.

Nontemporal Relationships

The transformation of nontemporal relationships between levels to the ER model is based on the following rule:

Rule 6: A nontemporal parent-child relationship is represented in the ER model as a binary relationship without attributes.

An example of this transformation for the relationship between the Product and Category levels in Fig. 5.7 is shown in Fig. 5.25a.[6]

To obtain the corresponding object-relational schema, we use the traditional mapping for binary one-to-many relationships and include a reference to the parent level in the child-level table. For example, the mapping of the Product level and the Product–Category relationship gives the relation shown in Fig. 5.25b. Since the levels are identified by surrogates, which are time-invariant, this mapping does not depend on whether the levels are temporal or not.

To define the Product table in SQL:2003, we need to create first a typed table Category with a surrogate in the CId attribute. This is shown in the following:

```
create type CategoryType as (
    CategoryName character varying(25),
    ... )
    ref is system generated;
create table Category of CategoryType (
    constraint CategoryPK primary key (CategoryName),
```

---

[5] If they are smaller than 4000 bytes or not explicitly specified as a large object (LOB).

[6] For simplicity, we present only one temporal attribute.

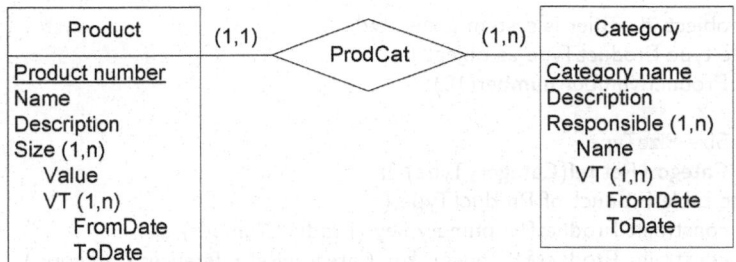

(a) ER schema

Product						
Pld	Product number	...	Size*			Category Ref
			Value	VT*		
				FromDate	ToDate	
1	QB876	...	10	05/2002	08/2002	C1
				08/2003	now	
			20	09/2002	07/2003	
2	QD555	...	18	05/2002	now	C2

(b) Object-relational representation

Fig. 5.25. Mapping a hierarchy with a nontemporal relationship

```
    ref is Cld system generated );
create type ProductType as (
    ProductNumber integer,
    . . .
    Size SizeCollType
    CategoryRef ref(CategoryType) scope Category references are checked )
    ref is system generated;
create table Product of ProductType (
    constraint ProductPK primary key (ProductNumber),
    ref is Pld system generated );
```

Note the CategoryRef attribute in ProductType, which is a reference type pointing to the Category table. This attribute can be used for the roll-up operation from the Product to the Category level.

The declarations in Oracle for representing the relationship between the Product and Category levels are very similar to those in SQL:2003:

```
create type CategoryType as object (
    CategoryName varchar2(25),
    . . . );
create table Category of CategoryType (
    constraint CategoryPK primary key (CategoryName))
```

```
        object identifier is system generated;
create type ProductType as object (
    ProductNumber number(10),
    ...
    Size SizeType,
    CategoryRef ref(CategoryType) );
create table Product of ProductType (
    constraint ProductPK primary key (ProductNumber),
    constraint ProductFK foreign key CategoryRef references Category )
    nested table Size store as SizeNT
    object identifier is system generated;
```

Temporal Relationships

Temporal relationships can link either nontemporal levels as in Fig. 5.9 or temporal levels as in Fig. 5.11. Further, the maximum cardinality of the child level can be equal to 1 or to n. Note that in order to have meaningful hierarchies, we assume that the maximum cardinality of the parent level is always equal to n.

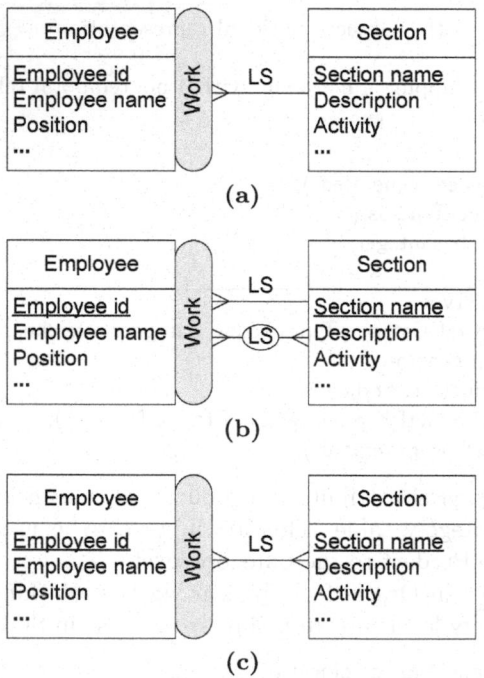

Fig. 5.26. Various cardinalities for temporal relationships linking nontemporal levels

Figure 5.26a shows a schema where the maximum child cardinality is equal to 1. In the figure, the instant and lifespan cardinalities are the same (i.e., one-to-many), allowing a child member to belong to at most one parent member during its entire lifespan; they indicate that an employee may work only in one section and that if he/she returns after a period of leave, he/she must be assigned to the same section.

There are two different cases where the maximum child cardinality is equal to n. In Fig. 5.26b, the instant and lifespan cardinalities are different, i.e., a child member is related to one parent member at every time instant and to many parent members over its lifespan. On the other hand, in Fig. 5.26c, the instant and lifespan cardinalities are the same, i.e., a child member is related to many parent members at every time instant.[7] The mappings of both cases are handled in the same way, since when the cardinalities are different we must keep all links according to the lifespan cardinality. The instant cardinality is then represented as a constraint stating that of all the links of a member, only one is current.

The following rule is used for mapping temporal relationships:

Rule 7: A temporal parent-child relationship is represented in the ER model as a binary relationship with an additional attribute for each of its associated temporality types; the latter are mapped according to Rule 1.

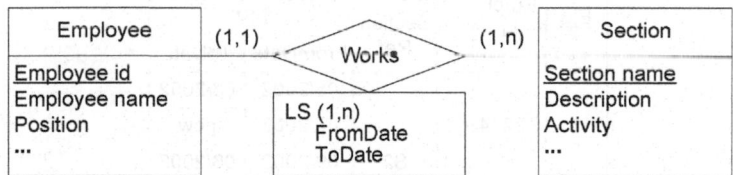

Fig. 5.27. Mapping the schema given in Fig. 5.26a into the ER model

An example of the application of this rule to the schema given in Fig. 5.26a is shown in Fig. 5.27. The lifespan of the relationship is represented as a multivalued composite attribute, since an employee can be hired several times in the same section, i.e., it corresponds to a set of intervals (Rule 1d). The mapping to the ER model of the schemas in Figs. 5.26b,c is similar except that the cardinalities for both roles of the relationship are (1,n).

There are several options for representing temporal relationships in the object-relational model. The first option consists in creating a separate table for the relationship; this is shown in Fig. 5.28a for the ER schema shown in Fig. 5.27. The Works table is composed of the surrogates of the Employee and Section levels, and the temporality of the relationship. Before the Works

---

[7] As we have seen in Chap. 3, this corresponds to a *nonstrict* hierarchy.

Works				
			LS*	
WId	Empl. Ref	Section Ref	FromDate	ToDate
1	E1	S1	05/2002	08/2002
			07/2003	now
2	E2	S1	05/2002	now

(a) One-to-many cardinalities

Employee					
			InSection		
				LS*	
EId	Empl. id	...	Section Ref	FromDate	ToDate
1	E2244	...	S1	05/2002	08/2002
				07/2003	now
2	E2345	...	S1	05/2002	now

(b) One-to-many cardinalities

Employee					
			InSection*		
				LS*	
EId	Empl. id	...	Section Ref	FromDate	ToDate
1	E2244	...	S1	05/2002	08/2002
				07/2003	now
			S2	09/2002	06/2003
2	E2345	...	S1	05/2002	now

(c) Many-to-many cardinalities

Fig. 5.28. Various object-relational representations of temporal links

table is defined in SQL:2003, the Employee and Section tables must have been previously declared as typed tables. The Works table is defined as follows:

```
create type WorksType as (
    EmployeeRef ref(EmployeeType) scope Employee references are checked,
    SectionRef ref(SectionType) scope Section references are checked,
    LS IntervalSetType )
    ref is system generated;
create table Works of WorksType (
    constraint WorksPK primary key (EmployeeRef)
    ref is WId system generated );
```

In order to use the above table for the cardinalities illustrated in Figs. 5.26b,c it is necessary to change the key of the table as being composed of both EmployeeRef and SectionRef. However, in the case of Fig. 5.26b, an integrity constraint must be defined to ensure that at each instant an employee is related to at most one section. Section 5.13.1 shows examples of such a temporal constraint.

The second option consists in representing a temporal relationship as a composite attribute in one of the level tables; this is shown in Fig. 5.28b for the relationship in Fig. 5.27. In the figure, the Employee table includes the InSection attribute, which stores the Section surrogates with their associated temporal spans. The corresponding SQL:2003 declaration is as follows:

```
create type InSectionType as (
    SectionRef ref(SectionType) scope Section references are checked,
    LS IntervalSetType );
create type EmployeeType as (
    EmployeeId integer,
    . . .
    InSection InSectionType )
    ref is system generated;
create table Employee of EmployeeType (
    constraint EmployeePK primary key (EmployeeId),
    ref is EId system generated );
```

If the cardinalities between the Employee and Section levels are many-to-many (Figs. 5.26b,c), the InSection attribute should be defined as InSectionType multiset allowing an employee to work in many sections; this is shown in Fig. 5.28c. However, in the case of Fig. 5.26c, an integrity constraint must ensure that at each instant an employee is related to at most one section, as was the case for Fig. 5.26a.

The second option above, i.e., the inclusion of an additional attribute in a level table, allows one to store all attributes and relationships of a level in a single table. This table thus expresses the whole semantics of a level. However, the choice from among the alternative object-relational representations may depend on query performance requirements and physical-level considerations for the particular DBMS used, such as join algorithms and indexing capabilities. For example, defining the InSection attribute as a nested table in Oracle will require a join of two tables, thus not offering any advantage with respect to the solution of using a separate table for the Works relationship.

The ER and object-relational representations for temporal levels linked by temporal relationships are the same as those described earlier for nontemporal levels and temporal relationships, since the surrogates of parent and child members are time-invariant. An example of a MultiDim schema and its object-relational representation corresponding to this case is given in Fig. 5.29.

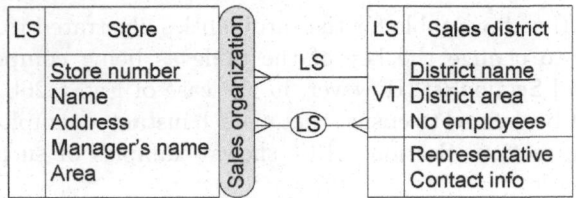

(a) MultiDim schema

Store								
Sld	LS*		Store number	...	SalesOrganization*			
	FromDate	ToDate			District Ref	LS*		
						FromDate	ToDate	
1	05/2002	now	QB876	...	A	05/2002	08/2002	
						10/2004	now	
					B	09/2002	09/2004	
2	05/2002	now	QD555	...	B	05/2002	now	
3	05/2002	08/2003	QE666	...	B	05/2002	now	
	09/2004	now						

SalesDistrict				
Dld	LS*		District name	...
	FromDate	ToDate		
A	05/2002	08/2003	Ixelles	...
	10/2004	now		
B	05/2002	now	Forest	...

(b) Object-relational representation

Fig. 5.29. Mapping a temporal relationship between temporal levels

5.11.4 Fact Relationships and Temporal Measures

Temporal measures may represent either events or states. In this section, we refer only to measures representing events whose valid time has a granularity of a month. Nevertheless, the results may be generalized straightforwardly to measures representing states, to other temporality types, and to other granularities.

The following rules are used for mapping fact relationships:

Rule 8: A fact relationship is represented in the ER model as an *n*-ary relationship.

Rule 9: A measure of a fact relationship is represented in the ER model in the same way as a temporal attribute of a level using Rule 5.

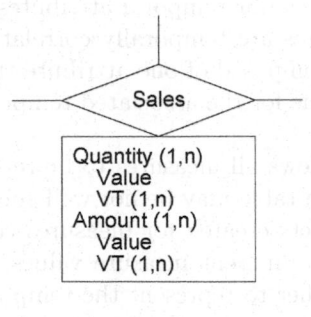

(a) ER representation

(b) Relational table for the Quantity measure

Sales				
Client fkey	Product fkey	Store fkey	Quantity	VT
C1	P1	S1	100	05/2002
C1	P1	S1	150	06/2002
C1	P1	S1	100	07/2002
C1	P1	S2	200	05/2002
C1	P1	S2	50	06/2002
C1	P1	S2	80	07/2002
...	...	...	...	...

Sales							
Sid	Client Ref	Product Ref	Store Ref	Quantity*		Amount*	
				Value	VT*	Value	VT*
1	C1	P1	S1	100	05/2002	10000	05/2002
					07/2002	15000	06/2002
				150	06/2002	18000	07/2002
2	C1	P1	S2	200	05/2002	20000	05/2002
				50	06/2002		07/2002
				80	07/2002	5000	06/2002
...	...	...	...	...	...	...	...

(c) Object-relational representation

Fig. 5.30. Mapping of the fact relationship shown in Fig. 5.5

Figure 5.30a shows the mapping to the ER model of the fact relationship with temporal measures shown in Fig. 5.5. Mapping this fact relationship to the relational model gives two tables. In Fig. 5.30b we show the table for the Quantity measure; the table for the Amount measure has a similar structure. The table has columns for the foreign keys of the participating dimensions, for the measure value, and for the associated temporal support. However, the relational schemas for measures can be simplified if additional information is available. For example, if all measures are temporally correlated, they can be represented in one table. In our example, this means that we add the Amount attribute to the table in Fig. 5.30b.

As shown in Fig. 5.30c, the object-relational representation creates a table for the fact relationship that includes as attributes the references to the surrogate keys of the participating levels. In addition, every measure is mapped into a new attribute in the same way as was done for temporal attributes of a level. As for the relational case, if all measures are temporally correlated, they can be represented in a single attribute, composed of one attribute that keeps the value for each measure and another one for the associated temporal support.

Even though the object-relational model allows all measures to be represented in a single table, in practice the resulting table may be not well suited for aggregation with respect to time. The objects created for measures contain a two-level nesting: one level to represent different measure values for the same combination of foreign keys and another to represent the temporal elements. Therefore, it is difficult to express aggregation statements related to time when one is accessing the second-level nesting.

When the object-relational representation is chosen, it is important to consider physical-level features of the particular object-relational DBMS used. For example, when nested varying arrays are used in Oracle, timestamps cannot be indexed, and comparisons of valid times must be done by programming. On the other hand, if two nested tables are used, indexing and comparisons are allowed, improving query formulation and execution. However, to access a measure and its corresponding valid time, the two nested tables must be joined, in addition to a join with the main table containing the foreign keys. Therefore, depending on the specific features of the target object-relational DBMS, the relational representation may be more suitable for representing in a more "balanced" manner all attributes that may be used for aggregation purposes.

5.12 Summary of the Mapping Rules

In this section, we recapitulate all of the mapping rules given in the previous sections for transforming the MultiDim model into the ER model.

Rule 1: A temporality type is represented in the ER model in one of the following ways:

> **Rule 1a:** An instant is represented as a monovalued attribute.
> **Rule 1b:** A set of instants is represented as a multivalued attribute.
> **Rule 1c:** An interval is represented as a composite attribute that has two attributes indicating the beginning and the end of the interval.
> **Rule 1d:** A set of intervals is represented as a multivalued composite attribute composed of two attributes indicating the beginning and the end of the interval.

Rule 2: A nontemporal level is represented in the ER model as an entity type.

Rule 3: A temporal level is represented in the ER model as an entity type with an additional attribute for each of its associated temporality types; the latter are mapped according to Rule 1.

Rule 4: A nontemporal attribute is represented in the ER model as a mono-valued attribute.

Rule 5: A temporal attribute is represented in the ER model as a multival-ued composite attribute that includes an attribute for the value and an additional attribute for each of its associated temporality types; the latter are mapped according to Rule 1.

Rule 6: A nontemporal parent-child relationship is represented in the ER model as a binary relationship without attributes.

Rule 7: A temporal parent-child relationship is represented in the ER model as a binary relationship with an additional attribute for each of its asso-ciated temporality types; the latter are mapped according to Rule 1.

Rule 8: A fact relationship is represented in the ER model as an n-ary relationship.

Rule 9: A measure of a fact relationship is represented in the ER model in the same way as a temporal attribute of a level using Rule 5.

5.13 Implementation Considerations

The mapping to an object-relational database provides structures for storing dimensions and measures that change over time. Therefore, users can have access to various versions of multidimensional data. This could help them to understand various analysis scenarios, for example an increase in the sales of some product after a change in its size, or a decrease in the total sales for some category of products due to the fact that the category had been split into two new categories. However, manipulation and aggregation of time-varying data is not an easy task. The solution could include management of changes at the schema level (for example adding a new level to a hierarchy or deleting measures) as well as at the instance level (for example including a new member in a level, changing an association between products and categories, or splitting categories). In our model, we consider changes at the instance level, since changes at the schema level require appropriate mechanisms for schema versioning or schema evolution.

Further, even though there exist several specialized prototype systems (as will be described in Sect. 5.14), in this section we shall refer to more gen-eral solutions that can be implemented in current DBMSs. We shall briefly refer to three aspects: (1) keeping the semantic equivalence between the con-ceptual and logical schemas, (2) aggregation of measures in the presence of time-varying levels and hierarchies, and (3) handling different temporal gran-ularities in dimensions and measures during aggregations.

5.13.1 Integrity Constraints

After the logical structures that implement a conceptual schema have been de-fined, it is necessary to complement these structures with additional integrity constraints that capture the particular semantics of an application domain. In the case of a temporal data warehouse, many integrity constraints must be implemented to enforce the inherent semantics of time.

Current DBMSs provide a limited set of predefined integrity constraints. These constraints include "not null" attributes, key attributes, unique attributes, referential integrity, and simple check constraints. All other integrity constraints must be explicitly programmed by the designer, and this is typically done using triggers. In this section, we discuss with the help of an example the issue of enforcing temporal integrity constraints. We use Oracle 10g as the target implementation platform. For a detailed explanation of all the temporal constraints, the reader is referred to [274].

The definition in Oracle of the object-relational structures corresponding to the schema in Fig. 5.29a is as follows:

```
create type IntervalType as object ( FromDate date, ToDate date );
create type IntervalSetType as varray(10) of IntervalType;
create type DistrictNameValueType as object (
    Value varchar(20),
    VT IntervalSetType );
create type DistrictNameType as table of DistrictNameValueType;
create type DistrictAreaValueType as object (
    Value number(10),
    VT IntervalSetType );
create type DistrictAreaType as table of DistrictAreaValueType;
create type NoEmployeesValueType as object (
    Value number(10),
    VT IntervalSetType );
create type NoEmployeesType as table of NoEmployeesValueType;
create type StoreType as object (
    Lifespan IntervalSetType,
    StoreNumber char[12],
    Name varchar2(30),
    Address varchar2(50),
    ManagerName varchar2(50),
    Area number(10) );
create table Store of StoreType
    nested table Lifespan store as StoreLifespanNT;
create type SalesDistrictType as object (
    Lifespan IntervalSetType,
    DistrictName DistrictNameType,
    DistrictArea DistrictAreaType,
    NoEmployees NoEmployeesType,
    Representative varchar2(35),
    ContactInfo varchar2(100) );
create table SalesDistrict of SalesDistrictType
    nested table Lifespan store as SalesDistrictLifespanNT
    nested table DistrictName store as DistrictNameNT
    nested table DistrictArea store as DistrictAreaNT
```

```
                nested table NoEmployees store as NoEmployeesNT;
        create table Store_SD as (
                StoreRef ref StoreType,
                SalesDistrictRef ref SalesDistrictType,
                FromDate date,
                ToDate date );
```

Note that in the above definitions, the parent-child relationship between stores and sales districts is implemented as a relational table (Store_SD). As we shall see next, this facilitates the implementation of temporal constraints.

In current object-relational DBMSs, integrity constraints can only be defined on the tables, not on the underlying types. For this reason, several constraints must be defined on the nested tables to verify the validity of lifespans. We give below the constraints for stores; similar constraints must be enforced for sales districts:

```
        alter table StoreLifespanNT
                add constraint FromDateUQ unique (FromDate);
        alter table StoreLifespanNT
                add constraint ToDateUQ unique (ToDate);
        alter table StoreLifespanNT
                add constraint ValidInterval check (FromDate<ToDate);
```

The above constraints ensure that in the lifespan of a single store, there are no two lines with the same value of FromDate or ToDate and that FromDate is smaller than ToDate. Nevertheless, these constraints do not ensure that all intervals defining the lifespan are disjoint. The following trigger ensures this constraint:

```
        create trigger StoreLifespanOverlappingIntervals
                before insert on Store
                for each row
        declare
                cnt number(7):=0;
        begin
                select count(*) into cnt
                from table(:new.Lifespan) l1, table(:new.Lifespan) l2
                where l1.FromDate<l2.ToDate and l2.FromDate<l1.ToDate;
                if ( cnt>1 ) then
                        raise application error(-20300, 'Overlapping intervals in lifespan')
                end if;
        end
```

In the above trigger, the condition in the where clause ensures that the two lifespan intervals defined by l1 and l2 overlap. Since this condition is satisfied if l1 and l2 are the same tuple, the cnt value must be greater than one for having two distinct intervals in the lifespan that overlap.

As already said in Sect. 5.5.3, temporal parent-child relationships must satisfy a constraint stating that the lifespan of a relationship instance must be covered by the intersection of the lifespans of the participating members. The following triggers ensure this constraint:

```
create trigger Store_SDLifespan1
    before insert on Store_SD
    for each row
declare
    cnt number(7):=0;
begin
    select count(*) into cnt
    from Store s
    where :new.StoreRef=s and not exists (
        select * from table(s.column_value.Lifespan) l
        where l.FromDate<=:new.FromDate and :new.ToDate<=l.ToDate );
    if ( cnt>0 ) then
        raise application error(-20301,
            'Lifespan of relationship is not contained in lifespan of child member')
    end if;
end;
create trigger Store_SDLifespan2
    before insert on Store_SD
    for each row
declare
    cnt number(7):=0;
begin
    select count(*) into cnt
    from SalesDistrict d
    where :new.SalesDistrictRef=d and not exists (
        select * from table(d.column_value.Lifespan) l
        where l.FromDate<=:new.FromDate and :new.ToDate<=l.ToDate );
    if ( cnt>0 ) then
        raise application error(-20302,
            'Lifespan of relationship is not contained in lifespan of parent member')
    end if;
end;
```

Further, in order to ensure the correctness of roll-up and drill-down operations, in multidimensional hierarchies it is also required that a child member must be related to a parent member throughout its lifespan, and vice versa. This amounts to saying that the lifespan of a child member is equal to the union of the lifespans of its links to all parent members, and vice versa. Computing this union corresponds to a temporal projection of the table Store_SD, where the results are coalesced. It should be noted that coalescing is a complex and costly operation in SQL. We compute next the intervals during which a store is associated with any sales district. Computing the corresponding information for a sales district is done in a similar way.

```
create view Temp(StoreRef, FromDate, ToDate) as
    select StoreRef, FromDate, ToDate from Store_SD
```

```
create view StoreLifespanLinks(StoreRef, FromDate, ToDate) as
    select distinct F.StoreRef, F.FromDate, L.ToDate
    from Temp F, Temp L
    where F.StoreRef=L.StoreRef and F.FromDate<L.ToDate
    and not exists ( select * from Temp M
        where M.StoreRef=F.StoreRef
        and F.FromDate<M.FromDate and M.FromDate<=L.ToDate
        and not exists ( select * from Temp M1
            where M1.StoreRef=F.StoreRef
            and M1.FromDate<M.FromDate and M.FromDate<=M1.ToDate ) )
    and not exists ( select * from Temp M2
        where M2.StoreRef=F.StoreRef
        and ( (M2.FromDate<F.FromDate and F.FromDate<=M2.ToDate)
        or (M2.FromDate<=L.ToDate and L.ToDate<M2.ToDate) ) )
```

Finally, the following trigger ensures that the lifespan of a store is equal to the union of the lifespans of its links, which is computed in the view Store-LifespanLinks:

```
create trigger StoreAlwaysLinkedSalesDistrict
    before insert on Store
    for each row
declare
    cnt number(7):=0;
begin
    select count(*) into cnt
    from table(:new.Lifespan) l1
    where not exists (
        select * from StoreLifespanLinks l2
        where l1.FromDate=l2.FromDate and l1.ToDate=l2.ToDate );
    if ( cnt>0 ) then
        raise application error(-20303,
        'Store is not related to a sales district throughout its lifespan')
    end if;
end;
```

Synchronization relationships constrain the lifespans of linked object types. For example, suppose that in Fig. 5.29 a synchronization relationship of type starts specifies that a Store member can be related to a Sales district member only if their lifespans both start at the same instant. A set of triggers must be generated to preserve the semantics of this relationship in the physical schema. One of them fires on insertions into Store_SD as follows:

```
create trigger StoreStartsSalesDistrict
    before insert on Store_SD
    for each row
declare
    cnt number(7):=0;
begin
    select count(*) into cnt
```

```
from Store s, SalesDistrict d, table(s.column_value.Lifespan) l1,
    table(d.column_value.Lifespan) l2
where :new.StoreRef=s and :new.SalesDistrictRef=d and not exists (
    select * from table(s.column_value.Lifespan) l3
    where l3.FromDate<l1.FromDate )
and not exists (
    select * from table(d.column_value.Lifespan) l4
    where l4.FromDate<l2.FromDate )
and l1.FromDate<>l2.FromDate;
if ( cnt>0 ) then
    raise application error(-20301,
        'Lifespans of parent and child do not start at the same instant')
    end if;
end;
```

The purpose of this trigger is to roll back the insertion of a new instance of the
parent-child relationship if the lifespans of the related store and sales district
do not start together.

As can be seen throughout this section, enforcing temporal integrity constraints is a complex and costly operation. In this section, we have given only
triggers that monitor the insertion operations; additional triggers must be introduced to monitor the update and delete operations. In addition, similar
triggers must be generated for all temporal levels and temporal parent-child
relationships in a schema. The problem of defining a set of triggers that ensures a set of temporal integrity constraints is an open area of research, both
in the database community and in the data warehouse community.

5.13.2 Measure Aggregation

In conventional data warehouses, measures are the only elements that change
over time. Therefore, when a user requests aggregated values with respect to
some time period, the measure values included in that time period are aggregated. However, the situation is different when this aggregation is performed
in the presence of time-varying dimension data. Consider the example shown
in Fig. 5.31a, where changes to levels and to the relationship between them are
stored. Figure 5.31b shows two tables of the corresponding relational schema.
Figure 5.31c shows examples of values for the association of stores with sales
districts and for the Quantity measure; in the figure, the time line has a granularity of a month. At the beginning of the period shown in Fig. 5.31c, the
three stores S1, S2, and S3 are associated with the sales district SD. After
three months, the organizational division changes and the sales district SD is
split into two districts SD1 and SD2, leading to a reassignment of stores to
sales districts: stores S1 and S2 are included in the sales district SD1, while
store S3 in the sales district SD2. The values of the Quantity measure for the
stores are also shown in Fig. 5.31c.

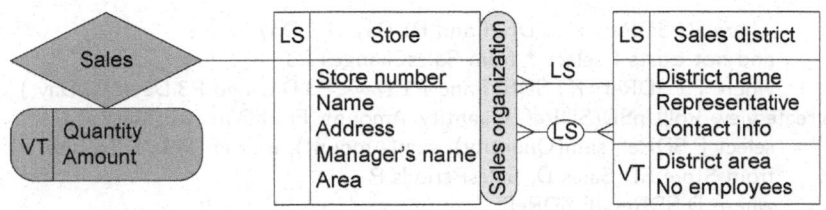

(a) Excerpt from the schema shown in Fig. 5.5

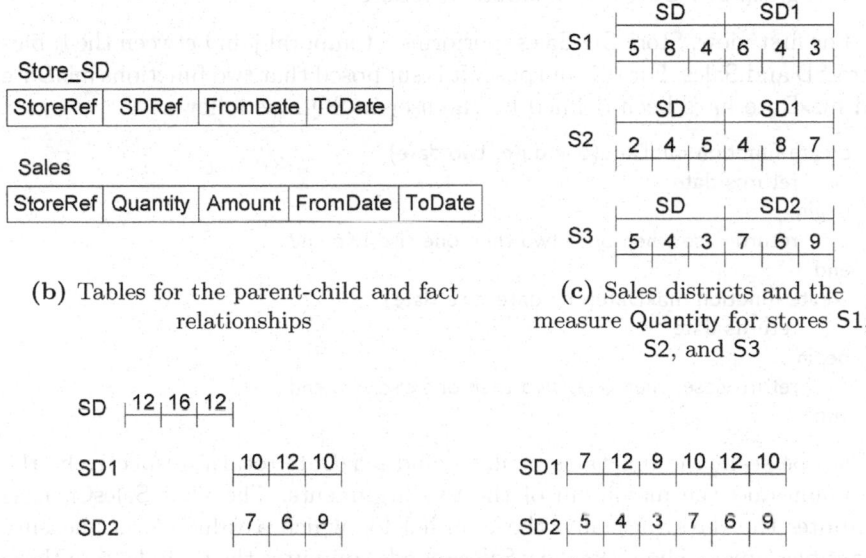

(b) Tables for the parent-child and fact relationships

(c) Sales districts and the measure Quantity for stores S1, S2, and S3

(d) Traditional temporal aggregation   (e) Eder and Koncilia's [61] aggregation

Fig. 5.31. Approaches to measure aggregation in the presence of temporal relationships

Let us suppose that a user wishes to roll up to the sales district level. The aggregation of measure values can be done in standard SQL using the approach given in [323]:

```
create view Store_SD_Sales(SDRef, Quantity, Amount, FromDate,ToDate) as
    select distinct D.SDRef, S.Quantity, S.Amount,
        maxDate(S.FromDate,D.FromDate), minDate(S.ToDate,D.ToDate)
    from Sales_SalesDistrict D, Sales S where D.StoreRef=S.StoreRef
    and maxDate(S.FromDate,D.FromDate)<minDate(S.ToDate,D.ToDate)
create view SalesChanges(SDRef, Day) as
    select distinct SDRef, FromDate from Store_SD_Sales
    union select distinct SDRef, ToDate from Store_SD_Sales
create view SalesPeriods(SDRef, FromDate, ToDate) as
    select P1.DNumber, P1.Day, P2.Day
    from SalesChanges P1, SalesChanges P2
```

```
        where P1.SDRef=P2.SDRef and P1.Day<P2.Day
        and not exists ( select * from SalesChanges P3
        where P1.SDRef=P3.SDRef and P1.Day<P3.Day and P3.Day<P2.Day )
create view RollUpSD(SDRef, Quantity, Amount, FromDate, ToDate) as
        select P.SDRef, sum(Quantity), sum(Amount), P.FromDate, P.ToDate
        from Store_SD_Sales D, SalesPeriods P
        where D.SDRef=P.SDRef
        and D.FromDate<=P.FromDate and P.ToDate<=D.ToDate
        group by P.SDRef, P.FromDate, P.ToDate
```

The first view, Store_SD_Sales, performs a temporal join between the tables Store_SD and Sales. For this purpose, it is supposed that two functions minDate and maxDate have been defined by the user in SQL, as follows.

```
create function minDate(one date, two date)
        returns date
begin
        return case when one<two then one else two end
end
create function maxDate(one date, two date)
        returns date
begin
        return case when one>two then one else two end
end
```

As can be seen, the functions minDate and maxDate return, respectively, the minimum and the maximum of the two arguments. The view SalesChanges computes the days (actually, the months) for which a value of the measure Quantity is given. Then, the view SalesPeriods computes the periods from these days. Finally, the view RollUpSD computes the sum of the measures for these periods.

Using the above approach, aggregation of the quantity measures shown in Fig. 5.31c gives the results shown in Fig. 5.31d. Note that during the first quarter the sales districts SD1 and SD1 do not exist, and therefore the aggregated values are unknown. The same situation occurs for the sales district SD after the splitting. This kind of aggregation is usual in temporal databases [148, 278, 323]. This approach is called "defensive" [62] in the temporal data warehouse community, since users must know that the sales district SD was split after the third month in order to make a correct interpretation of the results obtained.

On the other hand, in temporal data warehouses there is a clear tendency to use so-called "offensive" approaches, which provide users with various options for analyzing data that changes over time. An example based on the solution proposed by Eder and Koncilia [61][8] is shown in Fig. 5.31c. First, users must choose which version of the data they wish to analyze, i.e., they

---

[8] We omit some details of the proposed solution in order to facilitate understanding of the example.

must select a time instant before or after the splitting of the sales district. In the example, we suppose that the users have selected the version after this splitting. Therefore, the results of aggregation will refer to the sales districts SD1 and SD2 for all periods analyzed, as shown in Fig. 5.31e. The values of aggregated measures for the first quarter are based on "transformation" functions. These indicate how the measure originally assigned to the sales district SD should be distributed between the sales districts SD1 and SD2, which did not exist during this time period. In the example, we have chosen that 75% of this value will be assigned to sales district SD1 and 25% to sales district SD2. Note that this solution gives only approximate results, and it also requires knowledge about the application provided to the system in order to perform measure aggregation.

There are other approaches to temporal aggregation in data warehouses; these are discussed in Sect. 5.14. However, all these approaches, including the one described above, require in general customized systems, usually built on top of the DBMS, as well as extensions of query languages to manipulate time-varying data.

5.14 Related Work

The necessity to manage time-varying data in databases has been acknowledged for several decades (e.g., [67, 136]). However, no such consensus has been reached about representing time-varying multidimensional data. Publications related to temporal data warehouses have raised many issues, for example the inclusion of temporality types in temporal data warehouses (e.g., [3, 32]), temporal querying of multidimensional data (e.g., [195, 231]), correct aggregation in the presence of changes to data and structure (e.g., [63, 108, 195]), materialization of temporal views from nontemporal sources (e.g., [319]), multidimensional schema evolution (e.g., [25, 63, 195]), and implementation considerations in relation to a temporal star schema (e.g., [23]). Nevertheless, very little attention has been paid to conceptual modeling for temporal data warehouses and its subsequent logical mapping.

The publications reviewed next fall into four categories. First, publications that describe various temporality types that may be included in temporal data warehouses; second, conceptual models for temporal data warehouses; third, publications referring to a logical-level representation; and, finally, publications related to the handling of different granularities.

5.14.1 Types of Temporal Support

The inclusion of various temporality types in temporal data warehouses has been briefly mentioned in several publications. While most publications consider valid time [16, 23, 25, 42, 61, 63, 187, 204, 253, 319], none of them distinguish between lifespan and valid time support. As we saw in Sect. 5.5, this distinction is important, since it leads to several different constraints when levels form hierarchies.

With respect to transaction time, several approaches have been taken. Some approaches ignore transaction time [25, 194]. As we have already mentioned, the lack of transaction time precludes traceability applications. Pedersen et al. [231] briefly mention the possibility of including transaction time, but these authors do not provide an analysis of the usefulness of this temporality type in the data warehouse context. Some other approaches transform the transaction time from the source systems to represent the valid time [3, 187]. This is semantically incorrect, because data may be included in a database independently of its period of validity, for example data about previous or future employees might be added. Finally, some other approaches consider the transaction time generated in a temporal data warehouse in the same way as transaction time in a temporal database [150, 187, 260]. However, since data in a temporal data warehouse is neither modified nor deleted, the transaction time generated in a temporal data warehouse represents in fact the time when the data was loaded into the data warehouse.

Our proposal differs from those mentioned above in several respects: (1) we distinguish between lifespan support for levels and hierarchy links from valid time support for attributes and measures; (2) we include support for valid time, lifespan, and transaction time originating from a source system (if available); and (3) we include a new temporality type, i.e., loading time, that is generated in a temporal data warehouse. We have shown by means of examples the usefulness of these temporality types. To the best of our knowledge, only Bruckner and Min Tjoa [32] have discussed the inclusion of valid time, transaction time, and loading time. However, unlike our approach, these authors limit the use of these temporality types to active data warehouses and do not provide a conceptual model that includes these types.

5.14.2 Conceptual Models for Temporal Data Warehouses

Several models provide solutions for the problem of handling changes in multidimensional structures. These solutions fit into two categories [25]: schema evolution [22, 107, 108, 316] and historical models [24, 25, 42, 61, 63, 79, 194, 231]. In the former approach, a unique data warehouse schema is maintained and data is mapped to the most recent schema version. A set of operations that allow schema and instance modifications is usually provided. However, since only the last version of the schema is included, the history of the evolution is lost.

Historical models keep track of the evolution of the schema and/or instances, allowing the coexistence of different versions of the schema and/or instances. In general, the proposed models include temporal support for levels and for the links between them. Only a few models allow timestamping of instances. For example, Chamoni and Stock [42] allow timestamping of hierarchical assignments between level instances. The assignments are represented as a matrix, where the rows and columns correspond to the level instances and where the cells store the time of validity of hierarchical assignments between level instances. A similar approach that timestamps level instances and their hierarchical assignments was presented by Eder and Koncilia [61]. However, in the matrix of these authors, the rows and columns represent the old and new versions of instances and the cells include a value that is used for transforming measures from one version to another.

Most models are able to represent changes at the schema level; the changes at the instance level are managed as a consequence of schema modifications. These models focus mainly on the problem of data aggregation in the presence of different schema/instance versions.

Various mechanisms are used to create a new schema version. For example, Ravat et al. [254] defined a multiversion multidimensional model that consists of a set of star versions. Each star version is associated with a temporal interval that includes one version for a fact and one version for each dimension associated with a fact. Whenever changes occur at the schema level (to dimensions or to facts), a new star version is created. A similar approach was taken by Wrembel and Bebel [316], where each change at the schema or instance level results in the creation of a new schema version, even though that version has the same structure as the original one in the case of changes at the instance level. On the other hand, in the approach of Golfarelli et al. [79, 80], a schema is first transformed into a graph of functional dependencies, and then a new schema is created using schema modification operators.

Several models refer to the aggregation of measures in the presence of time-varying dimensional data, i.e., when queries refer to data included in several schema versions. Some publications require transformation functions between different schema/instance versions. For example, the augmented schema in [79, 80] contains elements in which the old and new schema versions differ. Users must then specify transformation actions between different versions. A set of mapping functions is also required in the solutions proposed by several other authors (e.g., [24, 25, 61, 63, 253, 254]). The model of Mendelzon and Vaisman [194, 195] extends the approach of Hurtado et al. [107, 108] by including timestamps for schema and instance versions. These authors also defined a specific language, called TOLAP (Temporal OLAP), that allows users to aggregate measures according to the dimension structures that existed when the corresponding measures were introduced. Another query language for temporal data was presented by Wrembel and Bebel [316]; this is based on publications of Bebel et al. [16] and Morzy and Wrembel [203, 204]. This language first decomposes a query into partial queries executed on a unique version, and then presents these partial results to users together with version and metadata information. In a second phase, integration of the partial results into a common set of data is performed, provided it is possible.

Some other publications relate to specific problems, such as maintaining changes in measures representing late registration events (i.e., events that need some confirmation in order to be valid) [260]. Since this work considered only valid time and transaction time for measures, changes to dimension data are not kept in that case.

In summary, the publications mentioned above have considered ways of providing temporal support for multidimensional models, allowing changes in dimension members, hierarchy links, and fact relationships to be expressed. Some of these publications also provide a query mechanism. However, none of the models propose a graphical representation of temporal multidimensional data that can be used for communicating between users and designers. Further, these models do not consider various other aspects considered in our work, for example a hierarchy that may have temporal and nontemporal levels linked by either temporal or nontemporal relationships, the inclusion of various temporality types for measures, and the problem of different temporal granularities between the source systems and the temporal data warehouse. In addition, these models do not provide an associated logical representation.

With respect to query and aggregation mechanisms, all of the above solutions proposed customized implementations. In contrast, we have shown that temporal data warehouses can be implemented in current DBMSs. In this way, we have provided a more general approach that does not require specific software for manipulating multidimensional data that varies over time. Since we are not currently considering schema versioning, our approach is closer to the solutions proposed for temporal databases, in particular with respect to temporal aggregation [202, 278, 285]. However, it is possible to extend our model with schema versioning. This extension can be done in a way similar to that proposed by Bebel et al. [16], Morzy and Wrembel [203, 204], and Wrembel and Bebel [316], where, in a first step, temporal aggregation is performed for each version of the schema/instances (corresponding to temporal aggregations based on instant grouping [278]), and then, in a second step, the users must decide whether the results can be integrated.

5.14.3 Logical Representation

Regarding the logical representation of temporal data warehouses, Bliujute et al. [23] introduced a temporal star schema that differs from the classical schema by the fact that the time dimension does not exist; instead, the rows in all tables of the schema are timestamped. These authors compared this model with the classical star schema, taking into account database size and performance. They concluded that the temporal star schema facilitates expressing and executing queries, it is smaller in size, and it does not keep redundant information.

Hurtado et al. [108] and Mendelzon and Vaisman [195] had given two logical representations for a conceptual multidimensional model; these representations are similar to the snowflake and star schemas, with the difference that each row is timestamped. Since this approach also refers to schema versioning, additional tables that contain information about schemas and their validity are required. On the other hand, the implementation proposed by Body et al.

[25] is based on a star schema, but incurs problems of data repetition for those instances that do not change between different versions.

Given the lack of a satisfactory solution to the problem of providing a logical representation of temporal data warehouses, we now briefly review logical models for temporal databases with the goal of adapting some of these ideas to the logical representation of temporal data warehouses.

Clifford et al. [48] distinguished between temporally grouped and temporally ungrouped historical data models. The former correspond to attribute-timestamping models using complex domains in non-first-normal form, while the latter correspond to tuple-timestamping models represented in first normal form. Although these two approaches model the same information, they are not equivalent: while a grouped relation can be ungrouped, there is not a unique grouped relation corresponding to an ungrouped one. Clifford et al. [48] and Wijsen [312] considered that temporally grouped models are more expressive. Wijsen [312] indicated that the approach given by Clifford et al. [48] has difficulties in managing time-varying data owing to the absence of an explicit group identifier. For this reason, our implementation is based on object-relational databases, which are able to represent temporally grouped models with an explicit group identifier (i.e., a surrogate).

One approach to logical-level design of temporal databases is based on mapping temporal conceptual models. While this is the usual practice for the design of conventional (i.e., nontemporal) databases, to the best of our knowledge only Detienne and Hainaut [56], Gregersen et al. [89], and Snodgrass [274] have proposed such an approach for obtaining a relational schema from a temporal conceptual model. In general, this approach consists in creating a table for each entity type that has lifespan support, a separate table for each timestamped monovalued attribute, and one additional table for each multivalued attribute, whether timestamped or not. This approach produces a significant number of tables, since entities and their time-varying attributes are represented separately. Thus, the resulting schema is not intuitive in relation to expressing the semantics of the modeled reality.

Another approach to logical-level design of temporal databases is to use normalization. Temporal functional dependencies have been defined, for example in [135, 310, 311, 312]. However, most of these approaches rely on the first normal form, which has well-known limitations in modeling complex data. Further, this normalization approach is complex and difficult for designers to understand. For this reason, we have chosen a mapping approach that transforms a temporal conceptual schema into a logical schema.

5.14.4 Temporal Granularity

There are many publications on temporal databases related to transformations between different granularities. In this section, we mention only some of them; more detailed references can be found, for example, in the publications of Bettini et al. [21] and Combi et al. [51]. Dyreson et al. [59] have defined mappings between different granularities as described in Sect. 5.8, while Bettini et al. [21] and Merlo et al. [196] referred to the problem of conversion between different temporal granularities and of handling data attached to these granularities. Bettini et al. [21] proposed calendar operations that allow one to capture the relationships that exist between temporal granularities. These authors defined point- and interval-based assumptions that can be used for data conversion between the same or different temporal granularities, respectively. Merlo et al. [196] considered the transformation of time-varying data for the generalization relationship in a temporal object-oriented data model. To ensure an adequate transformation between a type and its subtypes when they have different temporal granularities, these authors introduced and classified coercion functions.

Even though the problem of managing data with multiple temporal granularities has been widely investigated in the field of temporal databases, this is still an open research area in the field of temporal data warehouses. In particular, an analysis of whether the solutions proposed in the temporal-database community are applicable in the data warehouse context must still be done.

5.15 Summary

The joining together of two research areas, data warehouses and temporal databases, has allowed the achievements in each of these areas to be combined, leading to the emerging field of temporal data warehouses. Nevertheless, neither data warehouses nor temporal databases have a well-accepted conceptual model that can be used for capturing users' requirements. To help establish better communication between designers and users, we have presented a temporal extension of the MultiDim model in this chapter.

Our model provides several different temporality types: valid time, transaction time, and lifespan, which originate from the source systems, and loading time, generated in the temporal data warehouse. Levels may have temporal attributes and may also be temporal, i.e., they may associate a time frame with their members. Temporal relationships allow one to keep track of the evolution of the links between parent and child members. Our model allows one to represent instant and lifespan cardinalities, which indicate the number of members of one level that can be related to members of another level at any time instant and over the lifespan of the member, respectively. We have discussed three cases of temporal hierarchies, depending on whether the levels and/or the relationships between them are temporal or not. For temporal measures, we have presented several examples to justify the inclusion of the various temporality types supported by our model, and discussed how to aggregate measures when levels are related by temporal relationships. A temporal data warehouse must deal with different temporal granularities. This requires not only a mapping between the different granularities, but also adequate handling of measure aggregation. In this respect, we have discussed several solutions proposed for temporal databases that can also be used for temporal data warehouses.

In this chapter, we have also provided a mapping that allows the translation of the MultiDim model into the entity-relationship and object-relational models. We have shown examples of implementations in Oracle 10g, indicating the physical considerations that must be taken into account in the implementation of temporal data warehouses. Translating the constructs of the MultiDim model to the ER model allows a better understanding of their semantics. Further, the translation of the ER model into operational data models is well understood and can be done by the usual CASE tools. However, the ER model provides a less convenient conceptual representation of time-varying attributes, levels, and relationships than does the MultiDim model. The latter contains fewer elements, it clearly allows one to distinguish which data changes should be stored, and it leaves the more technical aspects outside the users' concern.

On the other hand, the mapping to the object-relational model is necessary to implement temporal data warehouses into current DBMSs. This mapping considers the semantics of the various elements of a multidimensional model, as well as the specificities of current DBMSs. The object-relational model allows temporal levels and hierarchies to be represented in a better way than in the relational model. In the former model, a level and its corresponding temporal attributes are kept together, whereas the relational model produces a significant number of tables, with well-known disadvantages in relation to modeling and implementation. Nevertheless, the relational model is more suitable for representing temporal measures because it facilitates aggregation procedures.

The proposed mapping may be customized according to the expected usage pattern, for example use with data-mining algorithms, and according to specific features of the target implementation system. For example, a user may choose a tool-specific method for multidimensional storage (for example, analytic workspaces in Oracle OLAP 10g) instead of relying on more general solutions such as those proposed in this chapter.

Providing temporal support in a conceptual multidimensional model allows temporal semantics to be included as an integral part of a data warehouse. Our multidimensional model provides a uniform way to represent changes and the time when they occur for all data warehouse elements. In addition, logical and physical models can be derived from such a conceptual representation. In this way, users are able to better represent the dynamics of real-world phenomena in a data warehouse, and thus expand their analysis capabilities.

6

Designing Conventional Data Warehouses

The development of a data warehouse is a complex and costly endeavor. A data warehouse project is similar in many aspects to any software development project and requires definition of the various activities that must be performed, which are related to requirements gathering, design, and implementation into an operational platform, among other things. Even though there is an abundant literature in the area of software development (e.g., [47, 245, 279]), few publications have been devoted to the development of data warehouses. Some of these publications [14, 113, 118, 145, 239] have been written by practitioners and are based on their experience in building data warehouses. On the other hand, the scientific community has proposed a variety of approaches for developing data warehouses [27, 28, 34, 41, 46, 78, 81, 113, 166, 201, 219, 235, 243]. Nevertheless, many of these approaches target a specific conceptual model and are often too complex to be used in real-world environments. As a consequence, there is still a lack of a methodological framework that could guide developers in the various stages of the data warehouse development process. This situation results from the fact that the need to build data warehouse systems arose before the definition of formal approaches to data warehouse development, as was the case for operational databases [165].

In this chapter, we propose a general method for conventional-data-warehouse design that unifies existing approaches. This method allows designers and developers to better understand the alternative approaches that can be used for data warehouse design, helping them to choose the one that best fits their needs. To keep the scope and complexity of this chapter within realistic boundaries, we shall not cover the overall data warehouse development process, but shall focus instead only on data warehouse design.

The chapter starts by presenting in Sect. 6.1 an overview of the existing approaches to data warehouse design. Then, in Sect. 6.2, we refer to the various phases that make up the data warehouse design process. Since data warehouses are a particular type of databases devoted to analytical processing [28], our approach is in line with traditional database design, i.e., it includes the phases of requirements specification, conceptual design, logical design, and physical design. In Sect. 6.3, we present a university case study that we then use throughout the chapter to show the applicability of the proposed method and to facilitate understanding of the various phases.

The subsequent sections are devoted to more detailed descriptions of each design phase. In Sect. 6.4, we propose three different approaches to requirements specification. These approaches differ in that users, source systems, or both may be the driving force for specifying the requirements. Section 6.5 covers the conceptual-design phase for data warehouses. With respect to database design, this phase includes some additional steps required for delivering conceptual schemas that represent users' requirements, but also take into account the availability of data in the source systems. Section 6.6 summarizes the three approaches to data warehouse development, highlighting their advantages and disadvantages.

Next, in Sects. 6.7 and 6.8, we provide insights into the logical and physical design phases for conventional data warehouses. Both of these phases consider the structure of the data warehouse, as well as the related extraction-transformation-loading (ETL) processes. Since we have already referred in Sect. 3.5 to the logical representation of data warehouses, in this chapter we present only a short example of a transformation of a conceptual multidimensional schema into a logical schema using the SQL:2003 standard. Then, we identify several physical features that can considerably improve the performance of analytical queries. We give examples of their implementation using Oracle 10g. Sect. 6.9 summarizes the proposed methodological framework for conventional-data-warehouse design. Finally, Sect. 6.10 surveys related work, and Sect. 6.11 summarizes this chapter.

6.1 Current Approaches to Data Warehouse Design

There is a wide variety of approaches that have been proposed for designing data warehouses. They differ in several aspects, such as whether they target data warehouse or data mart design, the various phases that make up the design process, and the methods used for specifying requirements. This section highlights some of the essential characteristics of the current approaches according to these aspects.

6.1.1 Data Mart and Data Warehouse Design

A data warehouse includes data about an entire organization that is used by users at high management levels in order to support strategic decisions. However, these decisions may also be taken at lower organizational levels related to specific business areas, in which case only a subset of the data contained in a data warehouse is required. This subset is typically contained in a **data mart** (see Sect. 2.9), which has a similar structure to a data warehouse but is smaller in size. Data marts may be physically collocated with the data warehouse or may have their own separate platform [113].

Similarly to the design of operational databases (Sect. 2.1), there are two major methods for the design of a data warehouse and its related data marts [145]:

- **Top-down design:** The requirements of users at different organizational levels are merged before the design process begins, and one schema for the entire data warehouse is built. Afterwards, separate data marts are tailored according to the particular characteristics of each business area or process.
- **Bottom-up design:** A separate schema is built for each data mart, taking into account the requirements of the decision-making users responsible for the corresponding specific business area or process. Later, these schemas are merged, forming a global schema for the entire data warehouse.

The planning and implementation of an enterprise-wide data warehouse using the top-down approach is an overwhelming task for most organizations in terms of cost and duration. It is also a challenging activity for designers [145] because of its size and complexity. On the other hand, the reduced size of data marts allows a business to earn back the cost of building them in a shorter time period and facilitates the design and implementation processes.

Leading practitioners usually use the bottom-up approach in developing data warehouses [113, 145]. However, this requires a global data warehouse framework to be established so that the data marts are built considering their future integration into a whole data warehouse. A lack of such a global framework may cause a data warehouse project to fail [219], since different data marts may contain the same data using different formats or structures.

Various frameworks can be applied to achieve this. For example, in the framework of Imhoff et al. [113] a global data warehouse schema is first developed. Then, a prototype data mart is built and its structure is mapped into the data warehouse schema. The mapping process is repeated for each subsequent data mart. On the other hand, Kimball et al. [145] have proposed a framework called the *data warehouse bus architecture*, in which dimensions and facts shared between different data marts must be conformed. A dimension is **conformed** when it is identical in each data mart that uses it. Similarly, a fact is **conformed** if it has the same semantics (for instance, the same terminology, granularity, and units for its measures) across all data marts. Once conformed dimensions and facts have been defined, new data marts may be brought into the data warehouse in an incremental manner, ensuring their compatibility with the already-existing data marts.

An approach to building a data warehouse by integrating heterogeneous data marts has been studied in [36, 37, 33, 296]. Cabibbo and Torlone [36] introduced the notion of dimension compatibility, which extends the notion of conformed dimensions in [145]. Intuitively, two dimensions belonging to different data marts are compatible when their common information is consistent.

Compatible dimensions allow the data of various data marts to be combined and correlated. These authors proposed two strategies for the integration of data marts. The first one consists in identifying the compatible dimensions and performing drill-across queries over autonomous data marts [36]. The second approach allows designers to merge the various data marts by means of materialized views and to perform queries against these views [37]. A tool that integrates these two strategies has been implemented [33, 296].

6.1.2 Design Phases

To the best of our knowledge, relatively few publications (e.g., [14, 46, 82, 113, 118, 145, 168, 244]) have proposed an overall method for data warehouse design. However, these publications do not agree on the phases that should be followed in designing data warehouses. Some authors consider that the traditional phases of developing operational databases (as described in Sect. 2.1), i.e., requirements specification, conceptual design, logical design, and physical design, can also be used in developing data warehouses or data marts. Other authors ignore some of these phases, especially the conceptual-design phase. Many publications refer to only one of the phases, without considering the subsequent transformations needed to achieve implementable solutions.

Some proposals consider that the development of data warehouse systems is rather different from the development of operational database systems. On the one hand, they include additional phases, such as workload refinement [82] or the ETL process [145, 168]. On the other hand, they provide various methods for the requirements specification phase, to which we refer in Sect. 6.1.3. Several publications mention the importance of creating metadata as a part of the data warehouse design process; this includes not only a description of the data warehouse model but also information about the source systems and the ETL processes [14, 41, 168, 219].

6.1.3 Requirements Specification for Data Warehouse Design

Several different approaches have been proposed for requirements specification, depending on whether the users, business goals, source systems, or a combination of these are used as the driving force. We give next a brief description of these approaches. More detailed descriptions are given in Sect. 6.10.

User-Driven Approach

This approach considers that the users play a fundamental role during requirements analysis and must be actively involved in the elucidation of the relevant facts and dimensions [76, 168]. Users from different levels of the organization are selected. Then, various techniques, such as interviews or facilitated sessions, are used to specify the information requirements [27, 113, 219, 303].

This approach is also called the demand-driven approach [314].

Business-Driven Approach

This approach considers that users are often not able to clearly formulate their particular demands. Therefore, the derivation of the structures of a data warehouse starts from an analysis of either business requirements or business processes. In the first case, the business requirements at the highest level of the organization are refined until the necessary multidimensional elements have been identified. Consequently, the specification obtained will include the requirements of users at all organizational levels, aligned with the overall business goals. On the other hand, the analysis of the business processes requires one to specify the various business services or activities that produce a particular output. The various elements participating in these activities may be considered as dimensions in the data warehouse schema. The metrics used by decision makers to evaluate business activities may be considered as measures.

This approach is also called the process-driven [166], goal-driven [78], or requirements-driven [243] approach.

Source-Driven Approach

In this approach, the data warehouse schema is obtained by analyzing the underlying source systems. Some of the proposed techniques require conceptual representations of the operational source systems [26, 34, 78, 81, 201], in most cases based on the ER model. Other techniques use relational tables to represent the source systems. These source schemas should exhibit a good degree of normalization in order to facilitate the extraction of facts, measures, dimensions, and hierarchies [78]. In general, the participation of users is not explicitly required [165]; however, in some techniques users need to either analyze the schema obtained to confirm the correctness of the derived structures [26], or identify some facts and measures as a starting point for the design of multidimensional schemas [82, 109, 201]. After the creation of an initial schema, users can specify their information requirements by selecting items of interest [118, 314].

This approach is also called the data-driven or supply-driven approach [78, 314].

Combined Approach

This approach is a combination of the business- or user-driven approach and the data-driven approach, taking into account what the business or the users demand and what the source systems can provide. In an ideal situation, these two components should match, i.e., all information that the users or the business require for analysis purposes should be supplied by the data included in the source systems. This approach is also called top-down/bottom-up analysis [28].

6.2 A Method for Data Warehouse Design

From the discussion in the previous section, we can see that the variety of existing approaches, in particular for the requirements specification phase, may be confusing for designers, even experienced ones. For this reason, we propose in this chapter a general method for data warehouse design that encompasses the various approaches. Because this framework allows one to describe the advantages and disadvantages of the various options precisely, designers will be able to choose the option that best fits their needs and the particularities of the data warehouse project at hand.

As in several of the proposals mentioned in the previous section, we consider that data warehouses are a particular type of databases dedicated to analytical purposes. Therefore, their design should follow the traditional database design phases, i.e., requirements specification, conceptual design, logical design, and physical design, as shown in Fig. 6.1. Nevertheless, there are significant differences between the design phases for databases and data warehouses; these are explained in later sections of this chapter. It is worth noting that although in Fig. 6.1 the various phases are depicted consecutively, in reality there are multiple interactions between them, especially if an iterative development process is adopted in which the system is developed in incremental versions with increased functionality.

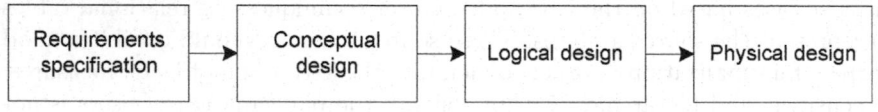

Fig. 6.1. Phases in data warehouse design

The first two phases, requirements specification and conceptual design, are the most critical ones, since they can considerably affect users' acceptance of the system [147]. Indeed, these two phases determine whether the relationship between the real world and the software world, i.e., between the users' needs and what the modeled system will offer, will be adequate. As will be described in the following sections, these two phases are also the ones that have the highest degree of variability, since the activities in these phases may be driven by analysis needs, by the information available in the source systems, or by a combination of both. The next phases, logical and physical design, are more technical phases that successively translate the conceptual schema obtained in the previous phase to target the implementation structures of a particular data warehouse tool.

Note that the phases in Fig. 6.1 do not depend on whether the top-down or the bottom-up approach is used, since, as explained in Sect. 6.1.1, the difference between the two approaches consists in whether the overall data warehouse schema is used to derive the schemas of the data marts or vice versa. Therefore, the phases in Fig. 6.1 may be applied to define either the overall data warehouse schema or the schemas of the individual data marts. The choice between the top-down and the bottom-up approach depends on many factors, such as the professional skills of the development team, the size of the data warehouse, the users' motivation for having a data warehouse, and the financial support, among other things. For example, if the user motivation is low, the bottom-up approach may deliver a data mart faster and at less cost. As a consequence, users can dynamically manipulate data using OLAP tools or create new reports; this may lead to an increase in users' acceptance level and improve the motivation for having a data warehouse. Nevertheless, recall from Sect. 6.1.1 that the development of these data marts should be done within a global enterprise framework in order to integrate them into a data warehouse. From now on, we shall use the term "data warehouse" to mean that the concepts that we are discussing apply also to data marts if not stated otherwise.

6.3 A University Case Study

To illustrate the proposed method, we shall use a hypothetical scenario of the design of a university data warehouse. Although this example is based on a real-world setting, the real application contains a vast amount of detailed information covering the many areas of activities of a typical university. For this reason, we have chosen to build a simplified example that allows us to focus on the methodological aspects and that should facilitate their understanding.

In recent decades universities everywhere have faced significant changes resulting from various factors, such as decreases in funding, political and organizational changes, globalization, and increased competition. In particular, owing to the globalization of higher education, students are looking abroad for the best educational opportunities, firms are scouring the world to set up research contracts, and highly skilled academics are seeking better conditions for research [208]. Attracting many students, research contracts, and well-prepared academics willing to develop research activities helps universities to improve their overall economic situation.

Therefore, the ranking of universities on various scales (worldwide, continental, national, regional, etc.) has become an important factor in establishing the reputation of a university at the international level. The rankings published by Shanghai Jiao Tong University since 2003 [119, 121, 123, 125] and by The Times since 2004 [286, 287, 288] have attracted wide attention worldwide. These two rankings are based on different methods. In addition, the rankings produced by Shanghai Jiao Tong University used one method for 2003 [120] and another for the period 2004–2006 [122, 124, 126], while in the rankings produced by The Times, the method has been slightly changed since the 2005 edition [115]. The authors of both rankings have recognized the methodological and technical problems that they found in producing them [115, 116, 167]. Nevertheless, these rankings may give important insights with respect to competing institutions.

For the purpose of our case study, we suppose that a university wants to determine what actions it should take to improve its position in the rankings. To simplify the discussion we consider only the Times ranking. The evaluation criteria in this ranking refer to the two main areas of activities of universities, i.e., research and education. However, a closer analysis shows that 60% of the criteria are related to research activities (peer review and citation/faculty scores) and 40% to the university's commitment to teaching. Therefore, we suppose that the decision-making users chose initially to analyze the situation related to research activities. To be able to conduct the analysis process, it was decided to implement a data warehouse system.

Universities are usually divided into faculties representing general fields of knowledge, for example medicine, engineering, and the sciences. These faculties comprise several different departments dedicated to more specialized domains; for example, the faculty of engineering may include departments of civil engineering, mechanical engineering, and computer engineering, among others. University staff (i.e., professors, researchers, teaching assistants, administrative staff, etc.) are administratively attached to departments. This traditional organizational structure is not well adapted to multidisciplinary research activities, which require expertise from several domains, possibly across different faculties. Autonomous structures called research centers support such multidisciplinary research. University staff from various faculties or departments may belong to these research centers. Research projects are conducted by one or several research bodies, which may be either departments or research centers.

The research department is the administrative body that coordinates all research activities at the university. It serves as a bridge between high-level executives (for example the Rector and the research council of the university) and researchers, as well as between researchers and external organizations, whether industrial or governmental. For example, the research department is responsible for the evaluation of research activities, for the development of strategic research plans, for promoting research activities and services, for managing intellectual property rights and patents, and for technology transfer and creation of spin-offs, among other things. In particular, the establishment of strategic research areas is based on the university's core strengths and ambitions, taking into account long-term potential and relevance. These areas are the focus of institutional initiatives and investments. On the basis of the institutional research strategy, faculties, departments, and research centers establish their own research priorities.

6.4 Requirements Specification

The requirements specification phase is one of the earliest steps in system development and thus entails significant problems if it is faulty or incomplete. Requirements analysis should attract particular attention and should be comprehensively supported by effective methods [314]. However, not much attention has been paid to the requirements analysis phase in data warehouse development [189]. In addition, the variety of the existing approaches to requirements specification has led to the situation that many data warehouse projects skip this phase; instead, they concentrate on technical issues, such as database modeling or query performance [220]. As a consequence, it is estimated that more than 80% of data warehouse projects fail to meet user needs [266] and do not deliver the expected support for the decision-making process.

Requirements specification determines, among other things, what data should be available and how it should be organized, as well as what queries are of interest. The requirements specification phase should lead the designer to discover the essential elements of a multidimensional schema, i.e., facts with associated measures, dimensions, and hierarchies [26, 27, 35, 109, 201, 238], which are required to facilitate future data manipulations and calculations. The requirements specification phase establishes a foundation for all future activities in data warehouse development [145]; it has a major impact on the success of data warehouse projects [266], since it directly affects the technical aspects, as well as the data warehouse structures and applications.

We present next a general framework for the requirements specification phase, which is based on existing approaches proposed by the research community. Although we separate the phases of requirements specification and conceptual design for readability purposes, in reality these phases often overlap. In many cases, as soon as some initial requirements have been documented, an initial conceptual schema starts to take shape. As the requirements become more complete, so does the conceptual schema [14].

We propose three approaches that consider as the driving force the business or user demands, the existing data in the underlying operational systems, or both. For each approach, we first present a general description, and then we refer in more detail to various steps and include examples to illustrate the process. We present the sequence of steps without indicating the various iterations that may occur between them. Our purpose is to provide a general framework to which details can be added and which can be tailored to the particularities of a specific data warehouse project.

6.4.1 Analysis-Driven Approach

General Description

In the analysis-driven approach, the driving force for developing the conceptual schema is business or user requirements. These requirements express the organizational goals and needs that the data warehouse is expected to address to support the decision-making process. Since users at several different management levels may require data warehouse support, the identification of key users is an important aspect. Stakeholders, users, business domain experts, and also an enterprise plan help the developer team to understand the purpose of having the data warehouse and to determine specific analysis needs. The information gathered serves as a basis for the development of the initial data warehouse schema.

Steps

The various steps in the analysis-driven approach to requirements specification are shown in Fig. 6.2. These are described as follows:

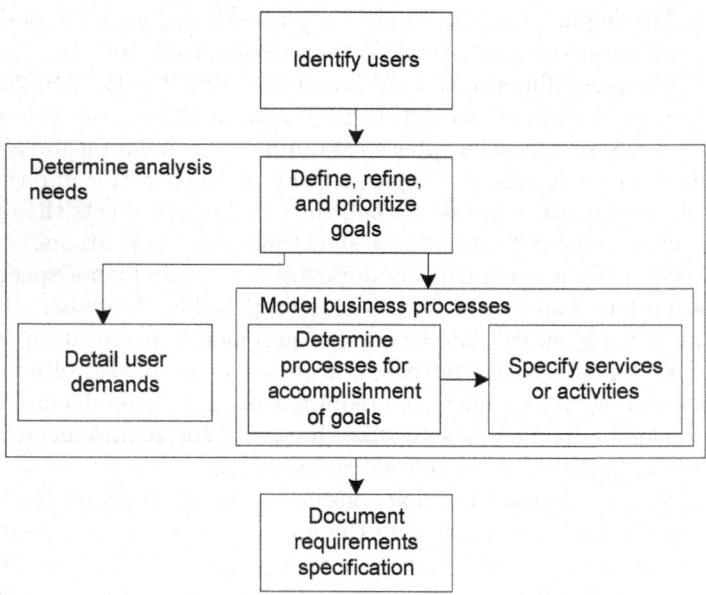

Fig. 6.2. Steps in requirements specification in the analysis-driven approach

- **Identify users:** Since a data warehouse provides an enterprise-wide decision-support infrastructure, users at several different hierarchical levels in the organization should be considered [166]. *Executive users* at the top organizational level may express their business needs by indicating what information is required, most probably in summarized form. These users help in understanding high-level objectives and goals, and the overall business vision [145]. *Management users* may require more detailed information and can refer to a more specific area of the organization. They can provide more insights into the business processes or the tactics used for achieving business goals. Finally, *professional users* can be responsible for a specific section or set of services and may demand specific information related to their area of interest. In addition, users representing different entities in an horizontal scheme of division of the organization (e.g., into departments) should also be considered. This will help in providing an overall view of the project and its scope. Therefore, even though the data warehouse project may start by focusing on a specific business area or process in the initial development, the identification of potential users should consider both vertical (hierarchical) and horizontal (departmental) division [145].

- **Determine analysis needs:** Determining analysis needs helps the developers to understand what data should be available to respond to users' expectations for the data warehouse system. Since we are focusing on multidimensional modeling, this phase should eventually identify facts with measures and dimensions with hierarchies; the order in which these elements are discovered is not important [14]. The process of determining analysis needs is a complex task that includes several steps:

  - **Define, refine, and prioritize goals:** The starting point in determining analysis needs is the consideration of business goals. Successful data warehouse projects assume that the goals of the company are the same for everyone and that the entire company will therefore be pursuing the same direction. Therefore, a clear specification of goals is essential to guide user needs and convert them into data elements [165] or to find critical business processes required for the accomplishment of goals. Since users at several different management levels participate in requirements specification, analysis needs may be expressed by considering both general and more specific goals. The specific goals should be aligned with the general ones to ensure a common direction of the overall development. The goal-gathering process can be conducted by means of interviews, facilitated sessions, or brainstorming based on various approaches, such as those specified in [28, 78, 145, 219]. The list of goals should be analyzed with regard to their similarities and the implications between them in order to reduce the number of goals to a manageable number [28]. For example, some goals may be considered as subgoals, and other goals may be combined because of their similarity or discarded because of their inconsistency. This analysis may require additional interaction with the users to establish the final list of goals.

    For every goal, the subsequent steps shown in Fig. 6.2 are performed. We propose two different approaches, based either on a more specific definition of user demands (left-hand path of Fig. 6.2) or on modeling of business processes (right-hand part of Fig. 6.2).

– **Detail user demands:** Additional interviews with specific users, focusing on more precise goal definitions, are conduced. These interviews allow designers to elicit the information needed for multidimensional schemas. At this stage, a list of queries or analytical scenarios can also help in defining what data is needed for analysis purposes. This list cannot be exhaustive, since the purpose of a data warehouse is to be able to answer queries created "on the fly". The person performing the interviewing process should be able to pose questions tailored to uncover data warehouse issues in more detail. Techniques other than interviews can also be used, such as workshops, questionnaires, or prototyping.

– **Model business processes:** The accomplishment of the goals is closely related to the business processes. Therefore, the relevant business processes should be determined for every specific goal. Since a business process is a group of services or activities that together create a result of value to a customer or a market [97], the identification of these services or activities takes place in the next step. Some activities or services may be complex and may require subsequent division into smaller tasks. Activities and services include data required for performing them, which may form part of the future multidimensional schema. Note that in this approach, even though we focus on processes, we do not ignore the presence of domain experts or users. These people may help in understanding the application domain and in choosing processes that correspond to the established goals. They also can indicate in a more precise manner what kinds of analysis scenarios are most important to them.

  Business processes can be considered implicitly and informally as described in [145], or a more formal business process model can be created, such as that proposed in [27].

The step of determining analysis needs should determine candidate facts, measures, dimensions, and hierarchies, as well as a preliminary list of possible queries or analytical scenarios that can be created. If possible, this step should also specify the granularities required for the measures, and information about whether they are additive, semiadditive, or nonadditive (see Sect. 2.9).

• **Document requirements specification:** The information obtained in the previous step should be documented. The documentation delivered is the starting point for the technical metadata and forms part of the business metadata (see Sect. 2.9). Therefore, this document should include all elements required by the designers and also a dictionary of the terminology, organizational structure, policies, and constraints of the business, among other things. For example, it could express in business terms what the candidate measures or dimensions actually represent, who has access to them, and what operations can be done. Note that we do not consider this document as a final specification of the requirements, since additional interactions may be necessary during the conceptual-design phase in order to refine or clarify some aspects. The preparation of the document for requirements specification can be done using recommendations given, for example, in [145, 161, 219, 220, 266, 303].

Example

In this section, we describe the application of the analysis-driven approach to produce a specification of requirements for our university data warehouse. We present each step and briefly describe it, referring to the particularities of our project. We do not include all details about how each step is developed, since these details could make the example unnecessarily long and difficult to understand.

Identify Users

In the first step, we identify users at various management levels who make strategic, tactic, or operational decisions related to the research process. In this example, three groups of users were established:

1. **Executive:** the rector, his advisors, and the research council responsible for establishing general policies for research at the university.
2. **Management:** representatives of the research department, including the sections responsible for the support of research activities and for administrative tasks, promotions, evaluation, and strategic analysis.
3. **Professional:** representatives of various research entities, including research centers and university departments.

Determine Analysis Needs

The next step, of determining analysis needs, starts with specification of the goals. Finding goals is a complex and creative task [27] and may require many sessions with users. In our example, we used only one general goal of improving the ranking of the university, considering the strategic research areas that had been established at the university. This goal can be decomposed into two subgoals related to improving the scores in two evaluation criteria of the Times ranking: the peer review and citation/faculty criteria. The peer review criterion (40% of the ranking score) is based on interviewing selected academics from various countries to name the top institutions in the areas and subjects about which they feel able to make an informed judgment [114]. The citation/faculty criterion (20% of the ranking score) refers to the numbers of citations of academic papers generated by each staff member [115].

Determining the activities that could improve these evaluation criteria required the participation of users at lower organizational levels, for example the management and professional levels. Interviews with users allowed us to conclude that, in the first step, information related to international projects, conferences, and publications was necessary to better understand the participation of the university's staff in international forums.

There are several sources of funding for research projects: from the university, from industry, and from regional, national, and international institutions. Independently of the funding scheme, a project may be considered as being international when it involves participants from institutions in other countries. International projects help to expand the circle of academics working in the same area, and thus could help to improve the peer review score. Further, participation in international conferences helps the university's staff to meet international colleagues working in the same or a similar area. In this way, not only can new strategic contacts be established (which may lead to international projects), but also the quality of the university's research can be improved. Knowledge about the international publications produced by the university's staff could also help to reveal the causes of decreasing scores in the citation/faculty criterion.

In the next step, we decided to conduct more sessions with the users to understand their demands in more detail. We did not create a university process model, since the two main processes, i.e., education and research, are very complex and may require sophisticated techniques to ensure their correct representation, such as that described in [27] for the business process model of the higher education process.

Since the accomplishment of goals requires decisions based on the analysis of existing situations, during the subsequent sessions, users expressed their analysis needs considering international publications, conferences, and projects. We give below some examples of these needs:

1. International publications:

   (a) Total number of international publications in various periods of time, considering calendar divisions of months, quarters, and years, as well as the academic calendar, represented by semesters and academic years.
   (b) Total number of international publications, taking into account various types of publications, i.e., journals, conference proceedings, and books, as well as the discipline to which they belong.
   (c) Total number of publications, classified by researcher, department, faculty, and discipline.
   (d) Total number of international publications, by publisher.

2. International conferences:

   (a) Costs related to participation of researchers in international conferences, classified according to various components, for example registration, travel, and lodging.

Table 6.1. Specification of dimensions and measures for the analysis scenarios in the example

Dimensions /measures	Hierarchy levels	Analysis scenarios												
		1a	1b	1c	1d	2a	2b	2c	3a	3b	3c	3d	3e	3f
Researcher	Researcher Department Faculty	–	–	✓	–	–	✓	–	✓	✓	✓	✓	✓	✓
Calendar time	Month Quarter Year	✓	–	–	–	–	✓	–	–	–	–	✓	–	✓
Academic time	Semester Academic year	✓	–	–	–	–	✓	–	–	–	–	–	–	–
Publication	Publication Journal Book Proceedings Publisher	✓	✓	✓	✓	–	–	–	–	–	–	–	–	–
Conference		–	–	–	–	✓	–	✓	–	–	–	–	–	–
Project	Project Department Faculty Research center	–	–	–	–	–	–	–	✓	✓	✓	✓	✓	✓
Discipline		–	✓	✓	–	–	–	–	–	–	–	–	–	✓
Role		–	–	–	–	–	–	✓	–	–	–	–	–	–
Registration cost		–	–	–	–	✓	✓	–	–	–	–	–	–	–
Travel cost		–	–	–	–	✓	✓	–	–	–	–	–	–	–
Lodging cost		–	–	–	–	✓	✓	–	–	–	–	–	–	–
Other expenses		–	–	–	–	✓	✓	–	–	–	–	–	–	–
Salary		–	–	–	–	–	–	–	–	–	✓	–	–	–
No. of hours		–	–	–	–	–	–	–	–	–	–	✓	–	–

(b) Cost of participation during various periods of time, considering both calendar and academic years, and also the organizational structure of departments and faculties.

(c) Role played in conferences, i.e., invited speaker, organizer, author, or member of the audience.

3. International projects:

  (a) Number of international projects in each department or research center.
  (b) Number of researchers involved in international projects.
  (c) Salary earned by researchers participating in international projects, considering individual researchers, as well as the departments, faculties, and research centers in charge of projects.
  (d) Number of hours per month, quarter, and year that researchers have dedicated to international projects.
  (e) Number of researchers with an affiliation in one department and participation in projects from other departments.
  (f) Numbers of projects and researchers, classified by discipline during various periods of time.

The above list of analysis scenarios required by the users indicates that there were three foci of analysis, which in a multidimensional model correspond to three fact relationships, i.e., international publications, conferences, and projects. Then, for each of these fact relationships, we established a list of dimensions (with their hierarchy levels if any) and measures. Next, we checked whether the analysis scenario could be created (i.e., whether we had all necessary elements) on the basis of the list of candidate dimensions and measures. A specification can be developed separately for each fact relationship or be shown in a summarized manner, as has been done in Table 6.1. The choice of which form is more suitable depends on the complexity of the system, the developers' skills, and the scope of the project.

Document Requirements Specification

The information compiled is included in the specification of the users' requirements. For example, it can contain summarized information as presented in Table 6.1 and also more descriptive parts that explain each component of the table.

The requirements specification document also contains the business metadata. For the university example, we had various ways to obtain this metadata, for example by interviewing users or university administration staff, or accessing the existing university documentation. Another possibility might be to use the data dictionary for US colleges and universities developed by the Consortium for Higher Education Software Services (CHESS) [291]. This provides definitions, codes, categories, and descriptions for many standard data elements related to universities, grouped into six categories: alumni, courses, facilities, finance, human resources, and students.

6.4.2 Source-Driven Approach

General Description

The source-driven approach relies on the data available in the source systems. It aims at identifying all candidate multidimensional schemas that can be realistically implemented on top of the available operational databases. These databases are analyzed exhaustively in order to discover the elements that can represent facts with associated measures, and dimensions with hierarchies. The identification of these elements leads to an initial data warehouse schema that may correspond to several different analysis purposes.

Steps

We describe next the steps in the source-driven approach to requirements specification. As with the analysis-driven approach above, we do not show the various iterations that may be required before the final data warehouse schema is developed. Each of the steps represents a complex task. In the following, we present only a brief description of these steps.

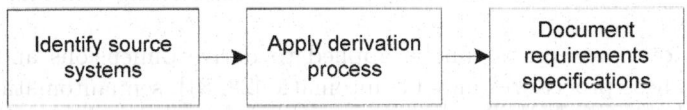

Fig. 6.3. Steps in requirements specification in the source-driven approach

- **Identify source systems:** The aim of this step is to determine the existing operational systems that can serve as data providers for the data warehouse. External sources are not considered at this stage; they can be included later on, when the need for additional information has been identified. In the presence of several operational systems, those that provide increased data quality [303] and stability of their schemas should be selected [83].

  This step relies on system documentation, preferably represented using the ER model or relational tables. However, in many situations this representation may be difficult to obtain, for example when the data sources include implicit structures that are not declared through the data definition language of the database, when redundant and unnormalized structures had been added to improve query response time, when the databases have been built by novice or untrained developers unaware of database theory and methods, or when the databases reside on legacy systems whose inspection is a difficult task [96]. In such situations, reverse engineering processes can be applied; these processes allow one to rebuild the logical and conceptual schemas of source systems whose documentation is missing or outdated [95]. Various methods can be used for database reverse engineering, such as those described in [96, 101, 102].

- **Apply derivation process:** Various techniques can be applied to derive multidimensional elements from operational databases [26, 28, 35, 81, 109, 201]. All these techniques require that the operational databases are represented using the ER model or relational tables.

  In general, in the first step of this process, the fact relationships and their associated measures are determined. This can be done by analyzing the existing documentation [26, 35, 201] or the structure of the databases [81]. Fact relationships and measures are elements that correspond to events that occur dynamically in the organization, i.e., that are frequently updated. If the operational databases are relational, they may correspond to tables and attributes, respectively. If the operational databases are represented using the ER model, facts may be entity or relationship types, while measures may be attributes of these elements. An alternative option may be to involve users who understand the operational systems and can help to determine what data can be considered as measures [109]. Identifying fact relationships and measures is the most important aspect of this approach, since these form the basis for constructing multidimensional schemas.

  Various procedures can be applied to derive dimensions and hierarchies. These procedures may be automatic [28, 81], semiautomatic [26], or manual [35, 109, 201]. The former two require knowledge about the specific conceptual models that are used for the initial schema and its subsequent transformations. The process of discovering a dimension or a leaf level of a hierarchy usually starts from identifying the static (not frequently updated) elements that are related to the facts. Then, a search for other hierarchy levels is conducted. For this purpose, starting with a leaf level of a hierarchy, every one-to-many relationship is revised.

  Unlike automatic or semiautomatic procedures, manual procedures allow designers to find hierarchies embedded within the same entity or table, for example to find city and province attributes in a store entity type. However, either the presence of system experts who understand the data in the operational databases is required or the designer must have good knowledge about the business domain and the underlying systems.

- **Document requirements specification:** As in the analysis-driven approach, the requirements specification phase should be documented. The documentation should describe those elements of the source systems that can be considered as fact relationships, measures, dimensions, and hierarchies. This will be contained in the technical metadata. Further, it is desirable to involve at this stage a domain expert who can help in defining business terminology for these elements and in indicating whether measures are additive, semiadditive, or nonadditive. The preparation of the document can be done using the insights contained in, for example, [109, 113, 303].

Example

To illustrate the source-driven approach, let us consider the ER schema in Fig. 6.4, containing information about academic staff and their teaching and research activities. To simplify the explanations, we have chosen a small excerpt from the overall operational database. Therefore, in this example we skip the step of identifying the source systems.

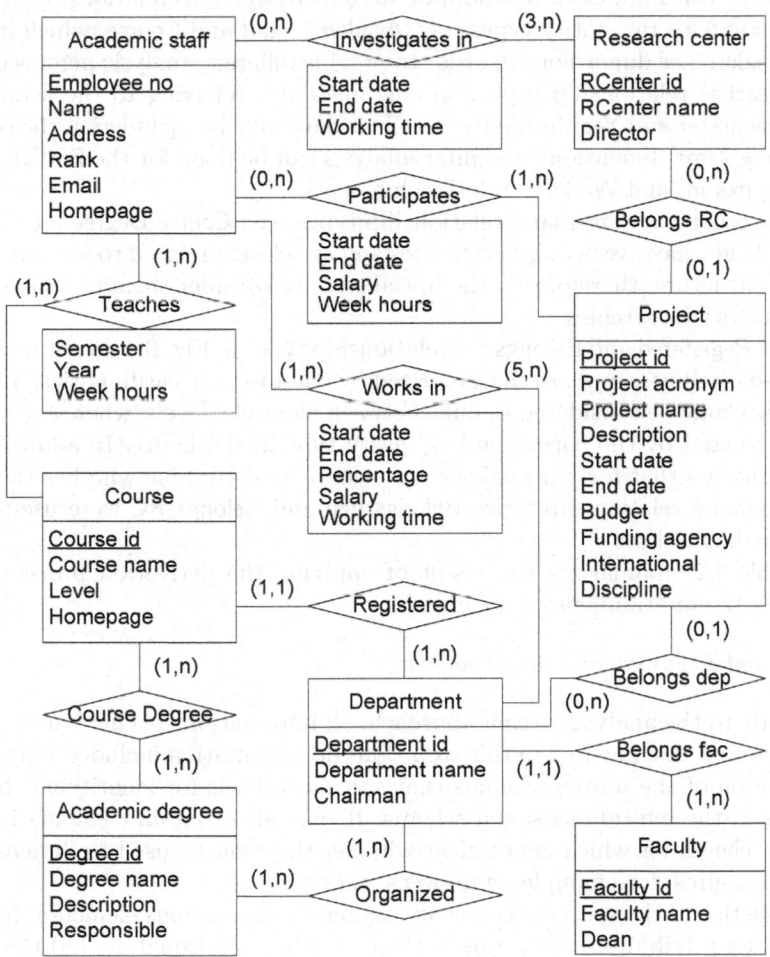

Fig. 6.4. Excerpt from the ER schema for the underlying operational database in the example

Apply Derivation Process

We chose a manual derivation process to provide a more general solution. However, the previously mentioned automatic or semiautomatic methods could also have been applied.

By analyzing the schema in Fig. 6.4, we can distinguish four many-to-many relationship types with attributes that represent numeric data: the relationship types Teaches, Participates, Investigates in, and Works in. These may be candidates for fact relationships in a multidimensional schema. For each of these candidate fact relationships, it is necessary to identify measures and dimensions. For example, the Teaches relationship type includes the Week hours attribute, which might be a candidate for a measure. This relationship type is associated with two entity types, i.e., Academic staff and Course, which might be considered as dimensions in order to provide different analysis perspectives. The Teaches relationship type also contains dates referring to the academic year (Semester and Year in the figure). This data may be included in the corresponding Time dimension. A similar analysis can be done for the Participates, Investigates in, and Works in relationship types.

The other many-to-many relationship types, i.e., Course Degree and Organized, do not have associated attributes and are closely related to the administrative structure; therefore, in the initial step we consider them as candidates for nonstrict hierarchies.

The Registered and Belongs fac relationship types in Fig. 6.4 do not have associated attributes and are characterized by one-to-many cardinalities. Therefore, they may be considered as links between hierarchy levels, where each level is represented by the corresponding entity type in the figure. In addition, it was identified that further analysis was needed to determine whether the two zero-to-many relationship types Belongs dep and Belongs RC were useful for defining hierarchies.

Table 6.2 summarizes the result of applying the derivation process and defining the multidimensional elements.

Document Requirements Specification

Similarly to the analysis-driven approach, all information specified in the previous steps is documented in this step. This documentation includes a detailed description of the source schemas that serve as a basis for identifying the elements in the multidimensional schema. It may also contain elements in the source schema for which is not clear whether they can be used as dimensions or as measures, for example, employees' ages.

Note that if the source schemas use names for the various elements (for instance, for attributes or relations in the case of the relational model) that are coded and difficult to understand, different names should be chosen for the elements of the multidimensional schema, specifying clearly the correspondences between the old and new names.

Table 6.2. Specification of multidimensional elements in the example

Fact relationships	Measures	Dimensions	Hierarchy levels and links
Teaches	Week hours	Academic staff Course Academic time	Course–Department Department–Faculty Course–Academic degree Academic degree–Department Department–Faculty
Investigates in	Working time	Academic staff Research center Calendar time	
Participates	Salary Week hours	Academic staff Project Calendar time	Project–Research center Project–Department Department–Faculty
Works in	Salary Working hours	Academic staff Department Calendar time	Department–Faculty

6.4.3 Analysis/Source-Driven Approach

The analysis/source-driven approach to requirements specification combines both of the previously described approaches, which may be used in parallel to achieve an optimal design. Therefore, two types of activities can be distinguished: one that corresponds to business demands (as described in Sect. 6.4.1) and another that represents the steps involved in creating a multidimensional schema from operational databases (as described in Sect. 6.4.2). Each type of activity results in the identification of elements for the initial multidimensional schema.

6.5 Conceptual Design

Independently of whether the analysis-driven or the source-driven approach has been used, the requirements specification phase should eventually provide the necessary elements for building the initial data warehouse schema. The purpose of this schema is to represent a set of data requirements in a clear and concise manner that can be understood by the users. This schema will serve as a basis for the analysis tasks performed by the users; it will also be used by the designers during future evolution of the data warehouse. Nevertheless, depending on the approach used for requirements specification, various aspects should be considered before a final conceptual schema is developed. In the following, we shall refer in more detail to the various steps of the conceptual-design phase and show examples of their execution.

We shall use the MultiDim model described in Chap. 3 to define the conceptual schemas. The notation of the model is given in Fig. 3.1 and in Appendix B. Nevertheless, other conceptual models that allow designers to define an abstract representation of a data warehouse schema can also be used.

6.5.1 Analysis-Driven Approach

General Description

Analysis of the users' requirements leads to the development of an initial multidimensional schema. However, in order to support organizational decisions, the various elements present in this schema should correspond to existing data. Therefore, it is necessary to verify that the data required by the users is available in the source systems prior to developing logical and physical schemas. During the process of verification, a detailed description of the required transformations between data in the source systems and elements of the data warehouse schema must created. In the case of missing data items, modification of the schema must be performed, considering the users' viewpoint. Modifications to the schema may lead to changes in the transformations.

Steps

The creation of a conceptual schema consists of three steps, specified below:

- **Develop the initial schema:** Well-specified business or user requirements lead to clearly distinguishable multidimensional elements, i.e., facts, measures, dimensions, and hierarchies. Therefore, a first approximation to the conceptual schema can be developed. It is recommended that a conceptual model is used to improve communication with nonexpert users. This schema should be validated against its potential usage during analytical processing. This can be done by first revising the list of queries and analytical scenarios and also by consulting the users directly. Designers should be cognizant of the features of the multidimensional model in use and pose more detailed questions (if necessary) to clarify any aspects that are unclear. For example, a schema may contain different kinds of hierarchies, as specified in Sect. 3.2, some dimensions may play different roles, and attributes and measures may be derived. Note that during this step, the refinement of the conceptual schema may require several iterations with the users.
- **Determine data availability and specify mappings:** The data contained in the source systems determines whether the proposed conceptual schema can be transformed into logical and physical schemas and be fed with the data required for analysis. All elements included in the conceptual schema are checked against the data items in the source systems.

This process can be time-consuming if the underlying source systems are not documented, are denormalized, or are legacy systems. The result of this step is a specification of mappings for all elements of the multidimensional schema that match data in the source systems. This mapping can be represented using descriptive methods as proposed in [145] or using UML diagrams as described in [170]. This specification includes also a description of required transformations, if they are necessary. Note that it is important to determine data availability at an early stage in data warehouse design to avoid unnecessary effort in developing logical and physical schemas for which the required data may not be available.

- **Develop final schema and refine mappings:** If data is available in the source systems for all elements of the conceptual schema, the initial schema may be considered as the final schema. However, if not all multidimensional elements can be fed with data from the source systems, a new iteration with the users to modify their requirements according to the availability of data is required. As a result, a new schema is developed and is presented to the users for acceptance. The changes to the schema may require modification of existing mappings.

During all steps of the conceptual-design phase, the specification of business and technical metadata is in continuous development. As was explained in Sect. 2.9, the technical metadata will include information about the data warehouse schema, the data source schemas, and ETL processes. For example, the metadata for a data warehouse schema model may provide information such as aliases used for various elements, abbreviations, currencies for monetary attributes or measures, and metric systems. Similarly, the elements of the source systems should be documented in an abstract manner. This may be a difficult task if conceptual schemas for these systems do not exist. However, as explained in Sect. 6.4.2, it is worth applying reverse engineering techniques to obtain such schemas in order to facilitate future evolution of the data warehouse.

The metadata for the ETL processes should consider several elements. For example, this metadata will include the mappings developed in the second step above, as well as general descriptions of the required transformations. Further, since a data warehouse should contain current data in order to better support decision processes, not only is an initial loading of data from the source systems into the data warehouse necessary, but also the frequency of subsequent data refreshment must be considered. For example, data in a fact table may be required on a daily or monthly basis, or after some specific event (for example, after finishing a project), while changes in dimension data may be included after their inclusion in the source systems. Therefore, users should specify a data refreshment strategy that will correspond to their analysis needs.

Example

Develop Initial Schema

Since the users had considered three different foci of analysis, related to publications, conferences, and projects, three multidimensional schemas were developed.

The multidimensional schema for publications is shown in Fig. 6.5. We shall now explain several aspects of this schema. Firstly, since all kinds of publications (including those in national forums) are important[1] for analyzing research activities, we did not create another dimension for publications. Instead, we included an additional hierarchy level Type, which allows users to select between international or national publications. Secondly, to be able to analyze publications according to the discipline to which they belong, we included this information as a separate level. Note that a nonstrict hierarchy Subject was created, since a publication may belong to several disciplines. Thirdly, a publication belongs to one of three different categories, i.e., journal, book, or conference proceedings. Therefore, we used a generalized hierarchy to represent publications. Fourthly, since academic semesters can start and finish in the middle of a month, we included two time hierarchies, i.e., Academic time and Calendar time. Finally, we included the nonstrict hierarchy Affiliation to represent the fact that researchers may be affiliated to several departments.

The above conceptual model can be used to analyze some of the queries described in Sect. 6.4.1. Some examples of these queries, issued by users at different organizational levels, are as follows:

- Professional users (for instance, research managers) may require the numbers of publications produced by researchers to see their productivity, or the numbers of publications in specific conferences that are considered as fundamental in a specific research area. These statistics can be given for various periods of time.
- Management-level users (for instance, the research department and its administrative units) may be interested in knowing how many publications were produced by a specific publisher in order to consider the university's subscription to that publisher's media. Further, analysis of publications by discipline can be done to promote specific research areas in which the university can offer its expertise, for instance in the form of consulting services or in the creation of spin-offs.
- Executive-level users (for instance, the Rector, his advisors, and the research council) may require information about the total number of publications by year to analyze whether international publications are increasing or decreasing.

---

[1] This information is required to accomplish the goal of increasing the overall research activity at the university.

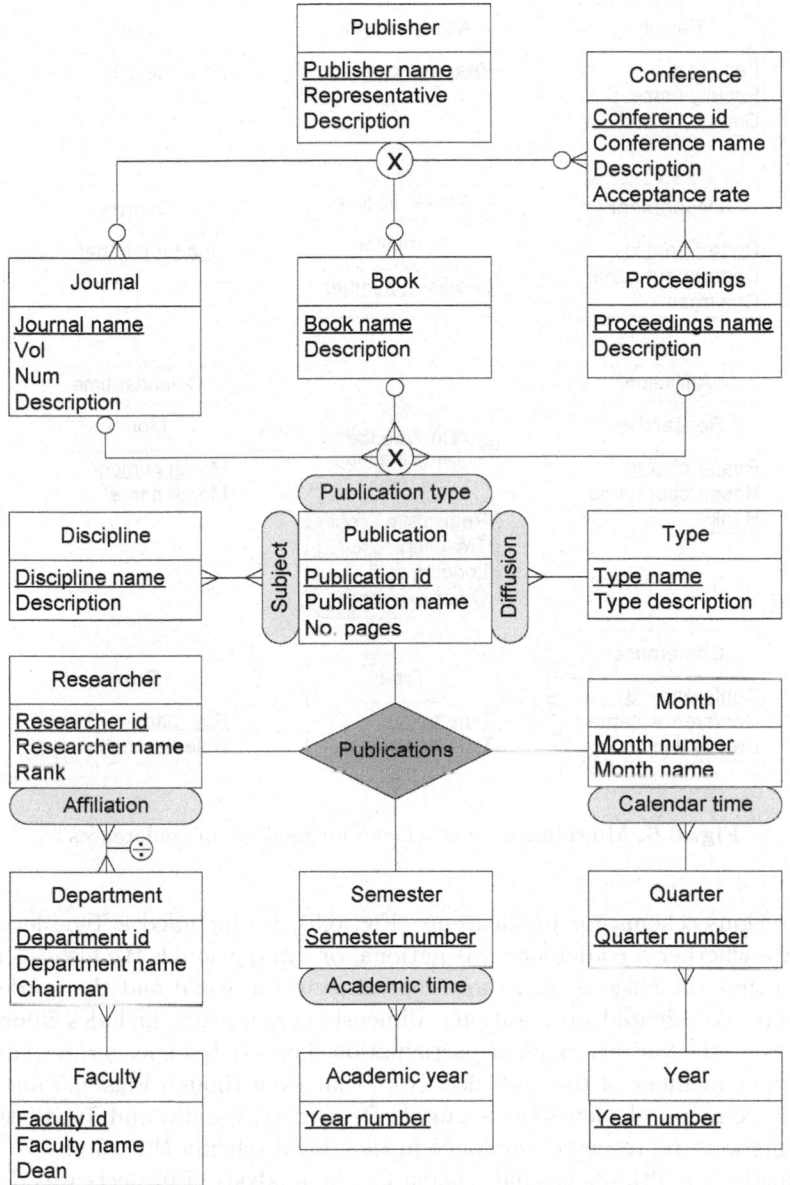

Fig. 6.5. Multidimensional schema for analysis of publications

Another multidimensional schema, for conferences, is shown in Fig. 6.6. In accordance with the users' requirements, we established several measures, i.e., Registration cost, Traveling cost, Lodging cost, and Other expenses. Similarly to

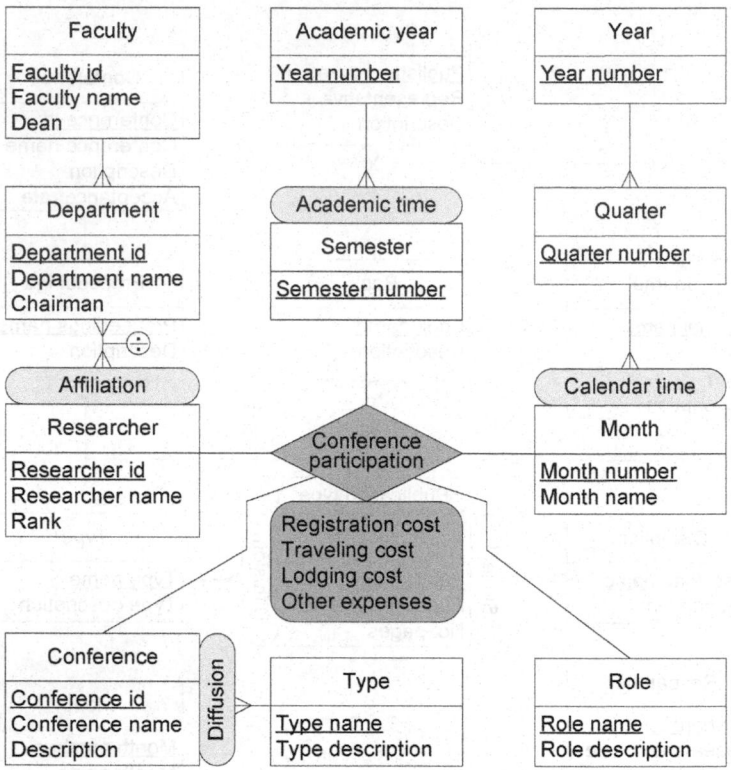

Fig. 6.6. Multidimensional schema for analysis of conferences

the previous schema for publications (Fig. 6.5), we included a Type level to indicate whether a conference was national or international. We also referred to two time dimensions, i.e., those representing the usual and the academic calendar. We created an additional dimension, Role, that includes information about the various kinds of participation, i.e., invited speaker, organizer, author, or member of the audience. Note that even though Figs. 6.5 and 6.6 contain common elements (for example Researcher, Month, and Type), these elements were represented only once in the global schema (Fig. 6.8).

Finally, a multidimensional schema for the analysis of projects was developed, as shown in Fig. 6.7. A generalized hierarchy was used to model the fact that projects can belong either to research centers or to departments. Similarly to the case of publications (Fig. 6.5), we used Type and Discipline levels to indicate whether a project was national or international and to which research area it belonged, respectively.

The three multidimensional schemas (Figs. 6.5, 6.6, and 6.7) were merged to form a data mart that contained data to support decisions related to

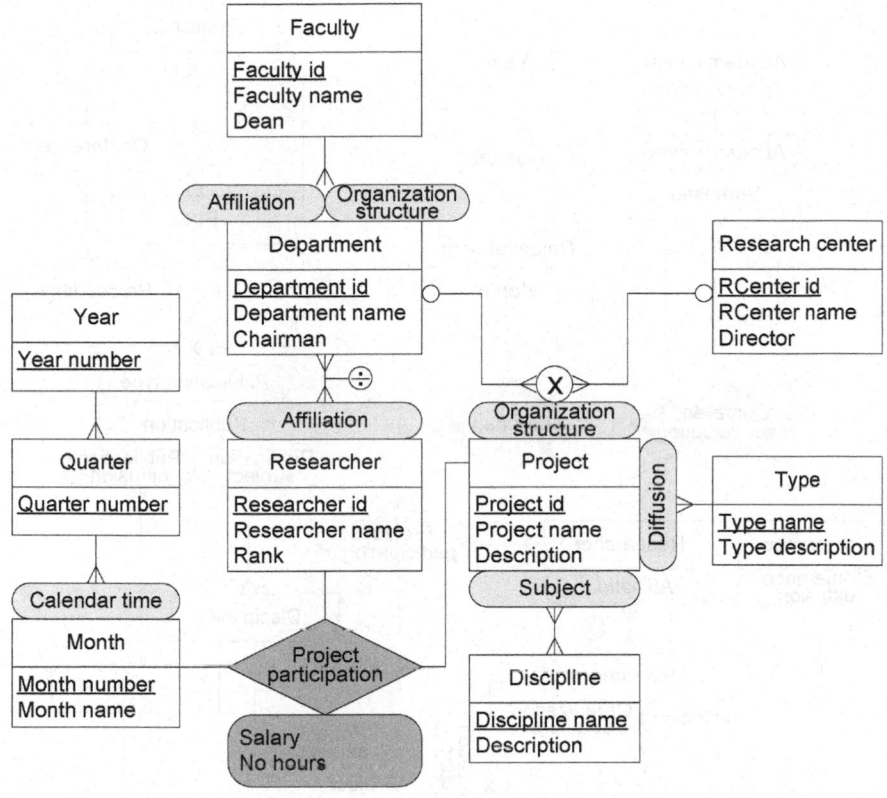

Fig. 6.7. Multidimensional schema for analysis of projects

improving the university's ranking score. The resulting schema is shown in Fig. 6.8. In this figure, only the level names are shown; attribute names are not included. Note that the resulting schema allows additional queries that may be issued using drill-across operations. As an example, information about the total of number of publications by research center or by area can be requested to analyze their international impact.

Since in real-world applications a data mart or a data warehouse schema may contain many more elements, it may become difficult to visualize all levels and fact relationships together. In order to know which levels are common to different fact relationships, we propose the use of a matrix such as that shown in Table 6.3. The columns of the matrix contain the names of the fact relationships, and the rows contain the names of levels comprising hierarchies or one-level dimensions. Every cell of the matrix containing the symbol ✓ indicates that the corresponding level is related (directly as a leaf level or indirectly, forming a hierarchy) to the corresponding fact relationship. This matrix is similar to that proposed by Kimball et al. [145] for **conformed**

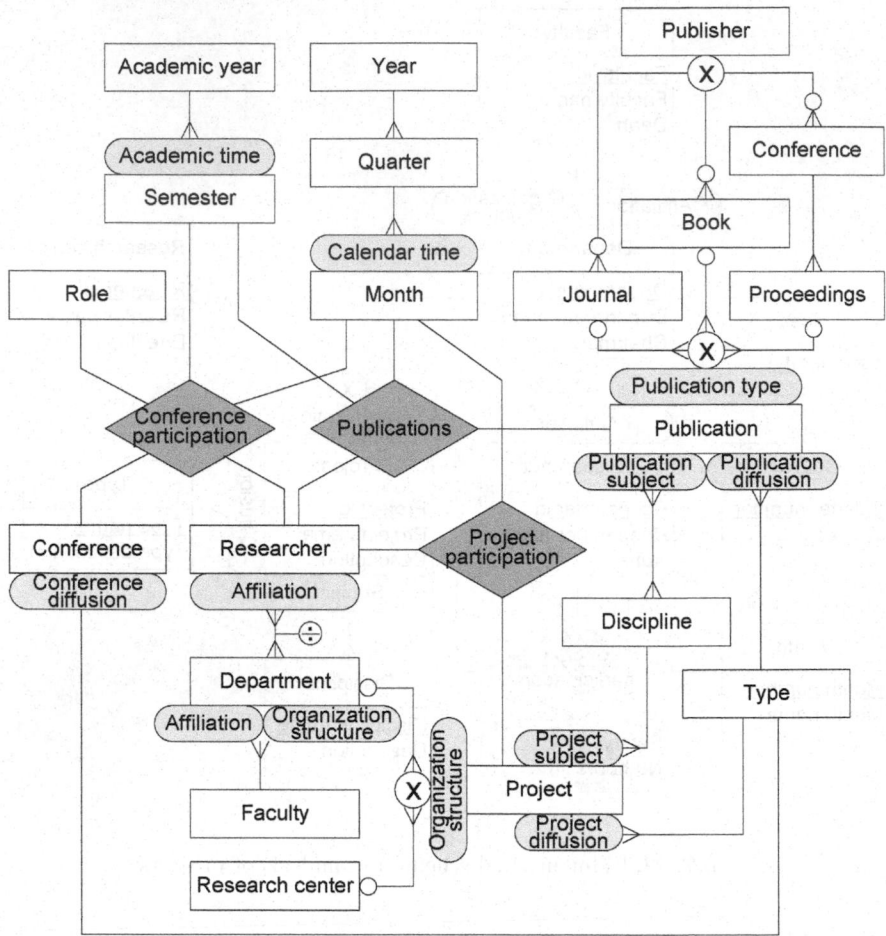

Fig. 6.8. A data mart schema obtained by merging the schemas shown in Figs. 6.5, 6.6, and 6.7

dimensions. However, we have extended the concept of conformed dimensions to **conformed levels**, since in our model levels can be reused between different hierarchies.

Check Data Availability and Specify Mappings

The next step in the proposed method is to check the availability of data in the source systems for all elements included in the data warehouse schema. We do not present here the structures of the underlying source systems, owing to their complexity. In the best scenario, both schemas are documented using a conceptual model, thus facilitating the task of specifying mappings. However,

Table 6.3. Specification of conformed levels

Fact relationship/ Dimension	Publications	Conference participation	Project participation
Researcher	✓	✓	✓
Department	✓	✓	✓
Faculty	✓	✓	✓
Publication	✓	—	—
Journal	✓	—	—
Discipline	✓	—	✓
Type	✓	✓	✓

in the absence of the conceptual representation of the source systems, their logical structures can be used instead.

Table 6.4. Data transformation between sources and the data warehouse

Source table	Source attribute	DW level	DW attribute	Transformation
Conference	Conference id	Conference	Conference id	—
Conference	Short name	—	—	—
Conference	Name	Conference	Name	—
Conference	Description	Conference	Description	—
Conference	Date	Calendar time Academic time	—	✓
Conference	International	Type	Type name	✓
Conference	City	—	—	—
. . .	. . .	. . .	. . .	. . .

Table 6.4 shows an example of a table that specifies the way in which source tables and attributes of the operational databases are related to the levels and attributes of the data warehouse. The last column indicates whether a transformation is required. For example, data representing the Conference id, Name, and Description of conferences in an operational database can be used without any transformation in the data warehouse for the corresponding attributes of the Conference level (Fig. 6.6). However, since the information indicating whether a conference is national or international is included as a flag in the operational database (the International attribute in Table 6.4), a transformation is necessary before its inclusion in the Type level in the data warehouse (Fig. 6.6).

In this example, considering all remaining elements of the multidimensional schemas in this way, the revision process revealed the following:

- For the multidimensional schemas for publications and conferences given in Figs. 6.5 and 6.6, we do not have information about time represented by academic cycles (semesters and academic years). However, we do have the exact dates of publications and conferences. Therefore, during the ETL process, this information can be included to specify the required time frame.
- The Type level, which indicates whether a publication, conference, or project is national or international, must also be derived from the operational databases. This information is included as a flag in the corresponding tables of the operational databases.
- The data required for the Discipline level exists in the operational database for the description of projects; however, it is not included in the database for publications. The latter contain keywords that can be used to derive the discipline. However, users must make a decision as to whether this derivation should be done automatically or manually, since automatic derivation may introduce misleading information when the same keyword belongs to several disciplines.
- In the operational databases, the information about journals and conference proceedings does not include publishers. Therefore, a decision must be made as to whether this level will be included only for books (for which it is present in the operational databases) or whether additional effort should be made to find this information and to include it in the data mart.

Note that not all of this information can be represented in Table 6.4. Additional documentation should be delivered that includes more detailed specification of the required mappings and transformations.

Develop Final Schema and Refine Mappings

Revision and additional consultation with users are required in order to adapt the multidimensional schema to the content of the data sources. When this has been done, the final schema and the corresponding mappings are developed. The metadata for the source systems, the data warehouse, and the ETL processes are also developed in this step. Besides the specification of transformations, the metadata includes abstract descriptions of various features mentioned earlier in this section. For example, for each source system, its access information must be specified (for instance login, password, and accessibility). Further, for each element in the source schemas (for instance the entity and relationship types), we specify its name, its alias, a description of its semantics in the application domain, etc. The elements of the data warehouse schema are also described by names and aliases and, additionally, include information about data granularity, policies for the preservation of data changes (i.e., whether they are kept or discarded), loading frequencies, and the purging period, among other things.

6.5.2 Source-Driven Approach

General Description

After the operational schemas have been analyzed, the initial data warehouse schema is developed. Since not all facts will be of interest for the purpose of decision support, input from users is required to identify which facts are important. Users can also refine the existing hierarchies, since some of these may be "hidden" in an entity type or a table. As a consequence, the initial data warehouse schema is modified until it becomes the final version accepted by the users. Note that in this stage it is straightforward to obtain a specification of the mappings, since the resulting multidimensional schemas are based on source data.

Steps

The conceptual-design phase can be divided into several steps, described below:

- **Develop initial schema:** Since the multidimensional elements have been identified in the previous requirements specification phase, the development of an initial data warehouse schema is straightforward. Similarly to the analysis-driven approach, we recommend that a conceptual model, such as that proposed in this book, is used in order to facilitate future communication with business users and evolution of the schema. The usual practice for this kind of schemas is to use names for the various schema elements that facilitate understanding by the users. However, in some cases users may be familiar with the technical names used in the source systems. Therefore, designers should develop a dictionary of names to facilitate communication with the users.
- **Determine user interest:** Until now, the participation of the users has been minimal, consisting of responding only to specific inquiries from the designer. In this step, users are incorporated in a more active role. These users usually belong to the professional or administrative level, since these kinds of users possess enough knowledge of the underlying systems to understand the proposed schema. The schema is examined in all its details in order to determine what kind of analysis can be done. However, the initial schema may require some modification for several reasons: (1) it may contain more elements than those required for the analysis purposes of the decision-making users; (2) some elements may require transformation (for example, attributes into hierarchies); and (3) some elements may be missing even though they exist in the source systems (for example, owing to confusing names or because no updating behavior is expected). Note that the inclusion of new elements may require further interaction with the source systems.

- **Develop final schema and mappings:** Users' recommendations about changes are incorporated into the initial schema, leading to a final conceptual schema that should be approved by the users. In this stage, an abstract specification of mappings and transformations (if required) between the data in the source systems and the data in the data warehouse is defined.

During all the above steps of the conceptual-design phase, a specification of the business, technical, and ETL metadata should be developed, following the same guidelines as those described for the analysis-driven approach.

Example

Develop Initial Schema

The previous analysis of the operational databases (Fig. 6.4) leads to the multidimensional schema shown in Fig. 6.9. Note that, alternatively, each fact relationship could have been represented separately, i.e., by including only one fact relationship and its corresponding dimensions with hierarchies.

The schema in Fig. 6.9 includes four fact relationships. These can be used for various analysis purposes. For example, the Teaching fact relationship allows users to analyze the various educational activities of academic staff members, while the Affiliation fact relationship can be useful for evaluating the overall involvement of staff members in their departments. Each fact relationship is associated with corresponding dimensions (for example, the Academic staff or the Project dimension). A hierarchy Register, formed by the levels Course, Department, and Faculty, was created on the basis of the relationship types with one-to-many cardinalities that exist between the corresponding entities.

Note that two many-to-many relationship types in the schema in Fig. 6.4, i.e., Course Degree and Organized, were included in the nonstrict hierarchy Program in the schema in Fig. 6.9. However, as explained in Sect. 3.2.2, other options for representing nonstrict hierarchies could have been considered. For example, Academic degree could have been included as an additional leaf level of the Teaching fact relationship, leading to a nonstrict hierarchy formed by the levels Academic degree, Department, and Faculty.

On the other hand, after analyzing the two zero-to-many relationship types Belongs RC and Belongs dep in Fig. 6.4, a decision was made to represent them as a generalized hierarchy called Organization structure; this allows users to analyze projects that belong to either research centers or departments. Note that, on the basis of the dates included in the operational schema, we have created two different time dimensions, representing the usual calendar (Date) and the academic calendar (Semester). Further, the Date dimension plays two different roles, indicating the starting and the ending time instants of the corresponding facts.

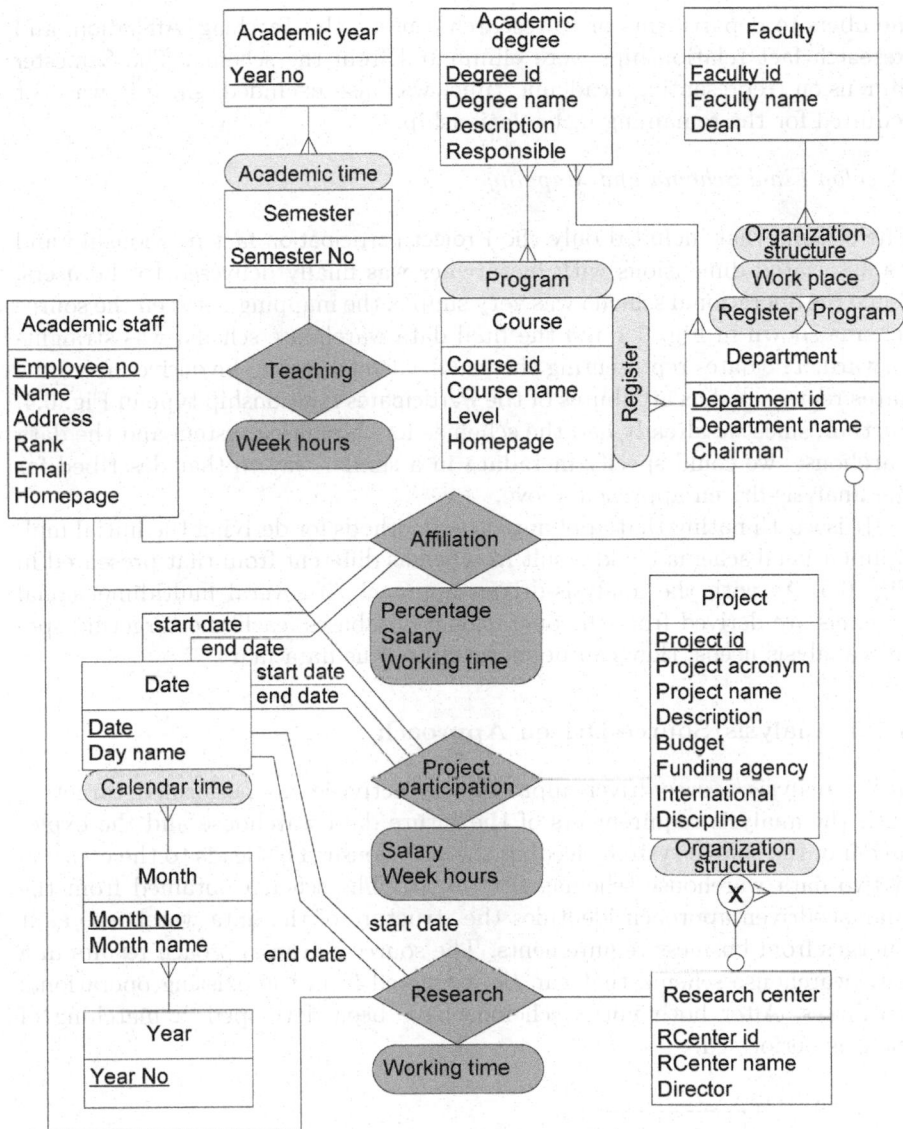

Fig. 6.9. Multidimensional schema derived from the ER schema of Fig. 6.4

Determine User Interest

The initial data warehouse schema as presented in Fig. 6.9 was delivered to the users, i.e., the representatives of the various research units. Since they were not interested in academic issues and the administrative assignment of staff members to departments or to research centers, the Teaching, Affiliation, and Research fact relationships were eliminated from the schema. The Semester dimension, representing academic time, was also excluded, since it was not required for the remaining fact relationship.

Develop Final Schema and Mapping

The schema that included only the Project participation fact relationship and its associated dimensions with hierarchies was finally delivered to the users. Since the operational schema was very simple, the mapping between the source schema shown in Fig. 6.4 and the final data warehouse schema was straightforward. The dates representing the usual calendar had to be derived from the dates represented as attributes of the Participates relationship type in Fig. 6.4. Further, since we already had the schemas for the source system and the data warehouse, we could specify metadata in a similar way to that described for the analysis-driven approach above.

It is worth noting that applying other methods for deriving the initial multidimensional schema could result in schemas different from that presented in Fig. 6.9. As with the analysis-driven approach, if several multidimensional schemas are derived from the operational databases, each one targeting specific analysis needs, they can be merged into one data mart.

6.5.3 Analysis/Source-Driven Approach

In the analysis/source-driven approach, two activities are performed, targeting both the analysis requirements of the future data warehouse and the exploration of the source systems feeding the warehouse. This leads to the creation of two data warehouse schemas (Fig. 6.10). The schema obtained from the analysis-driven approach identifies the structure of the data warehouse as it emerges from business requirements. The source-driven approach results in a data warehouse schema that can be extracted from the existing operational databases. After both initial schemas have been developed, a matching of them is performed.

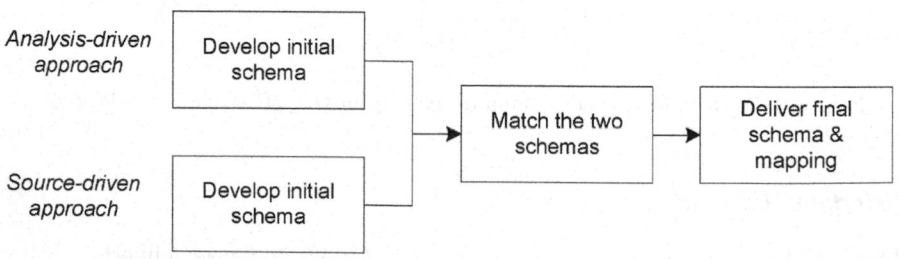

Fig. 6.10. Steps in conceptual design in the analysis/source-driven approach

This matching process is not an easy task. Several aspects should be considered, such as the terminology used, and the degree of similarity between the two solutions for each multidimensional element, for example between dimensions, levels, attributes, or hierarchies. Some solutions have been provided by Bonifati et al. [28] and Giorgini et al. [78]. However, these solutions are highly technical and may be too complex to implement.

An ideal situation arises when both schemas cover the same aspects of analysis, i.e., the user demands are covered by the data in the operational systems and no other data is needed to expand the analysis spectrum. In this case, the schema is accepted and mappings between elements of the source systems and the data warehouse are specified. Additionally, documentation containing metadata about the data warehouse, the source systems, and the ETL process, as described for the analysis-driven and source-driven approaches above, is developed.

Nevertheless, in real-world applications it is seldom the case that both schemas will cover the same aspects of analysis. Two situations may occur:

1. The users demand less information than what the operational databases can provide. In this case, it is necessary to determine whether users may consider new aspects of analysis or whether to eliminate from the schema those fact relationships that are not of interest to users. Therefore, another iteration of the analysis- and source-driven approaches is required. In this iteration, either new users will be involved who could be interested in the new possibilities provided by the source systems, or a new initial schema will be developed that eliminates from the schema some fact relationships and associated dimensions.

2. The users demand more information than what the operational databases can provide. In this case, the users may reconsider their demands and limit them to those proposed by the analysis-driven solution. Alternatively, the users may require the inclusion of external sources or legacy systems that were not considered in the previous iteration but contain the necessary data. Therefore, new iterations of the analysis- and source-driven approaches may again be needed.

The development of the initial schema according to the analysis- and source-driven approaches may be performed in a parallel or in a sequential manner, depending on the availability of users and the capacity of the development team.

Example

Let us suppose that, using the analysis-driven approach, we have created the schema for analysis of projects shown in Fig. 6.7. On the other hand, the source-driven approach gives the schema shown in Fig. 6.9. During the matching process, we can clearly see that the latter schema includes additional information not required by the users, i.e., they demand less information than the operational databases can provide. Therefore, during another iteration within the analysis-driven approach, a decision should be made whether analysis related to the Teaching, Affiliation, and Research fact relationships in Fig. 6.9 is important in accomplishing business goals. For example, decision makers might be interested in the Teaching fact relationship in order to target another goal, aimed at a more effective distribution of human resources leading to a decrease in the cost of courses.

Note some other similarities and differences between the two schemas. In Fig. 6.9, the Type and Discipline levels are not defined as hierarchies, since they are difficult to distinguish by only analyzing data. Further, the Time dimension in Fig. 6.7 has only one hierarchy, while the Date dimension in Fig. 6.9 contains two hierarchies, corresponding to the starting and ending times. Further, the measures Salary and No hours in Fig. 6.7 refer to monthly values for researchers participating in projects. On the other hand, these measures in Fig. 6.9 are represented by considering the starting and ending instants of the periods during which the specific measure values exist.

After additional iterations, a final schema will be developed. This includes the data delivered by the source-driven approach, with refinements necessary to better represent users' needs (for example, addition of a level Discipline). New mappings will also be developed. Further, all documentation will be created; this includes business and technical metadata for the source and data warehouse systems, as well as for the ETL process.

6.6 Characterization of the Various Approaches

In this section, we give an overall description of the three approaches to data warehouse development, describing various aspects that it is important to consider before choosing an approach for a specific data warehouse project.

6.6.1 Analysis-Driven Approach

The analysis-driven approach requires the intensive participation of users from different organizational levels. In particular, the support of executive-level users is important in order to define business goals and needs. The identification of key users for requirements specification is a crucial task. It is necessary to consider several aspects:

- The users targeted should be aware of the overall business goals and objectives to avoid the situation where the requirements represent the personal perceptions of the users according to their role in the organization or their specific business unit [314].
- Users who would dominate the requirements specification process should be avoided or tempered in order to ensure that the information needs of different users will be considered.

- Users should be available and should agree to participate during the whole process of requirements gathering and conceptual design.
- Users should have an idea of what a data warehouse system and an OLAP system can offer. If this is not the case, users should be instructed by means of explanations, demonstrations, or prototypes.

The development team requires highly qualified professionals. For example, a project manager should have very strong moderation and leadership skills [165]. A good knowledge of information-gathering techniques and/or business process modeling is also required. It is important that data warehouse designers should be able to communicate with and to understand nonexpert users in order to obtain the required information and, later on, to present and describe the proposed multidimensional schema to them. This helps to avoid the situation where users describe the requirements for the data warehouse system using business terminology and the data warehouse team develops the system using a more technical viewpoint that is difficult for the users to understand [31].

Several advantages of the analysis-driven approach are as follows:

- Since the requirements of the data warehouse are derived from a business perspective, they provide a comprehensive and precise specification of the needs of stakeholders from their business viewpoint [166].
- Effective participation of users ensures a better understanding of the facts, dimensions, and relationships existing between them [76].
- Using formal techniques for defining business process models provides a formal description of the informational requirements of the users [27].
- This approach may increase the acceptance of the system if there is continuous interaction with potential users and decision makers [165].
- External data may be specified for inclusion in the data warehouse [165].
- Long-term strategic goals can be specified [165].

However, some disadvantages of this approach can play an important role in determining its usability for a specific data warehouse project:

- The specification of business goals can be a difficult process, and its result depends on the techniques applied and the skills of the developer team.
- The specification of users' requirements that are not aligned with business goals may produce a very complex schema that does not support the decision processes of users at all organizational levels.
- Requirements specification based on business processes can become more complicated if these processes cross organizational boundaries [166].
- The duration of the project tends to be longer than when the source-driven approach is used [165]. Therefore, the cost of the project can also be higher.
- The users' requirements might not be satisfied by the information existing in the source systems [27].

6.6.2 Source-Driven Approach

In the source-driven approach, the participation of the users is not explicitly required [165]. They are involved only sporadically, either to confirm the correctness of the structures derived or to identify facts and measures as a starting point for creating multidimensional schemas. Typically, the involved users come from the professional or the administrative organizational level, since data is represented at a low level of detail.

On the other hand, this approach requires highly skilled and experienced designers. Besides the usual modeling abilities, they should additionally possess sufficient business knowledge in order to be able to understand the business context and its needs, relying mainly on operational data. They should also have the capacity to understand the structure of the underlying operational databases.

The source-driven method has several advantages:

- It ensures that the data warehouse reflects the underlying relationships in the data [201].
- The data warehouse contains all necessary data from the beginning [238].
- Developing data warehouses on the basis of existing operational databases simplifies the ETL processes [201].
- The enterprise data model implemented in the source systems may provide a more stable basis for design than user requirements, which may be subject to change [201].
- Minimal user time is required to start the project [238].
- The development process can be fast and straightforward if well-structured and normalized operational systems exist [165].
- If the operational databases are represented using the ER model or normalized relational tables, automatic or semiautomatic techniques can be applied.

However, it is important to consider the following disadvantages before choosing this approach:

- Since business or user requirements are not gathered before the data warehouse design process begins, the only business needs that can be captured are those reflected in the underlying source data models.
- The system may not meet users' expectations [314], since the company's goals and the users' requirements are not reflected at all.
- The resulting schema has a low level of granularity, since operational data sources are used [165].
- This approach cannot be applied when the logical schemas of the underlying operational systems are very large and hard to understand or the data sources reside on legacy systems whose inspection and/or normalization is not recommendable [78].
- Since it relies on existing data, this approach cannot be used when long-term strategic goals must be analyzed or specified [165].

- Creating a multidimensional schema based on existing data may make the inclusion of hierarchies difficult, since they may be "hidden" in various structures, for example generalization relationships.
- It is difficult to motivate end users to participate in the process, since they are not used to working with large data models developed for and by specialists [314].
- The derivation process can be confusing without knowledge of the users' needs, since the same data can be considered as a measure or as a dimension attribute.

6.6.3 Analysis/Source-Driven Approach

As this approach combines the analysis-driven and source-driven approaches, the recommendations regarding users and the development team given above should also be considered here.

The analysis/source-driven approach has several important advantages:

- It generates a feasible solution (i.e., the solution is supported by the existing data sources) that better reflects the users' goals [28].
- It may indicate missing data in the operational databases that is required to support the decision-making process.
- If the source systems offer more information than what the business users initially demand, the analysis can be expanded to include new aspects not yet considered.

However, this approach has the following disadvantages:

- The development process is more complicated, since two schemas are required, one obtained from the definition of the business requirements and another derived from the underlying source systems.
- The integration process to determine whether the data sources cover the users' requirements may need complex techniques [28, 78].

6.7 Logical Design

There are two aspects that it is important to consider during the logical-design phase: firstly, the transformation of the conceptual multidimensional schema into a logical schema; and secondly, specification of the ETL processes, considering the mappings and transformations indicated in the previous phase. We shall refer next to these two aspects.

6.7.1 Logical Representation of Data Warehouse Schemas

As explained in Sect. 2.7, the logical representation of a data warehouse is often based on the relational data model [83, 118, 146, 188] using specific structures called star and snowflake schemas. Many data warehouse applications also include precomputed summary tables containing aggregated data that are stored as materialized views. However, we do not consider such tables to be part of the core logical schema.[2]

After the conceptual-design phase has been completed, it is necessary to apply mapping rules to the resulting conceptual schema in order to generate a logical schema. It should be clear that the mapping rules depend on the conceptual model used [35, 82]. In Sect. 3.5, we described some general mapping rules that allow one to translate the MultiDim conceptual model into the relational model. In this section, we apply these rules to multidimensional conceptual schemas developed in the previous phase. In order to provide a more general solution, we use the SQL:2003 standard [192, 193] to define relational tables.

We present next the transformation of the schema shown in Fig. 6.6 into relational tables. First, considering users' analysis needs, query performance, and data reuse, we must decide whether a star or a snowflake representation should be chosen.[3] Since the levels of the Calendar time and Academic time hierarchies are not reused in other hierarchies, for performance reasons we denormalize these hierarchies and include them in a single table instead of mapping every level to a separate table. This is done as follows:

```
create table CalendarTime (
      CalTimeSysId integer primary key,
      MonthNumber integer,
      MonthName character varying(15),
      QuarterNumber integer,
      YearNumber integer );
create table AcademicTime (
      AcaTimeSysId integer primary key,
      SemesterNumber integer,
      YearNumber integer );
```

The Affiliation hierarchy in Fig. 6.6 contains three levels: Researcher, Department, and Faculty. Since the Department level is used in another multidimensional schema (Fig. 6.7), in order to reuse existing data we have decided to use the snowflake representation for this hierarchy, i.e., we represent each level in a separate table. We first declare the Faculty level, which is then referenced in the Department level:

```
create table Faculty (
      FacultySysId integer primary key,
      FacultyId integer unique,
      FacultyName character varying(35),
```

---

[2] We shall discuss view materialization in the physical-design phase.

[3] In Sect. 3.2.1, we have already stated the advantages and disadvantages of using a star (denormalized) or a snowflake (normalized) schema for representing dimensions with hierarchies.

```
        Dean character varying(35) );
create table Department (
        DepartmentSysId integer primary key,
        DepartmentId integer unique,
        DepartmentName character varying(35),
        Chairman character varying(35),
        FacultyFkey integer,
        constraint FacultyFK
            foreign key (FacultyFkey) references Faculty(FacultySysId) );
```

The Affiliation hierarchy is a nonstrict hierarchy (Sect. 3.2.2), i.e., it contains a many-to-many relationship with a distributing factor between the Researcher and Department levels. In order to represent this relationship, we use a bridge table as explained in Sect. 3.2.2. For that purpose, we create a Researcher table and an additional table called ResearcherDepartment, which references both the Researcher and the Department table and includes the distributing factor:

```
create table Researcher (
        ResearcherSysId integer primary key,
        ResearcherId integer unique,
        ResearcherName character varying(35),
        Rank character varying(35) );
create table ResearcherDepartment (
        ResearcherFkey integer,
        DepartmentFkey integer,
        Percentage integer,
        constraint ResearcherFK
            foreign key (ResearcherFkey) references Researcher (ResearcherSysId),
        constraint DepartmentFK
            foreign key (DepartmentFkey) references Department (DepartmentSysId),
        constraint ResDepPK primary key (ResearcherFkey,DepartmentFkey) );
```

In a similar way, we define tables for the other levels, i.e., for Type, Conference, and Role:

```
create table Type (
        TypeSysId integer primary key,
        TypeName character varying(35) unique,
        TypeDescription character varying(128) );
create table Conference (
        ConferenceSysId integer primary key,
        ConferenceId integer unique,
        ConferenceName character varying(35),
        Description character varying(128),
        TypeFkey integer,
        constraint TypeFK foreign key (TypeFkey) references Type (TypeSysId) );
create table Role (
        RoleSysId integer primary key,
```

```
RoleName character varying(50),
Description character varying(128) );
```

Finally, the fact relationship table is created. This contains all measures included in the conceptual schema, and references all participating dimensions. Note the inclusion of referential integrity constraints in order to ensure correct data insertion:

```
create table ConferenceParticipation (
    ResearcherFkey integer,
    DepartmentFkey integer,
    ConferenceFkey integer,
    CalTimeFkey integer,
    AcaTimeFkey integer,
    RoleFkey integer,
    RegistrationCost decimal(5,2),
    TravelingCost decimal(5,2),
    LodgingCost decimal(5,2),
    OtherExpenses decimal(5,2),
    constraint ResearcherFK
        foreign key (ResearcherFkey) references Researcher (ResearcherSysId),
    constraint DepartmentFK
        foreign key (DepartmentFkey) references Department (DepartmentSysId),
    constraint ConferenceFK
        foreign key (ConferenceFkey) references Conference (ConferenceSysId),
    constraint CalTimeFK
        foreign key (CalTimeFkey) references CalendarTime (CalTimeSysId),
    constraint AcaTimeFK
        foreign key (AcaTimeFkey) references AcademicTime (AcaTimeSysId),
    constraint RoleFK
        foreign key (RoleFkey) references Role (RoleTimeSysId),
    constraint ConfPartPK primary key (ResearcherFkey,DepartmentFkey,
        ConferenceFkey,CalTimeFkey,AcaTimeFkey,RoleFkey) );
```

The above tables use primary keys (for example, FacultySysId in the Faculty table and DepartmentSysId in the Department table), whose values must be supplied by users during the insertion process. However, in Sect. 3.5 we stated the advantages of using surrogates, i.e., system-generated keys. The latter can be obtained by using the facilities of object-relational databases and by declaring typed tables, as explained in Sect. 2.3.2. Recall that typed tables are tables that require some structured types for their definition. They contain an additional self-referencing column that stores a value that uniquely identifies each row. For example, we can define a structured type to represent the Faculty level as follows:

```
create type FacultyType as (
    FacultyId integer,
    FacultyName character varying(35),
    Dean character varying(35) )
    ref is system generated;
```

Then, we create a typed table Faculty of this structured type:

```
create table Faculty of FacultyType (
    constraint facultyIdUnique unique(FacultyId),
    ref is FacultySysId system generated );
```

The clause ref is FacultySysId system generated indicates that FacultySysId is a surrogate attribute automatically generated by the system. The Department table can be declared in a similar way. It will additionally include a reference to the surrogate of the Faculty table:

```
create type DepartmentType as (
    DepartmentId integer,
    DepartmentName character varying(35),
    Chairman character varying(35),
    CollRef ref(FacultyType) scope Faculty
    references are checked )
    ref is system generated;
create table Department of DepartmentType (
    constraint DepartmentIdUnique unique(DepartmentId),
    ref is DepartmentSysId system generated );
```

In this way, the Department table contains references to the corresponding rows of the Faculty table. This may help to avoid costly join operations between hierarchy levels.

However, as already mentioned in Sect. 4.8.1, the choice between the relational model or the object-relational model depends on application needs. In making this choice, it is also important to consider the physical-level design and the specific features of the target DBMS, as we shall explain in Sect. 6.8.

6.7.2 Defining ETL Processes

During the conceptual-design phase, we identify the mappings required between the sources and the data warehouse. We also specify some transformations that may be necessary in order to match user requirements with the data available in the source systems. However, before implementing the ETL processes, several additional tasks must be specified in more detail.

In the logical-design phase, all transformations of the source data should be considered. Some of them can be straightforward, for example the separation of addresses into their components (for example, street, city, ZIP code, etc.) or the extraction of date components (for example, month and year). Note that the transformation may depend on the logical model. For example, in the relational model each component of a department address will be represented as separate attributes, while in the object-relational model these components can be defined as elements of a user-defined type.

Other transformations may require further decisions, for instance whether to recalculate measure values in the ConferenceParticipation and Project-Participation tables to express them in euros or to use the original currency and include the exchange rate. It should be clear that in real situations, more complex data transformations may be required. Further, since the same data can be included in different source systems, the issue of inconsistencies may arise and an appropriate strategy for resolving them must be devised.

Moreover, developers should design the necessary data structures for all elements for which users want to keep changes. For example, some of the ideas about slowly changing dimensions [145], as described in Sect. 1.1.3, could be applied.

A preliminary sequence of execution for the ETL processes should also be determined. This is required to ensure that all data will be transformed and included, with its consistency being checked. For example, in Fig. 6.6, first the Faculty, then the Department, and finally the Researcher level should be included. Obviously, this order is required so as to conform to the referential integrity constraints. Similarly, for the fact table, the data for the dimensions must be populated first, before the data for the fact table is loaded. Further, depending on the particularities of the specific data warehouse project, it might be important to address additional issues other than those specified in this section.

It should be obvious that the ETL processes can be the most difficult part of developing a data warehouse; they can be very costly and time-demanding. To facilitate the design and subsequent execution of the ETL processes, various commercial tools can be used, such as Microsoft Data Transformation Services, IBM WebSphere Information Integration Solutions, or Oracle Warehouse Builder.

6.8 Physical Design

As with the logical-design phase, we should consider two aspects in the physical-design phase: one related to the implementation of the data warehouse schema and another that considers the ETL processes. For each of them, we refer only to general physical-design issues, since these aspects are heavily dependent on the particularities of the data warehouse project and on specific features of the target database management system.

6.8.1 Data Warehouse Schema Implementation

During the physical-design phase, the logical schema is converted into a tool-dependent physical database structure. Physical-design decisions should consider both the proposed logical schema and the analytical queries or applications specified during the process of requirements gathering. A well-developed physical design should help to improve access to data, query performance, data warehouse maintenance, and data-loading processes, among other things. Therefore, a DBMS used for building a data warehouse should include features that assist the implementers in various tasks, such as managing very large amounts of data, refreshing the data warehouse with new data from the source systems, performing complex operations that may include joins of many tables, and aggregating many data items. This depends on the facilities provided by the DBMS regarding storage methods, indexes, table-partitioning facilities, parallel query execution, aggregation functions, and view materialization, among other things.

We refer next to some of these features, showing their importance in the context of a data warehouse. We discuss them using Oracle 10g with the OLAP option [213] as an example of a DBMS.

Storage Method

The usual practice in developing a data warehouse is to use a relational database owing to the advantages such databases offer, which were discussed in Sect. 3.5. Relational tables, used as a basic unit of data storage, are mostly organized into star or snowflake schemas. Then, data can be queried using classical SQL functions and operators, for example sum, count, and group by, or more sophisticated ones specially defined for data warehouses, for example, cube and rollup. If the DBMS does not provide the required query and calculation capabilities, specialized OLAP systems may also be used.

Currently, there is a clear tendency for leading DBMS companies, such as Oracle and Microsoft, to offer an integrated environment that allows implementers to manage data that is represented in relational tables and in a tool-specific multidimensional format. For example, Oracle OLAP integrates the multidimensional storage within relational databases through **analytic workspaces**. The data is stored as a LOB (large object) table[4] and can be manipulated by the OLAP engine, thus providing more sophisticated analysis capabilities than those provided by SQL for relational databases. Within a single database, many analytic workspaces can be created. These workspaces require the definition of a logical multidimensional schema that, later on, is mapped to the physical data model and populated with data. Users may choose whether to perform an analysis with data that is entirely stored in analytic workspaces or is distributed between analytic workspaces and relational tables.

Further, Oracle OLAP introduces special objects called **dimensions**. Although it is not mandatory to define them, if the application uses a multidimensional schema (for example, a star schema), they can give several benefits in terms of query rewriting, materialized-view management, and index creation,[5] among other things. A dimension in this context is an abstract concept that requires the existence of a physical dimension table containing the corresponding data. For example, for the Diffusion hierarchy in Fig. 6.6, we first create tables for the Type and Conference levels as follows:

---

[4] This kind of table is used to store very large amounts of unstructured data.

[5] We shall refer to these concepts in subsequent sections.

```
create table Type (
    TypeSysId number(6) primary key,
    TypeName varchar2(35) unique,
    TypeDescription varchar2(128) );
create table Conference (
    ConferenceSysId number(15) primary key,
    ConferenceId number(10) unique,
    ConferenceName varchar2(35),
    Description varchar2(128),
    TypeFkey number(6),
    constraint TypeFK foreign key (TypeFkey) references Type (TypeSysId) );
```

Then, we can define the Conference dimension as follows:

```
create dimension ConferenceDim
    level ConferenceLevel is (Conference.ConferenceId)
    level TypeLevel is (Type.TypeName)
    hierarchy Diffusion (
        ConferenceLevel child of TypeLevel
        join key (Conference.TypeSysId) references Type );
```

The dimension ConferenceDim includes two levels, i.e., ConferenceLevel and TypeLevel, indicating which relational tables and attributes they correspond to, i.e., Conference.ConferenceId and Type.TypeName, respectively. The definition of the hierarchy Diffusion specifies the parent-child relationship that exists between levels. Since we are using normalized tables for the Conference and Type levels, the above declaration includes an additional statement referring to the attributes used for the join operation.

However, Oracle OLAP establishes a set of constraints that must be fulfilled to ensure that correct results can be returned from queries that use these dimensions. These constraints may sometimes be too restrictive; for example, they do not allow one to define nonstrict and generalized hierarchies.

In spite of the limitations that can be found in commercial systems, however, we can say that an integrated architecture for representing multidimensional and relational data facilitates data administration and management. Such an architecture provides more capability for performing various kinds of queries without the necessity of using separate software and hardware platforms.

Fragmentation

The terms "fragmentation", "partitioning", and "clustering" in the field of databases means that a table is divided into smaller data sets to better support the management of very large volumes of data. As explained in Sect. 2.8, there are two ways of achieving fragmentation: vertical and horizontal [218].

Vertical fragmentation allows the designer to group attributes of a relation into smaller records. For example, a dimension may be fragmented so as to have the name and city attributes in one partition and the remaining attributes in another partition. Therefore, if a query requests names, more records can be retrieved into main memory, since they contain fewer attributes and thus are smaller in size.

On the other hand, horizontal fragmentation divides a table into smaller tables that have the same structure but with fewer records. For example, if some queries require the most recent data while others access older data, a fact table can be horizontally partitioned according to some time frame, for example years [234].

Therefore, since smaller data sets are physically allocated to several partitions, these smaller data sets considerably facilitate administrative tasks (a new partition can be added or deleted), increase query performance when parallel processing is applied, and enable access to a smaller subset of the data (if the user's selection does not refer to all partitions).

Fragmentation techniques should be chosen during physical data warehouse design. However, to be able to use them, implementers must have a good knowledge not only of the meaning and consequences of having partitioned dimension and fact tables, but also of which specific method of partitioning may work better. For example, Oracle [213] provides four types of horizontal-partitioning methods: range, hash, list, and composite. Each of them has different advantages and design considerations. Oracle gives a detailed description of them and empirical recommendations about which situations a specific method would work better in.

We now give an example of the definition of a partitioned table using Oracle. Supposing that the university in our earlier example is involved in many projects, we create a table for the Project level in Fig. 6.9 with two partitions, considering whether or not the project is international:

```
create table Project (
        ProjectSysId number(15) primary key,
        ProjectId number(10) unique,
        ProjectAcronym varchar2(25),
        ProjectName varchar2(64),
        ...
        International char(1),
        ...)
partition by range International (
        partition IntProj values ('Y'),
        partition NatProj values ('N') );
```

As can be seen, partitioned tables address the key problem of supporting very large data volumes by allowing the implementer to decompose the data into smaller and more manageable physical pieces.

Indexing

The most relevant queries submitted to data warehouse systems are "star queries" [141]. Such queries include references to dimension and fact tables.

They impose restrictions on the dimension values that are used for selecting specific measures; these measures are then grouped and aggregated according to the user's demands. The major bottleneck in evaluating such queries is the join of the central (and usually very large) fact table with the surrounding dimension tables. Therefore, to deal with this problem, new kinds of indexes, the bitmap and join indexes (see Sect. 2.8), are used [43, 141, 209, 213].

A bitmap index can be seen as a matrix where each entry corresponds to an address (i.e., a rowid) of a possible record and each column to the different key values. If the bit is set, this means that the record contains the corresponding key value. A mapping function converts the bit position of an actual rowid so that the bitmap index provides the same functionality as a regular index. Bitmap indexes provide the greatest advantage for columns in which the ratio of the number of distinct values to the number of rows in the table is small [213]. For example, a gender column with only two distinct values (male or female) is optimal for a bitmap index and can be defined in Oracle as follows:

```
create bitmap index ResearcherBitmapIdx on Researcher(Gender);
```

Bitmap indexes speed up access to the fact table when a "star join" is performed [141, 213]; this filters out dimension records that satisfy some conditions before the dimension table itself is accessed. If the "where" clause contains multiple conditions, the filtering process is repeated for each condition. Then, bitmaps corresponding to the various dimension values represented in each condition can be combined using logical operators. The resulting bitmap can be used to extract tuples from the fact table [141].

Another kind of indexes that can be used to speed up star queries in Oracle 10g is the bitmap join index [213], which creates a bitmap index for the join of two or more tables, i.e., for values obtained from the joined tables (Sect. 2.8). Oracle provides various options for creating such indexes, for example by considering one or more columns of the same dimension table, several dimension tables, or a dimension snowflake structure.

Therefore, in order to improve system performance, implementers should consider various kinds of indexes according to the particularities of the data warehouse and the DBMS.

Materialization of Aggregates

Another way of increasing query performance in data warehouses is to pre-calculate expensive join and aggregation operations and store the results obtained in a table in the database [141]. This issue has been studied extensively, mainly in the form of **materialized views** and the problem of maintaining them (e.g., [91]), as was mentioned in Sect. 2.8. Therefore, since materialized views contain query results that have been stored in advance, long-running calculations are not necessary when users actually execute SQL statements; instead, access to a materialized view is performed.

For example, in the schema for the analysis of conferences (Fig. 6.6), we can create a materialized view that stores the aggregations of conference cost according to the conference type and academic year:

```
create materialized view ConferenceCostMV
    build immediate
    refresh complete
    enable query rewrite
    as select T.TypeName, A.YearNo,
        sum(CF.RegistrationCost), sum(CF.TravelingCost),
        sum(CF.LodgingCost), sum(CF.OtherExpenses),
        count(*) as cnt, count(CF.RegistrationCost) as cntRegCost,
        count(CF.TravelingCost) as cntTravCost,
        count(CF.LodgingCost) as cntLodgCost,
        count(CF.OtherExpenses) as cntOthExp
    from Conference C, Type T, AcademicTime A, ConferenceFact CF
    where C.TypeIdFkey = T.TypeSysId and
        CF.ConferenceIdFkey = C.ConferenceSysId and
        CF.AcaTimeIdFkey = A.AcaTimeSysId
    group by T.TypeName, A.YearNo;
```

This example creates a materialized view ConferenceCostMV that computes the total number[6] and the aggregated value of the conference costs (considering four measures included in the schema, i.e., Registration cost, Traveling cost, Lodging cost, and Other expenses), presenting them for each conference type and academic year. In this example, we suppose that the table representing the Academic time hierarchy is stored as a denormalized structure, i.e., all levels are included in the same table called AcademicTime.

Note that Oracle requires one to define when materialized views should be populated, i.e., when the data will be loaded into the table created by the materialized-view statement. In this example, the immediate method is chosen (build immediate). The refreshment method must be also specified, i.e., when the data in the materialized view is **updated**. Oracle provides several different options; the one chosen in the example (refresh complete) indicates that after the inclusion of new data in the data warehouse, the whole materialized view is recalculated.

To ensure that a materialized view will be accessed instead of the query that created it being executed, the query rewrite mechanism must be enabled (enable query rewrite). Afterwards, Oracle's optimizer is able to automatically recognize when an existing materialized view may be used to satisfy a user request. The optimizer transparently rewrites the request using the materialized view instead of going to the underlying detailed tables.

Nevertheless, in Sect. 2.8 we mentioned several problems related to the design and management of materialized views, such as the selection and updating of materialized views, and query rewriting in the presence of materialized views. In addition, not all possible aggregations can be precalculated and

---

[6] This column is required by Oracle.

materialized, since this could lead to the phenomenon called "data explosion", where the number of aggregates grows exponentially with the number of dimensions and hierarchies [270]. Even though much research has been focused on these problems in recent years, there is still a need to improve the proposed solutions.

In general, even though the management of materialized views is not an easy task, they may significantly improve query performance. The aggregates chosen should satisfy the largest number of queries. Current DBMSs facilitate the processes of creating, populating, refreshing, and using materialized views. However, the additional overhead related to storage capacity and the management of additional structures should be considered.

6.8.2 Implementation of ETL Processes

In the physical-design phase, all the required extraction, transformation, and loading processes must be defined in a more detailed, ready-to-execute manner. Note that this decomposition into three different processes is done only conceptually, since the same software component can be in charge of all steps.

The extraction process must be implemented not only for the initial loading of the data warehouse but also for the subsequent refreshment. For the latter, either a full or an incremental extraction method can be used. Using the former method, the complete content of the source systems is extracted and included in the data warehouse. On the other hand, in the incremental extraction method, only the data that has changed since the last refreshment is extracted. This requires a mechanism to identify which data has been modified. Several methods exist; in Sect. 5.3.1, we described various kinds of source systems and briefly referred to methods of extracting only data that had changed. For example, the entire table can be copied from a source system into intermediate storage and compared with the data from the previous extraction using the set difference operator usually provided by SQL. Another method could include additional structures, such as timestamps in the source systems to indicate the last data change. Some DBMSs can provide facilities for change detection. For example, Oracle 10g includes a "change data capture" mechanism that identifies and captures data that has been added, updated, or removed from an Oracle relational source system; these changes are stored in a relational table and are available for use by applications or users.

Data extracted from a source system may require transformation prior to loading into the warehouse. The transformation process may be a simple SQL instruction or may require complex programming effort. All transformations specified during the conceptual- and logical-design phases should now be implemented.

Further, additional storage may be needed for the transformation process in order to check the incoming data or to change it into the required output. As was explained in Sect. 2.9, this storage is commonly called the **data staging**

area [146]. For example, the inclusion in the data staging area of lookup tables for the Conference table might help to determine whether a conference name from a source system is valid. Therefore, considering the various extraction methods and transformation needs, a decision should be made as to whether a data staging area is required. In the affirmative case, this additional storage should be designed and created.

Finally, the process of loading data into the data warehouse may be implemented using mechanisms provided by the specific DBMS, such as Oracle's SQL*Loader or the export/import facilities of the DBMS.

All processes implemented must be scheduled and processed in a specific order. The result must be tracked, and, depending on the success or the failure of a process or some part of it, subsequent, alternative processes may be started. Therefore, it is important to include adequate exception handling in each component process, as well as in the whole ETL process.

Implementing the ETL processes may be a very complex, time-consuming, and costly task, for which there is not yet much methodological support. To succeed, it is important to have a well-documented specification of the mappings and transformations obtained from the previous phases of conceptual and logical design, a good knowledge of the DBMS used for implementing the data warehouse, knowledge of the physical structure of the underlying source systems, and high-quality programming skills.

Note that we have left out of the discussion any issues related to the quality of the source data, and its cleaning and integration. Even though these aspects are essential in the data warehouse context and are included in several publications (e.g., [130, 145, 303]), there are outside the scope of this book.

6.9 Method Summary

In this section we present a global overview of the proposed method for the design of conventional data warehouses. As explained in Sect. 6.2, the process of data warehouse design follows the phases used for traditional database design, i.e., requirements specification, conceptual design, logical design, and physical design. However, as we saw in Sect. 2.5 and throughout this chapter, data warehouses differ in many aspects from traditional databases. Therefore, the phases mentioned above must be adapted to the particular characteristics of data warehouses. Specifically, we considered several driving forces for the requirements specification phase and distinguished three approaches, called the analysis-driven, source-driven, and analysis/source-driven approaches (see Sect. 6.4). Then, the subsequent phase of conceptual design differs depending on the approach chosen for the requirements specification phase. In contrast, the logical- and physical-design phases remain the same for the three approaches. Figs. 6.11–6.13 summarize in a graphical form the phases and the corresponding steps for the three approaches.

6.9.1 Analysis-Driven Approach

In the analysis-driven approach, illustrated in Fig. 6.11, the requirements specification phase focuses on expressing the needs of the users in terms of business goals. Based on these goals, two different lines of action can be performed: either the elaboration of the users' needs in a more detailed manner, or the analysis of the business processes with the corresponding services or activities. The result of this phase is a precise specification of the facts, measures, and dimensions, and the description of business metadata (see Sect. 6.4.1).

Then, the conceptual-design phase (described in Sect. 6.5.1) starts by developing an initial schema that includes the facts and dimensions identified in the previous step. This schema is revised in order to verify that the operational systems contain the data required by the users. In the affirmative case, the corresponding schema is considered as final. If the operational systems cannot provide the data required by the users, the initial schema is modified according to the availability of data in the source systems. In this phase, it is necessary to specify the mapping and the corresponding transformations between the data elements in the source systems and the data elements in the data warehouse. This mapping is an important component of the documentation needed to develop the ETL processes.

The two subsequent phases are devoted to developing the logical and physical schemas, and to specifying and implementing the ETL processes. Even though these two phases may seem similar in Fig. 6.11, there are important differences in the aspects that are considered in these two phases, as described in Sects. 6.7 and 6.8.

6.9.2 Source-Driven Approach

The source-driven approach, illustrated in Fig. 6.12, focuses on the data available in the source systems. In the first step of the requirements specification phase, the source systems that can deliver data for analysis purposes must be identified. The goal of the second step is to derive from these source systems the facts, measures, and dimensions. As described in Sect. 6.4.2, this process can be performed manually on the basis of the existing documentation of the source systems; alternatively, various semi-automatic or automatic procedures can be applied. During the following step, the designers develop the documentation, which includes the description of all data elements in the source systems that can be considered as candidates for facts, measures, dimensions, and hierarchies. This documentation is used in the first step of the conceptual-design phase (see Sect. 6.5.2), where an initial multidimensional schema is developed. Since not all elements of this initial schema will be of user interest, he/she must determine the elements required for analysis purposes. Afterwards, the final schema and mapping is developed in a similar way to that described for the analysis-driven approach. Finally, the two subsequent phases of logical and physical design are also the same as for the analysis-driven approach described above.

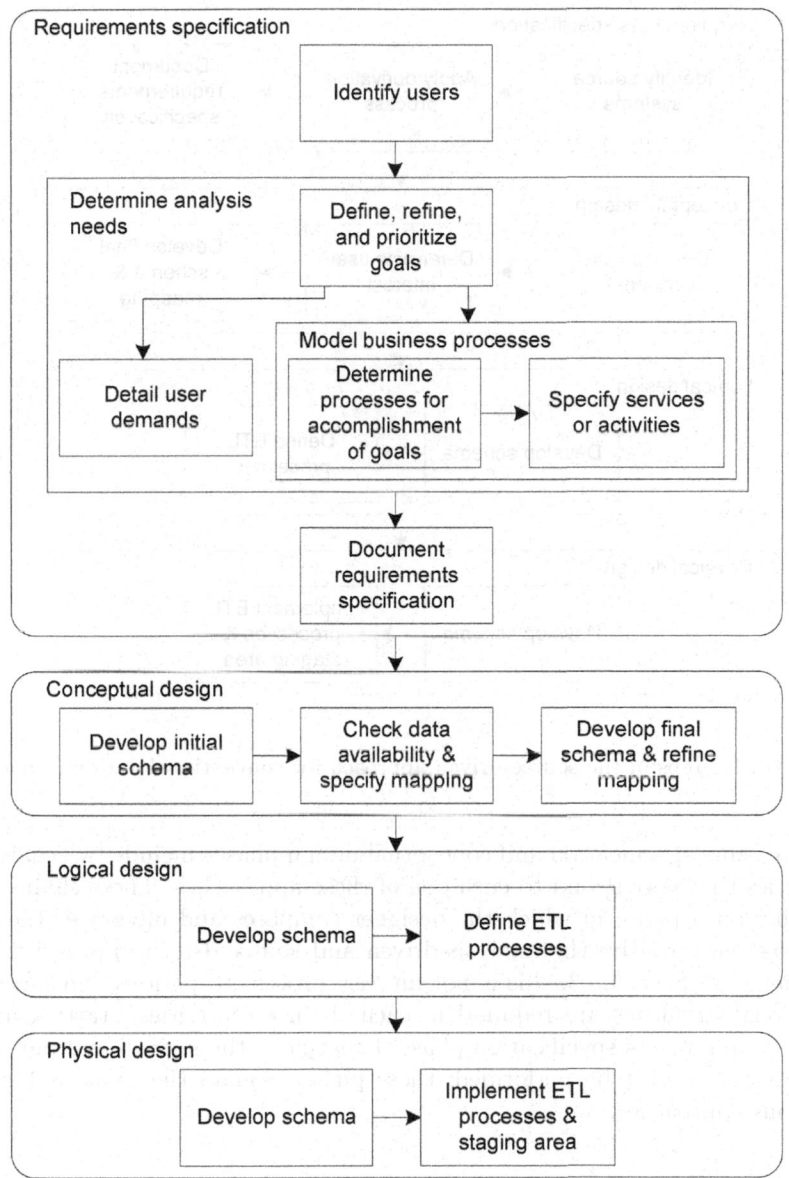

Fig. 6.11. Steps in the analysis-driven approach for conventional data warehouses

6.9.3 Analysis/Source-Driven Approach

As described in Sects. 6.4.3 and 6.5.3 and illustrated in Fig. 6.13, this approach combines the analysis-driven and source-driven approaches. Therefore, the

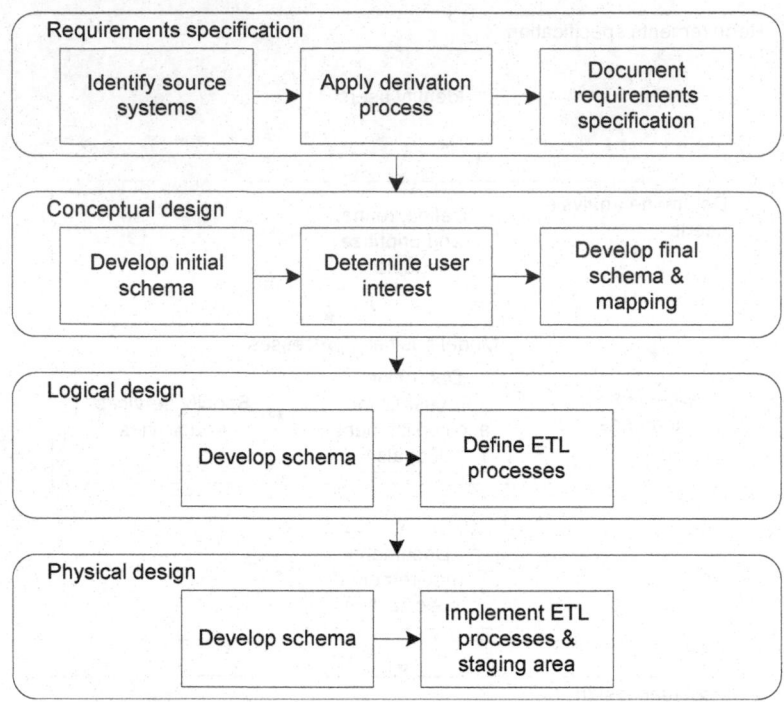

Fig. 6.12. Steps in the source-driven approach for conventional data warehouses

requirements specification and conceptual-design phases include two chains of activities that correspond to each one of these approaches. These chains join together on a phase in which the designer compares and integrates the two schemas delivered by the analysis-driven and source-driven approaches. As mentioned in Sect. 6.5.3, this is not an easy process to perform, and usually additional iterations are required in each chain of activities, starting again in the requirements specification phase. Eventually, the logical- and physical-design phases must be performed; these phases remain the same as for the previous approaches.

6.10 Related Work

In this section, we survey publications related to data warehouse design. As we have seen, the data warehouse design process is composed of several phases, i.e., requirements specification, conceptual design, logical design, and physical design. Since we have already reviewed publications related to conceptual, logical, and physical design of data warehouses in Sect. 3.8, we describe here

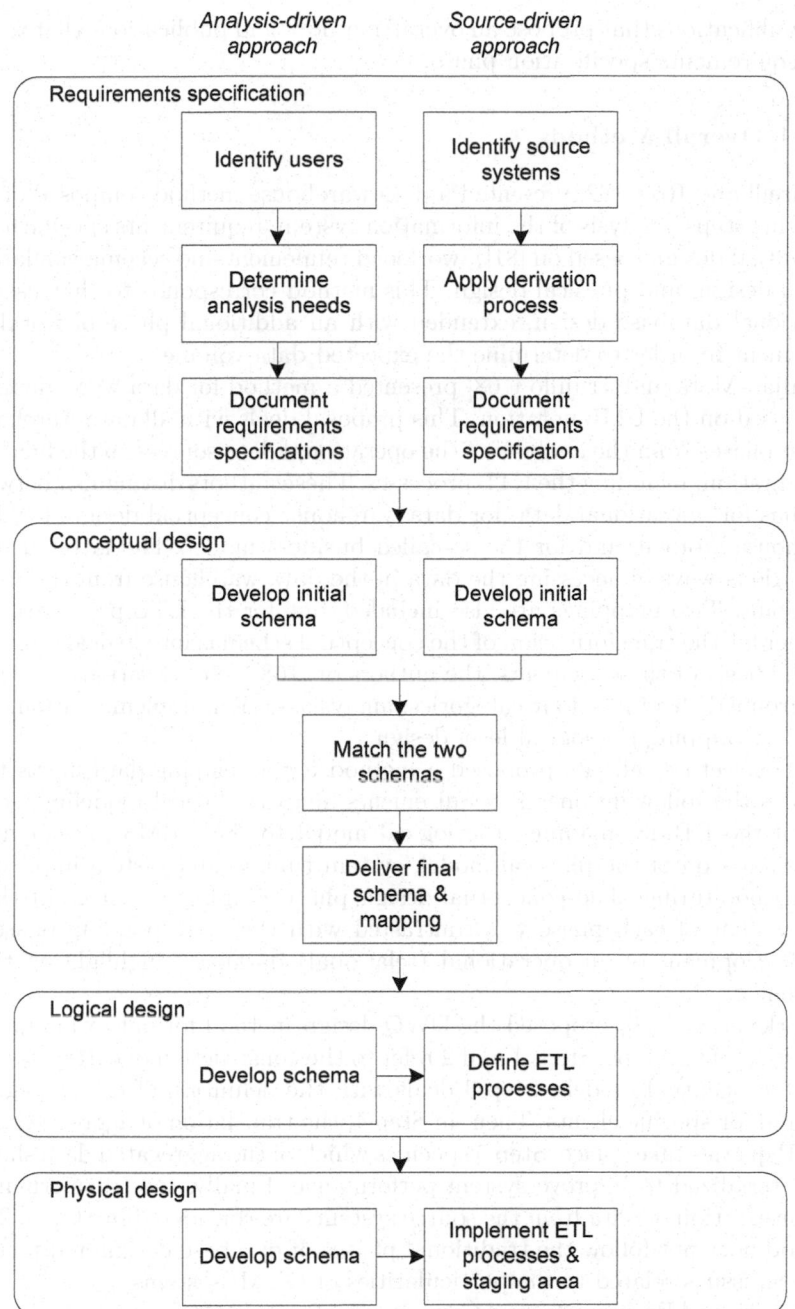

Fig. 6.13. Steps in the analysis/source-driven approach for conventional data warehouses

only publications that propose an overall method, and publications that target the requirements specification phase.

6.10.1 Overall Methods

Golfarelli and Rizzi [82] presented a data warehouse method composed of the following steps: analysis of the information system, requirements specification, conceptual design (based on [81]), workload refinement and schema validation, logical design, and physical design. This method corresponds to that used in traditional database design, extended with an additional phase of workload refinement in order to determine the expected data volume.

Luján-Mora and Trujillo [168] presented a method for data warehouse design based on the UML notation. This proposal deals with all data warehouse design phases from the analysis of the operational data sources to the final implementation, including the ETL processes. These authors distinguish between schemas for operational data, for data warehouse conceptual design, for data warehouse storage, and for the so-called business model. The latter defines the various ways of accessing the data in the data warehouse from the users' viewpoint. Two mappings are also included: one for the ETL processes, and another for the transformation of the conceptual schemas into logical schemas. On the basis of these elements, the authors of [168] defined various activities and grouped them into four categories: analysis, design, implementation, and test, i.e., skipping the logical-level design.

Chenoweth et al. [46] proposed a method for developing data marts that includes the following phases: requirements analysis, logical modeling, selection of the DBMS, mapping the logical model to the DBMS, development and evaluation of the physical model, system tuning, and system implementation/monitoring. The conceptual-design phase is missing in this method. The content of each phase was contrasted with the corresponding phase in the development of an operational (relational) database, highlighting their differences.

Jarke et al. [130] proposed the DWQ design method for data warehouses, consisting of six steps. Steps 1 and 2 refer to the construction of enterprise and source conceptual models. Step 3 deals with the definition of the aggregates required for specific clients. Then, in Step 4, the translation of aggregates into OLAP queries takes place. Step 5 specifies which of the aggregated data should be materialized to improve system performance. Finally, the transformation and integration of data from the source systems are considered in Step 6. This method does not follow the traditional phases of database design and focuses more on issues related to the particularities of OLAP systems.

Pereira and Becker [235] and Carneiro and Brayner [41] proposed a method for the development of data warehouse pilot projects targeted toward professionals experienced in the development of traditional databases. The approach of Pereira and Becker [235] is based on the life cycle proposed by Kimball et al. [145] and includes several additional phases, such as experimentation and prototyping. On the other hand, the approach of Carneiro and Brayner [41] starts by building a prototype and then, on the basis of a spiral model, builds a more complete version of the data warehouse at each iteration. The idea of including a pilot project in the data warehouse life cycle allows the internal team to gain experience in constructing a data warehouse; it also reduces the risk of failure.

6.10.2 Requirements Specification

As was mentioned in Sect. 6.1, several different approaches may be used in the requirements specification phase. In this section, we describe these approaches in more detail, grouping them into source-driven, user-driven, business-driven, and combined approaches. We do not present the advantages and disadvantages of each of these methods, since these have already been described in Sect. 6.4.

Source-Driven Approach

Böhnlein and Ulbrich-vom Ende [26] proposed a method for deriving initial logical data warehouse structures from the conceptual schemas of operational systems. This is done in three stages. The first stage defines measures based on the chain goals–services–measures. After determination of the measures, the derivation of dimensions is performed. This derivation is based on existence dependencies that are visualized using SERM (Structured Entity-Relationship Model). Hierarchies may also be derived by considering the cardinalities between entity types. The final step consists in specifying integrity constraints on the dimension hierarchies. Since the process of defining hierarchies is not always successful and requires creativity and considerable knowledge of the application domain, the authors of [26] consider that the derivation of multidimensional models should not be automated.

Golfarelli et al. [81] presented a graphical conceptual model for data warehouses called the Dimensional Fact Model and proposed a semiautomated process for building conceptual schemas from operational ER schemas. The first phase collects the documentation about the underlying operational systems. On the basis of this documentation, the designer and the end users determine the facts and the workload that the system is required to handle. Then, a semiautomated process is applied to find other multidimensional elements. For each fact, an attribute tree is built. The next two steps, of pruning and graphing the attribute tree, allow one to exclude attributes not required for analysis purposes. Finally, navigation through the schema along one-to-many relationships allows the dimensions and their hierarchies to be determined.

Moody and Kortink [201] proposed a method for developing dimensional models from enterprise data models represented using the ER model. This is done by a four-step approach that starts with classifying entities to represent facts, measures, dimensions, and hierarchies. For the latter, one-to-many relationships between functionally dependent entities are considered. Then, various multidimensional schemas can be created, such as star and snowflake schemas. The final step performs an evaluation and refinement of the proposed schema to produce a final data mart with a simpler structure.

Cabibbo and Torlone [34] presented a design method that starts from an existing ER schema. These authors assumed that this schema describes a "primitive" data warehouse that contains all operational information that can support business processing, but is not yet tailored to the activity of analysis. Cabibbo and Torlone derived a multidimensional schema and provided an implementation of it in terms of relational tables and multidimensional arrays. The derivation of the multidimensional schema is performed into several steps. In the first step, facts along with their measures have to be selected, and afterwards, dimensions for facts are identified by navigating the schema. Then, in the second step, the initial ER schema is restructured in order to express facts and dimensions explicitly, thus arriving at a multidimensional representation.

Hüsemann et al. [109] used the ER schemas of transactional applications to determine multidimensional requirements. In this approach, business domain experts select strategically relevant operational attributes, which are classified as dimensions or measures. The resulting requirements are presented in a tabular list of attributes along with their multidimensional purpose. Supplementary information (integrity constraints and additional derived attributes) can be added informally in a textual appendix. The authors of [109] focused on conceptual design based on this information, using functional and generalized multidimensional dependencies [162]. However, use of these kinds of dependencies precludes the inclusion of several kinds of hierarchies proposed in this book, such as generalized and nonstrict hierarchies.

Inmon [118] also considered that the data warehouse environment was data-driven. According to that author, the requirements should be understood after the data warehouse has been populated with data and has been used by a decision-support analyst. The data model is derived by transferring the corporate data model into a data warehouse schema and adding performance factors.

User-Driven Approach

Paim et al. [219] proposed a method for requirements analysis for data warehouse systems that consists in the following phases: requirements management planning, requirements specification, and requirements validation. The elicitation of requirements is done on the basis of communication with stakeholders using techniques such as interviews, prototyping, and scenarios. This requires the presence of application domain users. Paim and Castro [220] extend this method by including nonfunctional requirements, such as performance and accessibility.

Trujillo et al. [297] presented a conceptual model based on UML diagrams for designing data warehouses. This model can be used to create a conceptual schema within a method proposed for data warehouse design [168]. This method includes several steps that are user-driven, such as requirements specification, conceptual modeling, and data mart definition.

Freitas et al. [76] considered that to correctly understand facts, dimensions, and the relationships between them, effective user participation is required. These authors presented MD2, a tool based on the dimensional data-modeling approach, which facilitates the participation of users in several steps of the development of data warehouse applications.

Imhoff et al. [113] proposed a method that first develops a "subject area data model", which represents the subject of analysis using the ER model. This model can be developed on the basis of users' requirements obtained by applying various techniques, such as interviews or facilitated sessions.

Vaisman [303] proposed a method for the elicitation of functional and nonfunctional requirements that integrates the concepts of requirements engineering and data quality. This method refers to the mechanisms for collecting, analyzing, and integrating requirements. Users are also involved in order to determine the expected quality of the source data. Then, data sources are selected using quantitative measures to ensure data quality. The outcome of this method is a set of documents and a ranking of the operational data sources that should satisfy the users' information requirements according to various quality parameters.

Business-Driven Approach

Böhnlein and Ulbrich-vom Ende [27] derived data warehouse structures from business process models, taking account of the fact that users are often not able to clearly formulate their demands. These authors assumed that a data warehouse is designed according to a set of relevant business subjects and that each of these subjects is centered around a business process. They derived initial data warehouse structures on the basis of the Semantic Object Model (SOM) method used for business engineering. The SOM method helped them to obtain a "conceptual object schema", which was used later to identify facts, measures, and dimensions. Since they did not use any conceptual multidimensional model, they presented a logical model based on star schemas.

Bruckner et al. [31] introduced three different abstraction levels for data warehouse requirements, considering business, user, and system requirements. The business requirements represent high-level objectives of the organization, identifying the primary benefits that the data warehouse system will provide to the organization. The user requirements describe the task that the users must be able to accomplish with the data warehouse system. These requirements have to be in concordance with the business requirements. Finally, the detailed system requirements refer to data, functional, and other kinds of requirements.

Mazon et al. [189] presented an approach for including business goals in data warehouse requirements analysis. These requirements are then transformed into a multidimensional model. These authors used the i* framework [320], which is based on two kinds of models: a strategic dependency model, which describes the dependency between actors in an organizational context, and a strategic rationale model, which is used to understand the interests of actors and the way in which they might be addressed. They adapted these models to the data warehouse context, giving specific guidelines for building them. Then, other guidelines were given for mapping these models to a conceptual multidimensional model based on UML notation.

Giorgini et al. [78] proposed a goal-oriented approach for data warehouse requirements analysis using the Tropos method. The latter is an agent-oriented software development method, also based on the i* framework. The starting point is an explicit goal model of the organization. Goals are considered at various levels of management, leading to creation of "organizational" and "decisional" models. These models are refined in order to obtain a definition of facts, measures, and attributes. The highly technical requirement models are then mapped to the conceptual multidimensional model of Golfarelli et al. [81].

Kimball et al. [145, 146] based their data warehouse development strategy on choosing the core business processes to model. Then, business users are interviewed to introduce the data warehouse team to the company's goals and to understand the users' expectations of the data warehouse. Even though this approach lacks formality, it has been applied in many data warehouse projects.

Combination of Approaches

Bonifati et al. [28] presented a method for the identification and design of data marts. This method consists of three general parts: top-down analysis, bottom-up analysis, and integration. The top-down analysis emphasizes the users' requirements and requires precise identification and formulation of goals. On the basis of these goals, a set of ideal star schemas is created. On the other hand, the bottom-up analysis aims at identifying all the star schemas that can be implemented using the available source systems. This analysis requires the source systems to be represented using the ER model. The final integration phase allows designers to match the ideal star schemas with realistic ones based on the existing data.

Phipps and Davis [238] proposed the creation of a data warehouse conceptual schema using first the source-driven approach. Their automated technique produces candidate data warehouse conceptual schemas based on the existing data. The user-driven approach is then applied, since these authors consider that relying only on data sources may result in data warehouse schemas that are inadequate with respect to user needs. These needs are represented as queries that users wish to perform. Then, the question of whether the schemas obtained using the source-driven approach can answer user queries is reexamined.

Winter and Strauch [314] proposed a method that first determines the information requirements of data warehouse users and then compare these requirements with aggregate data from the source systems. The information requirements that are not covered are homogenized and prioritized. Then, the selection of more detailed data from the source systems is considered. The resulting multidimensional schema is evaluated. If the user requirements are still not met, other iterations between the various phases begin.

6.11 Summary

In this chapter, we have presented a general method for the design of conventional data warehouses. Our proposal is close to the classical database design method and is composed of the following phases: requirements specification, conceptual design, logical design, and physical design.

For the requirements specification phase, we have proposed three different approaches. The analysis-driven approach focuses on business or user needs. These needs must be aligned with the goals of the organization in order to ensure that the data warehouse system provides the necessary support for the decision processes at all organizational levels. However, the final data warehouse schema includes only those users' requirements for which corresponding source data exists. The second approach, called the source-driven approach, develops the data warehouse schema on the basis of the structures of the underlying operational databases, which are usually represented using the ER or the relational model. Users' needs are considered only within the scope of the available operational data. The third approach, the analysis/source driven approach, combines the first two approaches, matching the users' needs with the availability of data. The requirements specification phase is refined until the various elements of the multidimensional model can be distinguished, such as facts, measures, dimensions, and hierarchies. Then, the conceptual schema can be built, and evaluated by users.

The next phases of our method correspond to those of classical database design. Therefore, a mapping of the conceptual model to a logical model is specified, followed by the definition of physical structures. Note that the physical-level design should consider the specific features of the target DBMS with respect to the particularities of data warehouse applications. We have expanded these phases with the refinement of the ETL processes and the design of the data staging area, if that is necessary.

The proposed method has several advantages. First, it does not depend on a particular conceptual model, on a particular target implementation platform, or on any particular logical or physical organization of the database.

It also refers to the overall life cycle of data warehouse development, including the conceptual-design phase ignored by most practitioners and by some scientific publications. Additionally, it classifies and organizes the variety of current approaches to requirements specification into a coherent whole. It generalizes the proposed approaches to make them easy to understand without losing their semantic differences, and it provides clear guidelines that state in which situations each method may give better results. Therefore, the proposed method allows designers to choose a suitable approach according to the particularities of the data warehouse project.

Moreover, including the ETL processes (i.e., mappings and transformations) in the conceptual-design phase and extending the specifications of these processes in the logical-design phase allow developers to define and refine such processes without being burdened with the implementation details of the underlying database management systems. The subsequent phase of physical design defines the necessary mechanisms for the implementation of these processes.

Designing Spatial and Temporal Data Warehouses

Although spatial and temporal data warehouses have been investigated for several years, there is still a lack of a methodological framework for their design. This situation makes the task of developing spatial and temporal data warehouses more difficult, since designers and implementers do not have any indication about when and how spatial and temporal support may be included. In response to this necessity, in this chapter we propose methods for the design of spatial and temporal data warehouses.

In Sect. 7.1 we refer briefly to the current methods for designing spatial and temporal databases. We present in Sect. 7.2 a risk management case study and revisit in Sect. 7.4 the university case study that we introduced in the previous chapter. We use these case studies to show the implications from a methodological perspective of including spatial or temporal support in a data warehouse.

The methods for the design of spatial and temporal data warehouses are given in Sects. 7.3 and 7.5, respectively. Our proposal is based on the method for designing conventional data warehouses described in the previous chapter, which includes the phases of requirements specification, conceptual design, logical design, and physical design. However, in this chapter we revisit these phases, extending them by the inclusion of spatial and temporal support.

Moreover, we propose modifications to the three approaches to the requirements specification and conceptual-design phases, i.e., the analysis-driven, source-driven, and analysis/source-driven approaches. For each of them, we provide two options: the inclusion of spatial or temporal support either at the beginning of the requirements-gathering process or at a later stage, once a conceptual schema has been developed. We refer to these options as early or late inclusion of spatial and temporal support. In this way, we provide various alternatives for the designers of a spatial or temporal data warehouse; they should base their choice on users' knowledge of spatial or temporal features and the availability of spatial or temporal data in the source systems. For the late inclusion of spatial or temporal support, the requirements specification and conceptual-design phases intertwine significantly. Therefore, to facilitate reading, we present these two phases together. Further, since we already referred to the logical and physical design of spatial and temporal data warehouses in Chaps. 4 and 5, these aspects are only covered briefly in this chapter. Section 7.6 summarizes the proposed methodological framework for the design of spatial and temporal data warehouses. Finally, Sect. 7.7 surveys related work, and Sect. 7.8 concludes this chapter.

7.1 Current Approaches to the Design of Spatial and Temporal Databases

Unlike the case of conventional databases, there is not yet a well-established method for the design of spatial and temporal databases. In general, the four phases described for conventional-database design, i.e., requirements specification, conceptual design, logical design, and physical design, are considered for both spatial and temporal databases (e.g, [132, 268, 274, 315]).

Two main approaches to the design of spatial databases can be distinguished. In one approach, the spatial elements are included in the initial conceptual schema (e.g., [10, 268, 315]). The other approach starts with the development of a nonspatial schema, which is augmented afterwards with spatial elements (e.g., [143, 227]). In both cases, the spatially extended conceptual schema is then translated into logical and physical schemas using mapping rules. Nevertheless, owing to the lack of a well-accepted conceptual model for spatial-database design, in many situations the phase of conceptual design is skipped, starting the design process with the logical schema.

Similar approaches can be used for temporal-database design. In the first approach (e.g., [86, 89]), temporal support is included in the conceptual schema and this schema is mapped later into logical and physical schemas. Another approach (e.g., [56, 274]) initially ignores all temporal aspects when one is developing a conceptual schema. After the full design is complete, the conceptual schema is augmented with temporal elements. Similarly, the logical design is developed in two stages. First, a nontemporal conceptual schema is mapped into a nontemporal relational schema using common mapping strategies, such as those described in [66]. In the second stage of logical design, each of the temporal elements is included either as part of the relational tables or as integrity constraints.

7.2 A Risk Management Case Study

In this section, we introduce a hypothetical case study that will be used to illustrate the design of spatial data warehouses. This case study was inspired by an application for monitoring natural risks used at Cemagref, a public research center in Grenoble, France; this application has been described extensively in [227].

The problem at hand is to provide administrative offices with risk evaluation information on which to substantiate the granting or denying of permission to build houses and facilities in a mountainous area that is exposed to natural risks. A correct risk evaluation is fundamental, as permission may be legally challenged if a building is damaged or a person is injured by some natural hazard in a zone whose risk level was not correctly estimated.

Risk evaluation is based on knowledge of past natural hazards that have occurred within the monitored area. The events of major interest are avalanches, landslides, and land erosion. Considering only avalanche events, the hazard that recurs most often in the area studied, more than 70 000 observations made since 1920 by various people and various means have been collected. For each hazard event there may be several observations, both during the occurrence of the hazard and afterwards to precisely determine its consequences.

The process of risk evaluation is a long and complex one. The first step is to collect all knowledge about hazard events of a similar nature that have happened in a given zone to obtain a global understanding of the dynamics of natural hazards in that zone. Information at this level is grouped into hazard maps,[1] which describe the nature of the hazards taken into consideration, their geographical extent, various physical parameters, and the consequences of these events on humans and buildings. A given geographical area may be covered by two or three overlapping hazard zones if the area has been affected by hazards of different nature (for example, avalanches and landslides).

Hazard maps are used as input in the development of preliminary risk maps, called synthesis maps,[2] whose scope is to synthesize the hazards of any nature in a given zone to provide an initial appreciation of the level of hazard risk in that zone. Risk here is classified according to a hierarchy that defines three or four risk levels, such as strong, average, weak, and irrelevant. To produce these maps, various elements are taken into account, such as the nature of the events that have occurred (for example, avalanche, landslide, or erosion), the probability of a new occurrence, and the expected importance of new events. The development of the maps may call for additional field investigation by experts, either to complement the available data with additional measurements, or to validate or correct data that is imprecise or of insufficient reliability.

The final output of the application is the production of two kinds of maps described next. Risk maps[3] display three kinds of zones within the area covered. Red zones are those of significant risk, where no building is allowed. Blue zones are medium-risk zones, where only buildings that conform to given specifications are allowed. White zones are those considered as no-risk zones, where buildings are allowed with no constraint related to risk. These risk zones do

---

[1] *Carte de phénomènes naturels* in French.
[2] *Carte d'aléas* in French.
[3] *Plan de prévention des risques* in French.

not have to obey cadastral boundaries. On the other hand, land use maps[4] decompose an area into several zones (for example, dwellings, green spaces, and industrial areas), for which different regulations on land use and planning rights hold. Land use maps are obviously established on the basis of risk evaluation information. Hence, each land use zone conveys a specification of whether it is allowed to build new buildings in the zone. Land use zones must follow land plot boundaries.

As can easily be deduced from the above description, risk management is a permanent task, in which new events must be taken into account in order to reassess decisions taken in the past. To illustrate the design of spatial data warehouses, we suppose that a reclassification of the various risk zones and land use zones must be performed owing to an increasing number of hazard events.

7.3 A Method for Spatial-Data-Warehouse Design

In this section, we propose a method for designing spatial data warehouses which is based on that for conventional-data-warehouse design described in the previous chapter. Therefore, it includes the phases of requirements specification, conceptual design, logical design, and physical design. We revisit these phases, describing how to take spatial support into account. It is worth noting that our method is independent of the conceptual model used. Thus, various spatial multidimensional models can be used, including the MultiDim model proposed in Chap. 4.

The design of spatial data warehouses requires that the designers are familiar with spatial data management and with concepts related to spatial data warehouses, as explained in Chap. 4 of this book. Designers must also be able to work with users who have different degrees of knowledge about spatial data, i.e., from experts who manipulate spatial data in their everyday work to novices who are just discovering the advantages that the representation of spatial data can give them in supporting decision-making process.

7.3.1 Requirements Specification and Conceptual Design

As was the case for conventional data warehouses, spatial data warehouses can be designed on the basis of the analysis requirements of the users (or the business), the data available in the source systems, or both. This leads to three approaches for spatial-data-warehouse design, referred to as the **analysis-driven**, **source-driven**, and **analysis/source-driven** approaches.

Another aspect that must be determined is when to consider spatial support during the design process. We propose two different options, where spatial support is considered either in the early steps of the requirements specification

---

[4] *Plan d'occupation du sol* in French.

phase or in later steps. We call these two options **early** and **late inclusion of spatial support**, respectively. The choice between these two options depends on the users' knowledge of spatial data features and the presence of spatial data in the source systems. For example, if the users are proficient with concepts related to spatial data analysis and manipulation, early inclusion of spatial support may be chosen. The advantage of this solution is that it focuses on the right issues from the beginning of the design process. On the other hand, if the users are acquainted with design techniques for conventional (i.e., nonspatial) databases using the ER model or UML, late inclusion of spatial support may be chosen, since this may allow an initial solution to be built quickly, on top of which spatial support may be considered. Similarly, if the spatial data is not yet available in the internal source systems and must be acquired from outside the organization's boundaries, late inclusion of spatial support may be the recommended solution.

Note that the choice between early or late inclusion of spatial support is independent of which of the three approaches above is used. For example, a particular data warehouse project might choose a source-driven approach with early inclusion of spatial support in order to integrate existing spatial applications into a decision-support infrastructure that is used by experts cognizant of spatial databases.

Analysis-Driven Approach

In the analysis-driven approach to spatial-data-warehouse design, the requirements specification phase is driven by the analysis needs of the users or the business. Figure 7.1 shows the steps required in this approach.[5]

Early Inclusion of Spatial Support

In this case we assume that the users are familiar with concepts related to spatial data, including its manipulation and some kinds of spatial analysis. Therefore, from the beginning of the requirements specification process, the users may be able to express what spatial data they require in order to exploit the features of multidimensional models and to perform various kinds of spatial analysis, as described in Chap. 4. For this kind of users, the design process may be performed following the same steps as those for conventional-data-warehouse design, as shown in Fig. 7.1.

As can be seen in Fig. 7.1a, in the first step of the requirements specification phase, users at various management levels are identified to ensure that the requirements will express the goals of the organization. These users will help the developer team to understand the purpose of having a spatial data

---

[5] In this figure and the following ones, we mention both spatial and temporal support. The reason for this is that, in our method, the steps for building spatial and temporal data warehouses are the same. The method for building temporal data warehouses is given in Sect. 7.5.

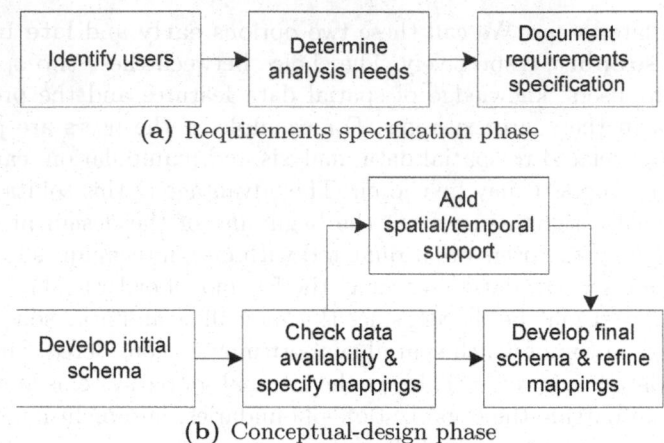

(a) Requirements specification phase

(b) Conceptual-design phase

Fig. 7.1. Steps of the analysis-driven approach for spatial and temporal data warehouses

warehouse and to determine the analysis needs, which are collected in the second step. The information gathered and the corresponding metadata are documented in the third step and serve as a basis for the next phase.

The conceptual-design phase (Fig. 7.1b) starts with the development of the initial spatial-data-warehouse schema. Note that this schema already includes spatial elements, since we assume that the users are able to refer to spatial data when expressing their specific analysis needs. Therefore, we follow the lower path of Fig. 7.1b. In the following step, it must be determined whether the data is available in the source systems, and the corresponding mappings with data warehouse elements are established. Note, however, that external sources may be needed if the required spatial support does not exist in the source systems. During the last phase, the final schema is developed; it includes all data warehouse elements, for which the corresponding data exists in the source systems (whether internal or external). Additionally, the corresponding mappings between the two kinds of systems is delivered.

We now illustrate this approach with our risk management application. In order to determine the analysis requirements, a series of interviews was conducted, targeting users at different organizational levels, i.e., from senior risk experts to field surveyors who inspect damage after a hazard event. From these interviews it was established that owing to the increasing number of hazard events, a reclassification of the various risk zones and land use zones had to be performed. The various analysis scenarios that were elicited were as follows:

1. The evolution in time of the extent and frequency of hazard events for the various types of risk zones (red, blue, and white) in different land plots.

2. The number of buildings affected by hazard events in different risk zones during different periods of time.
3. The evolution in time of the damage and number of victims caused by avalanches, considering the division of the area into districts and counties.
4. The repair cost of buildings located in white and blue risk zones that were affected by different types of hazard events.
5. The changes in the total area of the various kinds of risk zones according to their distribution within land plots.

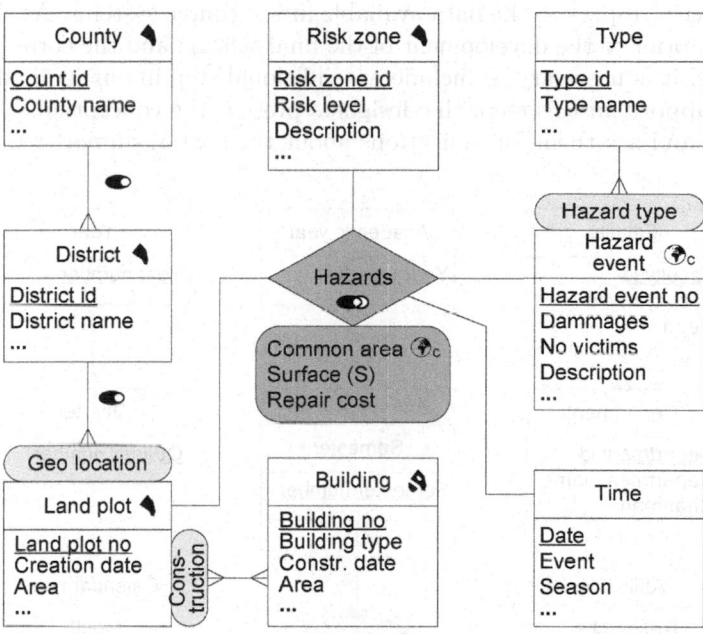

Fig. 7.2. Spatial data requirements expressed in a multidimensional schema

An analysis similar to that presented for the example in Sect. 6.4.1 was performed; the only difference lies in the fact that in this case spatial data must be taken into account. The resulting multidimensional schema is shown in Fig. 7.2. Afterwards, for each element of this schema, the source systems were inspected to find the corresponding data. However, the source systems did not include the information about the repair cost. Further, this information could not currently be obtained from external systems, since various private companies were involved in repairing the damage that had occurred. Therefore, the Repair cost measure had to be excluded from the schema. The two remaining measures, i.e., Common area and Surface, had to be calculated

during the extraction-transformation-loading (ETL) process on the basis of
the data present in the source systems.

Late Inclusion of Spatial Support

It may be the case that the users are not familiar with spatial data management
or that they prefer to start by expressing their analysis needs related
to nonspatial elements and include spatial support later on. In this case the
requirements specification and conceptual-design phases proceed as for a conventional
data warehouse, ignoring spatial features until the initial schema is
checked with respect to the data available in the source systems. As shown in
Fig. 7.1b, prior to the development of the final schema and the corresponding
mappings, it is necessary to include an additional step in our method to add
spatial support. In this step, the designers present the conceptual schema to
the users and ask them for indications about the spatial support required.

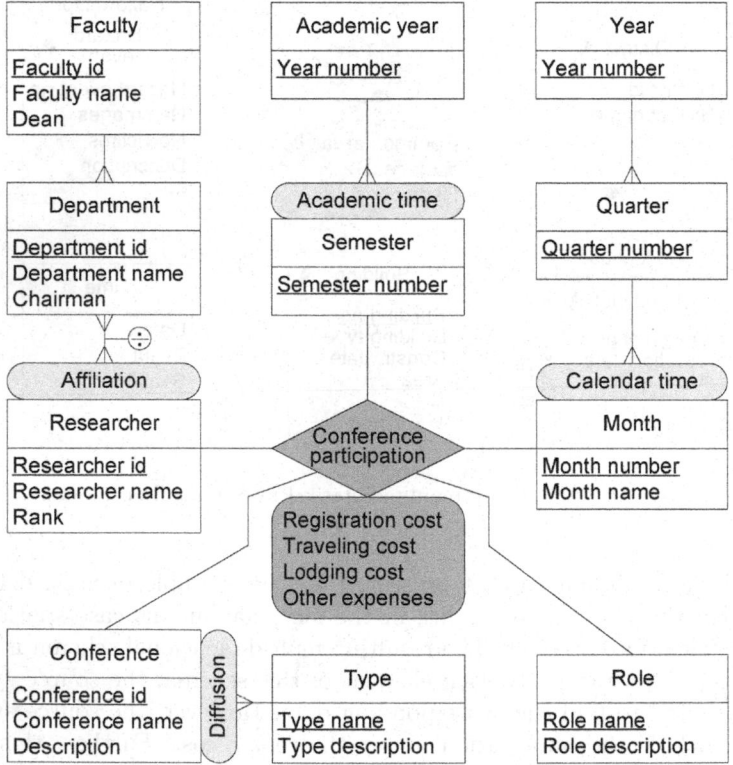

Fig. 7.3. Multidimensional schema for the inclusion of spatial elements

Suppose now that the MultiDim model is being used as a conceptual model for designing a spatial data warehouse. In this case, as described in Chap. 4, in the first step the designers may consider each level and decide whether that level, some of its attributes, or both should be represented spatially. Then, if a hierarchy includes two related spatial levels, the topological relationship between them is specified. If a fact relationship relates two or more spatial dimensions, the designers can help the users to determine whether some topological relationship between these dimensions may be of interest. In the affirmative case, a specific topological relationship should be included in the fact relationship to indicate a predicate for a spatial join operation. Finally, the inclusion of spatial measures may be considered. Note that the elements of the multidimensional schema could be analyzed in a different order, depending on the designers' skills and their knowledge about spatial data warehouses, and the particularities of the conceptual model used.

Similarly to the previous case, the step of checking data availability may require access to external sources, since spatial data may not be present in the underlying source systems. The final schema developed also includes modified mappings.

As an example, we can use the schema in Fig. 7.3 (a repetition of Fig. 6.6) that was developed for a conventional data warehouse in Sect. 6.5.1. When this schema was shown to the users, they decided to include information related to the location of conferences. They required visualization on maps not only of the cities where the conferences took place, but also their countries and continents. Therefore, a new hierarchy Geo location was added to the Conference dimension, as shown in Fig. 7.4. Traversing this hierarchy would help users to analyze the costs for conferences related to specific geographic areas.

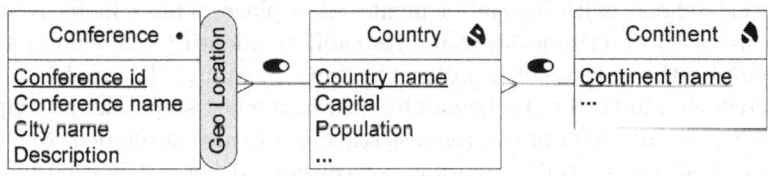

Fig. 7.4. A spatial hierarchy for the Conference dimension in Fig. 7.3

The spatial elements of the schema were checked against the data available in the source systems. Since the operational data did not include spatial components, external sources were used to obtain the corresponding information for the Conference dimension.

Source-Driven Approach

As explained in Sect. 6.4.2, this approach relies on the data in the source systems. Similarly to the analysis-driven approach, the spatial support may be included either early or late in the design process. Since the operational databases are the driving force in this approach, the choice between early or late inclusion of spatial requirements depends on whether these databases are spatial or not.

Early Inclusion of Spatial Support

If the source systems include spatial data, steps similar to those for conventional-data-warehouse design can be applied (Fig. 7.5a and the lower path of Fig. 7.5b). The requirements specification phase starts with the identification of the source systems; it aims at determining existing operational systems that may serve as data providers for the spatial data warehouse. External sources are not considered at this stage. In the second step, these sources are analyzed to discover multidimensional schema elements. However, a special derivation process needs to be designed for this purpose, since to the best of our knowledge, no semiautomatic or automatic procedure has been proposed in the literature for deriving the schema of a spatial data warehouse from the schemas of source systems. Therefore, this derivation process should be conducted manually and must rely on the designers' knowledge of the business domain and of spatial-data-warehouse concepts. The derivation process leads to the identification of elements for the multidimensional schema. This information, as well as metadata, is documented in the third step of the requirements specification phase.

In the first step of the conceptual-design phase, the development of a conceptual schema with spatial elements takes place. This schema is shown to the users to determine their interest and to identify elements that are important for the purpose of analysis. The users' recommendations for changes will be reflected in the final schema obtained in the last step, where mappings between the source and data warehouse schemas are also developed. Note that the step of adding spatial support is not considered here; it is used only for the late inclusion of spatial support, described later in this section.

Let us consider the spatially extended ER schema shown in Fig. 7.6. This schema is an excerpt from the operational database of the risk management application used as a case study in this chapter. As already said, this application is used for granting or denying permission to build houses and facilities in areas that are exposed to natural risks. In the discussion of the analysis-driven approach above, we have already described various elements of the schema, such as land plots, risk zones, and hazard events. This schema includes additional information about the owners of land plots, which may be private or public. It also includes information about land use zones, which provides a classification of zones, such as industrial, agricultural, forestry, and housing zones, as well as the assignment of land plots to land use zones.

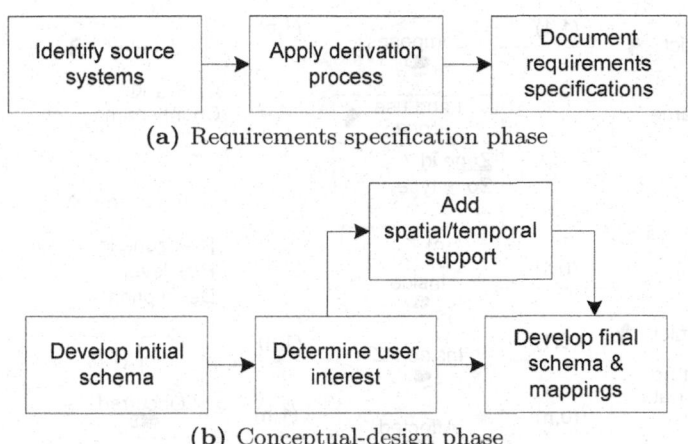

(a) Requirements specification phase

(b) Conceptual-design phase

Fig. 7.5. Steps of the source-driven approach for spatial and temporal data warehouses

To be able to identify multidimensional elements and to derive the corresponding data warehouse schema, we used a procedure similar to that described for the source-driven approach for conventional data warehouses. In order to find candidate fact relationships, we first considered the many-to-many relationship types: Affected, Belongs, Contains, and Occurred. Two of them, i.e., Affected and Belongs, contain numerical attributes: Percentage and Purchase price, respectively. Therefore, we chose these two relationship types as candidates for fact relationship types. The Contains relationship type could be included as a hierarchy level for the Land plot entity type, thus allowing users to analyze risks related to buildings. In analyzing the Occurred and Inside RZ relationship types, we realized that, to facilitate multidimensional analysis, we could include the Risk zone entity type as a new dimension for Affected, which is a candidate for a fact relationship type.

Two relationship types that were characterized by one-to-many cardinalities (i.e., Inside and Composes) were chosen as candidate levels for hierarchies. The subtypes of these two generalization/specialization relationship types contain only descriptive attributes. Therefore, to facilitate the analysis process, we decided to create an additional hierarchy level for each generalization/specialization relationship type. For the Hazard event entity type, this new level included members such as "avalanche", "landslide", and "erosion", and for the Owner entity type these members were "public" and "private".

The resulting schema is given in Fig. 7.7. Note that this schema also includes a time dimension related to the Risks and Ownership fact relationships. For the latter, the time dimension plays two different roles, indicating the period of time when a land plot belongs to a specific owner. The members of this

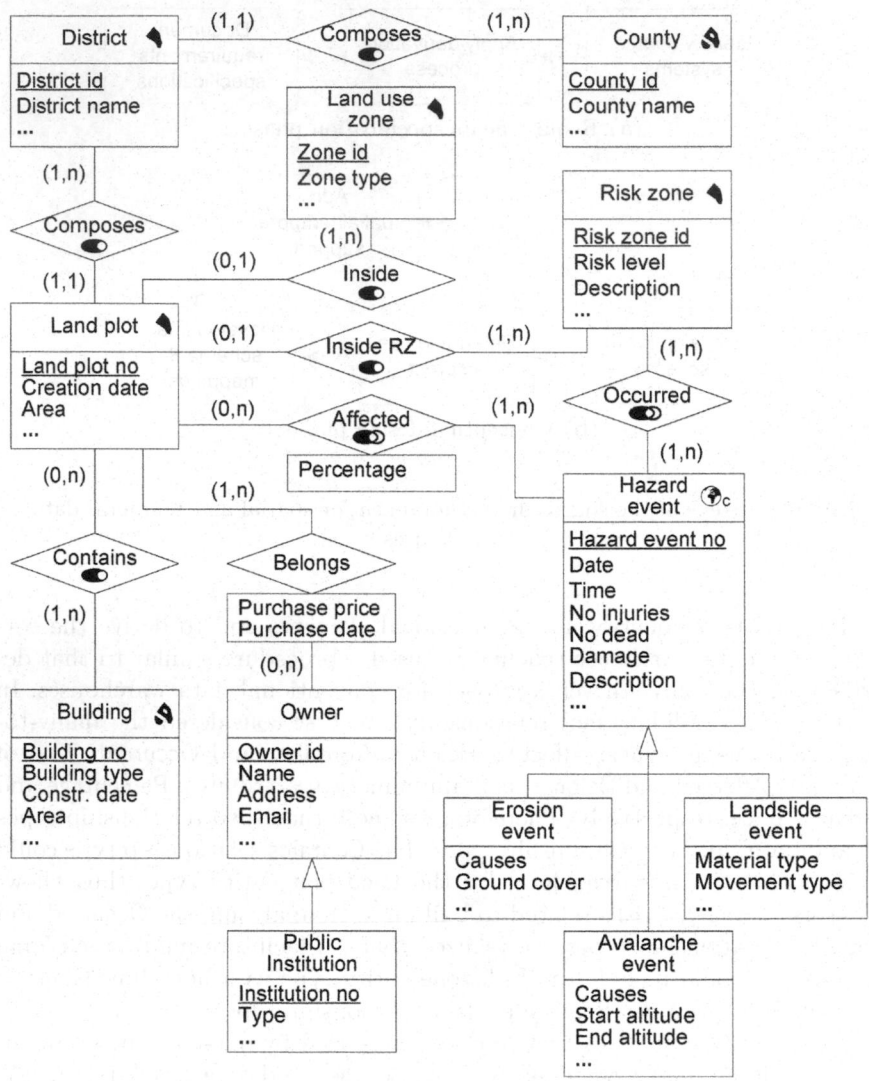

Fig. 7.6. Excerpt from the ER schema of the risk management application

time dimension were derived from the source data on the basis of attributes of type "date".

The schema in Fig. 7.7 was shown to the users. Since they were not interested in analysis related to the owners of land plots, the Ownership fact relationship and the corresponding dimensions (i.e., Time and Owner) were eliminated. The corresponding mapping and transformations were developed and delivered to the users.

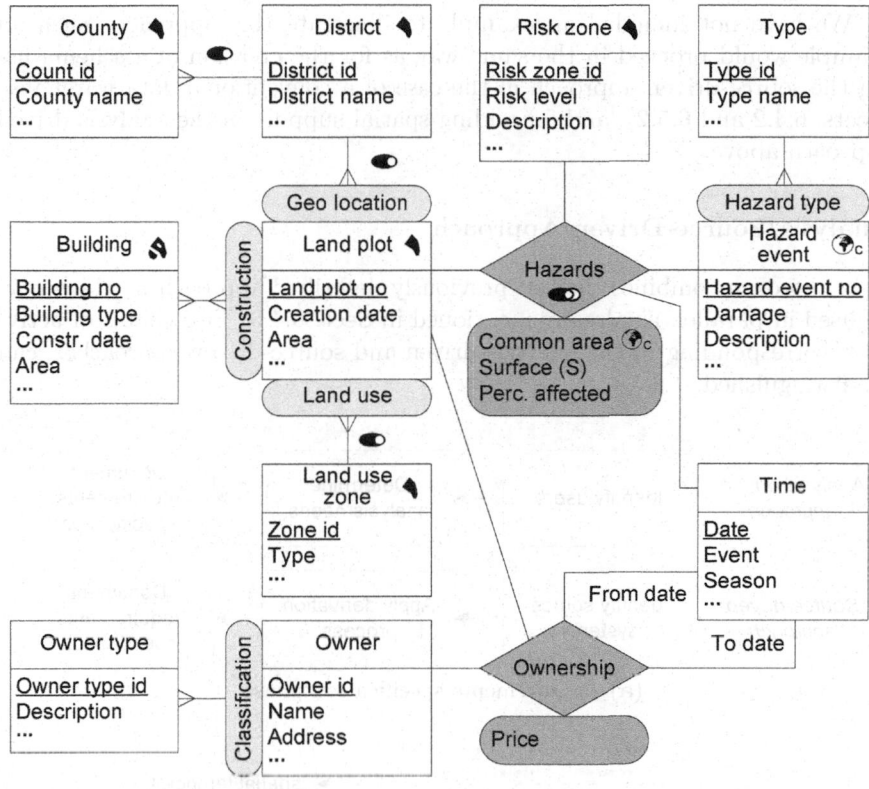

Fig. 7.7. Spatial-data-warehouse schema derived from the schema in Fig. 7.6

Late Inclusion of Spatial Support

Another situation arises when the source systems do not include spatial data, or when they do contain spatial data but the derivation process is complex and the designers prefer to focus first on nonspatial elements and to address spatial support later on. In these cases, the design process proceeds as for a conventional data warehouse, but at the end of the process a new step of adding spatial support is carried out (Fig. 7.5a and the upper path of Fig. 7.5b). Note that this support is considered only for the previously chosen elements of the multidimensional schema.

If the MultiDim model is used to represent the conventional-data-warehouse schema, the analysis of which elements can be spatially represented can be conducted in a way similar to that in the analysis-driven approach above. If spatial support is not provided by the underlying operational systems, external sources may deliver the required spatial data. The corresponding mapping should be included as part of the final schema.

We have not included an example to illustrate this approach. Such an example would proceed in the same way as for the creation of a schema using the source-driven approach in the case of a conventional data warehouse (Sects. 6.4.2 and 6.5.2), and for adding spatial support in the analysis-driven approach above.

Analysis/Source-Driven Approach

This approach combines the two previously described approaches, which may be used in parallel, as already mentioned in Sect. 6.4.3. Two chains of activities, corresponding to the analysis-driven and source-driven approaches, can be distinguished.

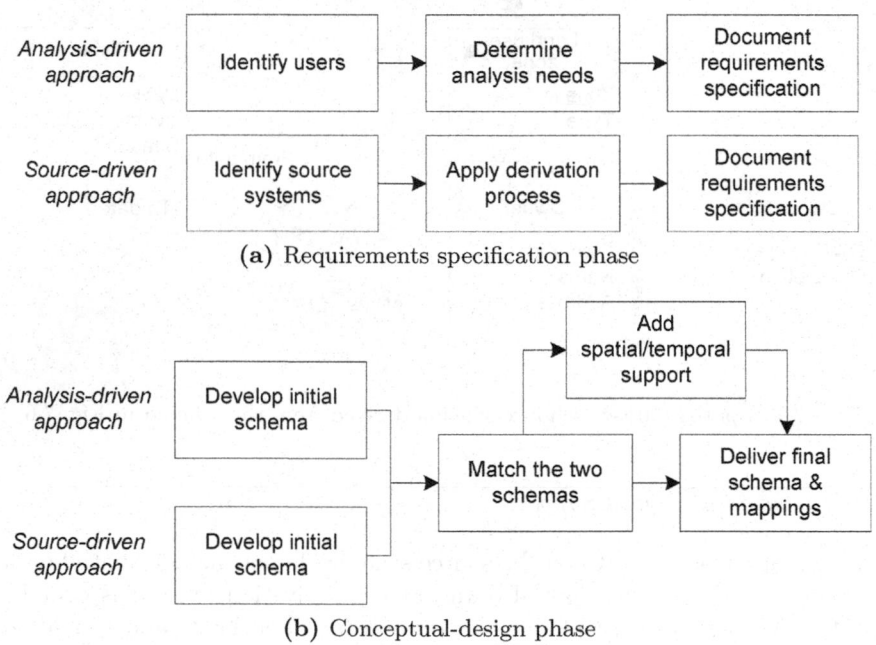

(a) Requirements specification phase

(b) Conceptual-design phase

Fig. 7.8. Steps of the analysis/source-driven approach for spatial and temporal data warehouses

Figure 7.8 shows the steps of the analysis/source-driven approach for spatial data warehouses. Similarly to the previous cases, we propose two different solutions, the choice of which depends on whether the users are familiar with concepts related to spatial data and whether the source systems include spatial data.

Early Inclusion of Spatial Support

If the users have knowledge about spatial data and if the source systems include such data, spatial support can be considered from the beginning of the requirements specification phase. Both the requirements specification phase and the conceptual-design phase are performed as described above for the analysis-driven and source-driven approaches. However, it is worth mentioning that the matching process that integrates the schemas obtained from the analysis and source chains can be a difficult task. Indeed, such a process is already complex for conventional data warehouses (Sect. 6.4.3) and in the present case it must additionally consider spatial elements. If the result of this matching process is satisfactory, the final schema is delivered. Otherwise, additional iterations may be necessary, as indicated for conventional-data-warehouse design (Sect. 6.4.3).

Late Inclusion of Spatial Support

On the other hand, if the users are not familiar with concepts related to spatial data or if the source systems do not include it, all steps up to the process of matching the schemas obtained from the analysis and source chains are the same as for a conventional data warehouse (Sect. 6.4.3). A new step is then included to add spatial support (upper path of Fig. 7.8b). During this step, the resulting multidimensional schema is analyzed for the purpose of inclusion of spatial support in various elements. Similarly to the previous approaches, external sources may be needed to obtain spatial data.

7.3.2 Logical and Physical Design

The logical and physical design of a spatial data warehouse should consider the various aspects mentioned in Sects. 6.7 and 6.8 for conventional data warehouses, which refer to the mapping of a conceptual schema into a logical and a physical schema, and to the ETL process.

Schema Definition

Two approaches exist for implementing spatial data warehouses. In one approach, an integrated architecture uses an object-relational DBMS with a spatial extension. In the other, a dual architecture uses separate management systems for spatial and nonspatial data, which is the case in most GISs. In Sect. 4.1 we described in detail the differences between these two kinds of architectures, and in Sect. 4.8 we presented our rationale for implementing a data warehouse using an integrated architecture and the object-relational model.

One of the advantages of the object-relational model is that it represents the geometry of spatial objects as an attribute of a spatial data type. However, before implementing this model into a specific DBMS, a decision must be made regarding two aspects: (1) whether a vector or a raster data model is more suitable for storing individual spatial objects, and (2) whether a spaghetti, network, or topological model should be used for storing a collection of spatial objects. In Sect. 4.1, we referred to these models, presenting their characteristics and their advantages and disadvantages. Further, in Sect. 4.8.2 we explained our rationale for using vector and spaghetti data models when implementing a spatial data warehouse using Oracle Spatial 10g.

We present below an example of how the Geo location hierarchy in Fig. 7.2 can be implemented using Oracle Spatial:

```
create type CountyType as object (
      Geometry mdsys.sdo_geometry,
      CountyId number(10),
      CountyName varchar2(50) );
create table County of CountyType (
      constraint countyPK primary key (CountyId) )
      object identifier is system generated;
create type DistrictType as object (
      Geometry mdsys.sdo_geometry,
      DistrictId number(10),
      DistrictName varchar2(50) );
create table District of DistrictType (
      CountyRef not null,
      constraint districtPK primary key (DistrictId),
      constraint districtFK foreign key (CountyRef) references County )
      object identifier is system generated;
create type LandPlotType as object (
      Geometry mdsys.sdo_geometry,
      LandPlotNo number(10),
      CreationDate date,
      Area number(10) );
create table LandPlot of LandPlotType (
      DistrictRef not null,
      constraint landPlotPK primary key (LandPlotNo),
      constraint landPlotFK foreign key (DistrictRef) references District )
      object identifier is system generated;
```

The other elements present in Fig. 7.2, i.e., hierarchies, levels, the fact relationship, and the measure, can be mapped in a similar way.

Additionally, during the translation of the conceptual schema into the logical and physical schemas, various aspects described in Sect. 4.9 should be considered, such as the various options for including a spatial attribute in a level or implementing triggers to check the topological relationship between hierarchy levels.

Since spatial data warehouses also contain thematic data, developers should take account of various physical issues that we have described for conventional-data-warehouse design (Sect. 6.8). These include indexing, data fragmentation, and materialization of aggregates, among other things; these aspects are essential for improving system performance and facilitating data management.

Oracle Spatial provides **spatial indexes** to optimize the execution time of spatial queries. These include finding objects that are spatially related, i.e., objects such that there is some topological relationship different from "disjoint" between them. Oracle Spatial uses R-tree indexing, which is similar to B-tree indexing, but instead of elements being represented as basic data types, the elements in R-trees can be n-dimensional objects located in space. For example, a spatial index for the LandPlot table can be implemented as follows:

```
create index LandPlotIdx on LandPlot (Geometry)
    indextype is mdsys.spatial_index;
```

In Oracle Spatial, spatial indexing is mandatory for data included as parameters for some operators, such as sdo_relate and sdo_within_distance. It is also necessary for executing spatial join operations, as described in Sect. 4.5. For example, to select those land plots that intersect risk zones with a risk level equal to 3, the geometries of land plots and of risk zones must be spatially indexed. Then, a query can be issued as follows:

```
select ref(LP), LandPlotNo
from LandPlot LP, RiskZone RZ, table(sdo_join('LandPlot',
    'Geometry', 'RiskZone', 'Geometry', 'mask=anyinteract')) j
where j.rowid1 = LP.rowid and j.rowid2 = RZ.rowid and RZ.RiskLevel = 3;
```

Further, if the table containing spatial data is partitioned on some scalar data, for example, on CreationDate in the LandPlot table, the spatial index can also be partitioned. This is done automatically by Oracle Spatial if the last line of the definition of the spatial index includes an additional keyword local.

Implementation of the ETL Process

As we mentioned in Sects. 6.7.2 and 6.8.2 for conventional-data-warehouse design, the implementation of the ETL process can be a difficult task, since several different aspects must be considered, such as the transformations required, the sequence of execution of the various processes, and refreshment frequency and methods.

Additionally, other difficulties related specifically to spatial data may arise. As described by Rigaux et al. [256], spatial data may be stored using a variety of formats that not only are different for vector and raster data but also are specific to the software used for managing spatial data. Further, differences may also exist in the reference systems and in the precision used for the same spatial data stored in several systems. Since spatial data warehouses obtain data from source systems (whether internal or external), these differences must be resolved during the ETL process. Some software companies are already addressing the issue of the variety of formats used for spatial data. For example, PCI Geomatics [229] offers an ETL tool for populating Oracle 10g GeoRaster databases from source systems that supports over 100 different raster and vector formats.

Several options exist for loading spatial data into a data warehouse, depending on the particularities of the target DBMS. For example, Oracle Spatial 10g provides two methods for loading data [216]. Bulk loading is used to import large amounts of ASCII data into an Oracle database and requires the creation of a control file; this includes a definition of how the various components of the mdsys.sdo_geometry data type are represented in the file. On the other hand, when small amounts of spatial data are to be loaded into a data warehouse, an insert operation such as that presented in Sect. 4.9.1 can be applied.

7.4 Revisiting the University Case Study

Let us consider again the university case study discussed in the previous chapter, but focusing now on education activities. In this case, the users are interested in establishing a better distribution of resources among faculties according to the number of students enrolled. This may improve the Times ranking criterion related to the staff-to-student ratio [288], which represents 20% of the overall score. Therefore, an analysis related to student enrollment, courses offered, and teaching duties is essential for supporting decisions aimed at accomplishing these goals.

As described in Sect. 6.3, universities are divided into faculties (for instance, the Faculty of Engineering), which comprise various departments (for instance, the Department of Computer Engineering). University staff (i.e., professors, researchers, teaching assistants, administrative staff, etc.) are affiliated administratively to departments. While most staff members are affiliated to a single department, staff members may be shared between departments, possibly from different faculties (for instance, a professor may be appointed 60% in the Faculty of Engineering and 40% in the Faculty of Sciences). In addition, there may also be some staff members who have been appointed to part-time positions (for instance, an external professor teaching a particular course). Therefore, the employment figures of faculties are expressed in full-time equivalent (FTE) terms, representing the number of full-time employees who could have been employed if the jobs done by part-time employees had been done by full-time employees. The salary expenses of appointed staff are allocated within the global budget of the university. In addition, temporary staff may be hired by departments and faculties according to their own financial resources.

Faculties offer various academic degrees (for instance, Master of Science on computer engineering), and these degrees may be organized by several departments (for instance, Master of Science on bioinformatics is jointly organized by the Computer Science and the Biology departments). Academic degrees are composed of several years (for instance, two years for Master of Science on computer engineering), which in turn are composed of several courses. The same course may be given as part of several degrees, possibly organized by different faculties (for instance, a statistics course may be given as part of both the computer engineering and the psychology degree). Each faculty is responsible for the organization of the set of courses it offers, in particular for appointing the academic staff involved in each course, and for the putting together of courses into academic degrees. When students enroll for an academic degree, they must take various courses determined by their corresponding degree.

The university allocates an annual endowment for educational purposes to each faculty. Further, as already said, the university allocates a number of personnel (expressed in FTE) to each faculty. The distribution of both of these resources among faculties is reexamined on a yearly basis and depends on multiple factors, of which the number of students enrolled in each faculty is the most important one. There is significant variation in student enrollment across the years, both in the total number of students and in the faculties or degrees in which they are enrolled. There are many reasons for this variation, which include the labor market supply and demand, and socioeconomic and financial factors. As a consequence, faculties must adapt their numbers of staff members in order to face this variation in student enrollment. For this reason, upon departure of a staff member, his or her position is reexamined in order to determine which department or faculty it will be reallocated to.

7.5 A Method for Temporal-Data-Warehouse Design

In this section, we present a method for temporal-data-warehouse design similar to that described in Sect. 7.3 for spatial-data-warehouse design, i.e., it includes the phases of requirements specification, conceptual design, logical design, and physical design. As in the spatial case, our method is independent of the conceptual model used for representing temporal multidimensional schemas.

Designing temporal data warehouses requires designers with a good knowledge of the inherent semantics of temporal aspects. In particular, they should be able to understand the various temporality types and their implications in the data warehouse context. They should also be capable of explaining to users technical concepts related to time-varying multidimensional elements, such as the possibility of representing changes in dimension data, the inclusion of the various kinds of temporal support for measures, and the management of different time granularities.

7.5.1 Requirements Specification and Conceptual Design

As was the case for the design of conventional and spatial data warehouses, we distinguish three approaches to requirements specification and conceptual design for temporal data warehouses: the analysis-driven, source-driven, and analysis/source-driven approaches. We also consider two different possibilities for addressing temporal features during the development process, depending on whether the users are familiar with concepts related to temporal data and whether the source systems are able to provide temporal support. Even though temporal DBMSs are not yet available, we assume that the source systems may provide some temporal support, which is managed explicitly by application programs.

The steps in requirements specification and conceptual design are similar to those described earlier for spatial data warehouses and illustrated in Figs. 7.1, 7.5, and 7.8. We briefly describe each one of these approaches next.

Analysis-Driven Approach

This approach relies on the requirements of the users (or the business) for developing a conceptual schema. Figure 7.1 shows the steps required in this approach, considering whether temporal support is included at an early or late stage of the temporal-data-warehouse design process.

Early Inclusion of Temporal Support

When the users are able to express their needs for temporal support for the various elements of a multidimensional schema, the phases of the analysis-driven approach to temporal-data-warehouse design are similar to those for conventional-data-warehouse design. However, each step must consider additionally the temporal data required.

The requirements specification phase starts with identifying users at various management levels and determining their analysis needs. Designers should focus not only on distinguishing multidimensional elements, as described in Sect. 6.4.1, but also on specifying the elements for which users need to store the history of changes. In the last step, all requirements are documented and serve as a basis for the next phase.

The conceptual-design phase starts with developing a first version of the temporally extended multidimensional schema. The next step checks the availability of data, i.e., the existence of corresponding data in the source systems, considering both nontemporal and temporal elements. We have indicated in Sect. 5.3.1 what types of temporal support can be obtained for different types of source systems. However, if the users' requirements for temporal data cannot be satisfied by the data available in the source systems, it should be decided whether to modify the source systems to be able to track future changes to data, to prepare suitable intermediate storage (for example, a data staging area as described in Sect. 6.8.2) and processes to capture future changes, or to represent the changes directly in the data warehouse. The final step delivers a multidimensional schema that can be implemented with the data available; the correspondences between elements of the multidimensional schema and the data in the source systems are also specified.

In our university case study, we considered the two goals expressed in Sect. 7.4, i.e., to establish a better distribution of financial and personnel resources among faculties according to the numbers of students in each faculty, and to improve the Times ranking, in particular the criterion related to the staff-to-student ratio. After interviewing users from the Education Department, and representatives of faculties, departments, and the various academic degrees, we were able to refine the analysis needs and to establish scenarios that could better support the decision process.

Users were interested in analyzing the evolution in time of the number of students enrolled in the university. The number of students had to be analyzed according to courses, and according to academic degrees and their component years, and according to departments and faculties. This would allow users to establish which courses or degrees were successful, since they were taken by many students, and those that did not attain a minimum quota of students. Some other important indicators in this respect were the numbers of academic staff in the various departments, and the total numbers of lecture and lab hours they provided. This would allow users to determine the adequacy of the number of staff members with respect to the number of courses and degrees offered by the departments. Therefore, the following analysis scenarios were elicited:

1. The percentage of the number of students choosing a course with respect to the number of students enrolled during a year in which the course is given.
2. The average number of hours provided by professors in different departments.
3. The total number of hours offered by departments with respect to the number of professors on their staff.
4. The number of hours provided by professors in departments which are not the one to which they are affiliated.
5. The staff-to-student ratio in different departments.

In addition, the users expressed a particular interest in keeping track of changes in various elements, since that would allow them to better understand the evolution of the numbers of students taking various courses. In this way, users might be able to analyze whether changes that have been made have helped to the university to approach the accomplishment of its goals. The changes that the users wished to track were the following:

1. Opening and closing of academic degrees.
2. Opening and closing of courses.

3. Evolution of the courses that made up academic degrees.
4. Evolution of the persons responsible for academic degrees.

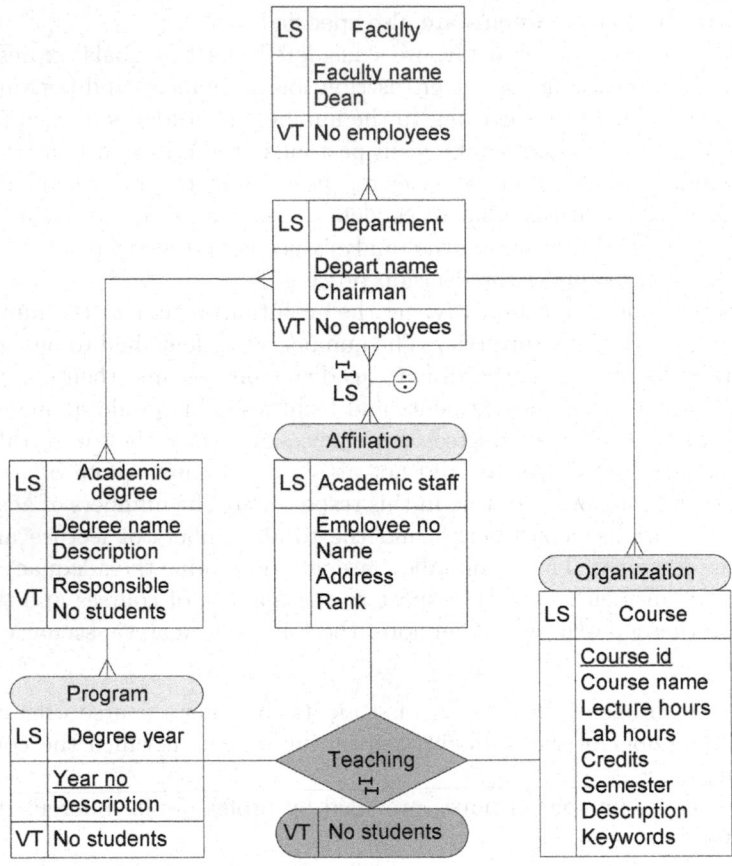

Fig. 7.9. Resulting schema with early inclusion of temporal support

These specifications helped in the identification of the various multidimensional elements and in the inclusion of temporal support for some of them. The resulting multidimensional schema is shown in Fig. 7.9. In the next step, the source systems were checked to determine the availability of data for the above analysis requirements. This revealed that the operational systems did not contain all of the historical information required by the users. Therefore, it was necessary to create additional processes to capture in the data warehouse the changes made to the operational databases. The final schema and the corresponding mappings were delivered to the users.

Late Inclusion of Temporal Support

In many cases, users and/or designers may prefer to first consider the analysis requirements without taking temporal data into account, leaving the inclusion of temporal support to a later stage when a working conceptual schema is already available. In this situation, the design process proceeds as for a conventional data warehouse until the step in which the conceptual model is compared with the data sources to check data availability (Fig. 7.1a and the upper path in Fig. 7.1b). At this point, a new step is included to add temporal support and produce the final schema and corresponding mappings.

The inclusion of temporal data requires us to consider the various temporality types that may be associated with the elements of a multidimensional schema. If the MultiDim model is used, we may focus first on lifespan support. Therefore, every level and parent-child relationship are verified for the inclusion of lifespan support. After that, the attributes of the levels are verified to determine whether they should include valid time support. If for some elements of the schema the users need to analyze changes that have occurred in the source data and/or to know when data was introduced into the data warehouse, the transaction time and/or loading time, respectively, should be included for those elements.

Nevertheless, a different approach should be taken for measures. In current nontemporal multidimensional schemas, the evolution of measures is stored by associating a time dimension with the corresponding fact relationship. In contrast, recall that in temporal multidimensional models, measures represent time-varying attributes of fact relationships. Therefore, the members of the time dimension of a nontemporal multidimensional schema should be transformed into temporal support for measures.[6]

Similarly to the previous case of early inclusion of temporal support, during the step that checks data availability the designers should verify whether the temporal support required for analysis can be implemented on the basis of the existing source systems. If this is not the case, additional structures should be created to capture future data changes.

Note that the inclusion of temporal support after a conceptual schema has been developed is a common practice in temporal-database design, as explained in Sects. 7.1 and 7.7. We believe that this approach may be simpler to use than the early inclusion of temporal support, since the latter approach requires one to simultaneously consider two different aspects: multidimensional elements and their temporal support.

We shall not give an example of applying late inclusion of temporal support. This amounts to following the analysis-driven approach for conventional-data-warehouse design, after which the conceptual multidimensional model is presented to the users in order to establish which elements they wish to keep the history of changes for.

---

[6] For more details, refer to Sect. 5.7.

Source-Driven Approach

The source-driven approach relies on the source data to create a data ware-house schema. As was the case for spatial data, the choice between early or late inclusion of temporal support depends on whether temporal support is provided by the source systems.

Early Inclusion of Temporal Support

If the source systems include temporal data, the three steps in Fig. 7.5a and the first two steps in Fig. 7.5b can be used to develop a conceptual multidimensional schema with temporal support. These steps are similar to the ones described for conventional data warehouses, with the difference that temporal support is considered in every step from the beginning.

After the source systems have been selected in the first step, the derivation process performed in the second step aims at identifying the various elements of the multidimensional model. Since, to our knowledge, there are no proposals for such a derivation process, this must be conducted manually by designers, relying on their experience and their knowledge of operational databases and aspects of temporal support. This process can be a complex task; it requires one to differentiate the elements in the source systems that correspond to measures, dimensions, and hierarchies and also to analyze whether the changes to these elements are included in the source systems.

An example of an operational schema that includes temporal information is given in Fig. 7.10. It corresponds to the schema in Fig. 6.4, augmented with attributes that allow changes to data to be kept. For example, the Academic staff entity type has a multivalued composite attribute Rank with components Name, Start date, and End date. This attribute keeps information about the academic ranks assigned to academic staff during various periods of time. Similar structures are included for the Director attribute in the Research center entity type and for the Name attribute in the Department entity type. Note that attributes indicating various periods of time (i.e., Start date and End date) for an instance as a whole are included in the Course, Research center, and Project entity types, as well as in the Investigates in, Participates, and Works in relationship types.

We have already described in Sects. 6.4.2 and 6.5.2 the transformation of this schema (without temporal support) into a multidimensional schema. The derivation process applied now is very similar, considering the attributes that have an associated time frame in the operational databases as temporal attributes in the data warehouse. For example, as can be seen in the resulting schema in Fig. 7.11, the multivalued complex attribute Rank in the ER schema becomes an attribute Rank with valid time support in the multidimensional schema. Similar transformations have been applied for the attributes Director in Research center and Name in Department. The lifespan support in the multidimensional schema results from transforming the Start date and End date

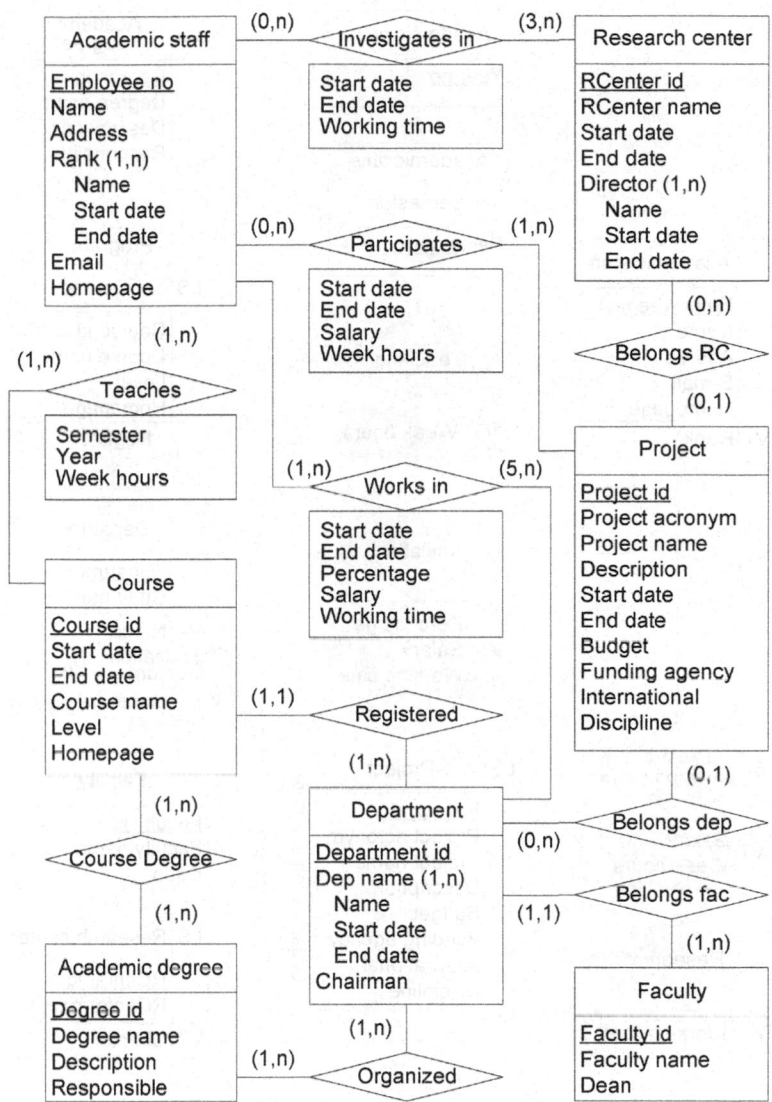

Fig. 7.10. Excerpt from an operational ER schema with temporal data

attributes included as simple attributes of the corresponding entity types (for example Course).

Since measures in temporal data warehouses are handled as temporal attributes of their corresponding fact relationships, we do not require the presence of the time dimension in the schema, as already explained in Sect. 5.3. However, academic years differ from a standard time dimension, because they

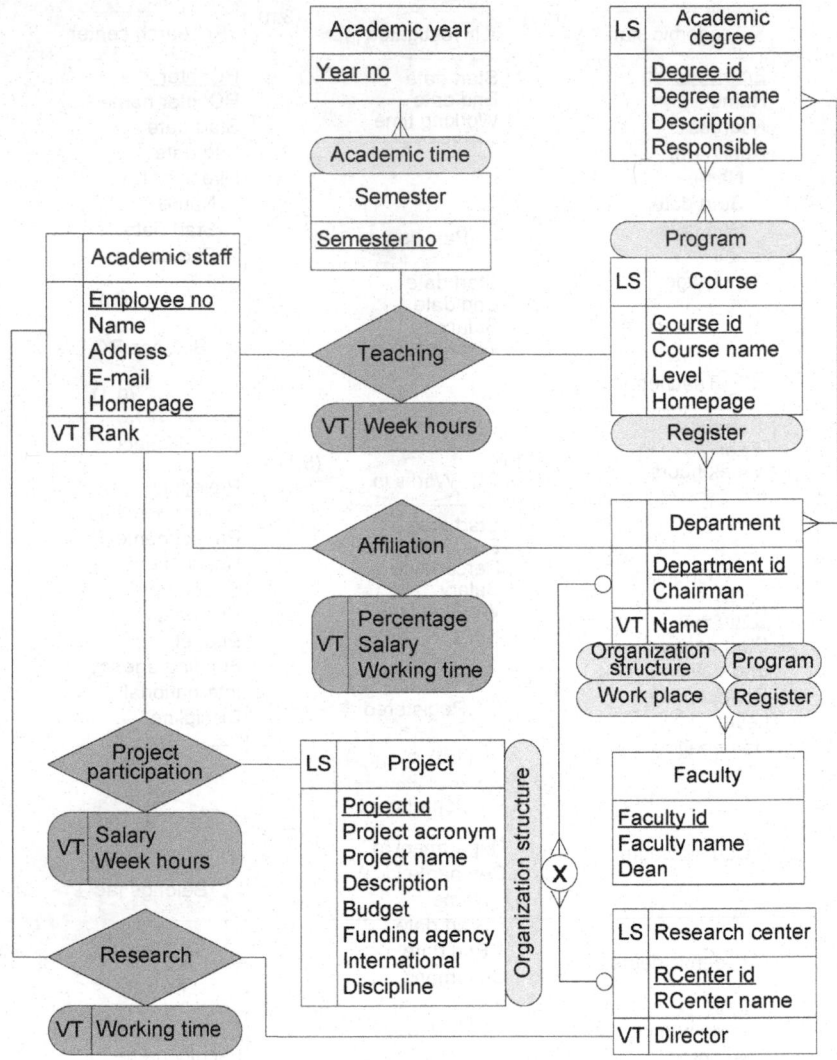

Fig. 7.11. Temporal-data-warehouse schema derived from the schema in Fig. 7.10

do not start and end at a regular point in time. Therefore, to facilitate the analysis, we have decided to include this time dimension in the schema. It is used for selection of members (slice-and-dice operations) but not for indicating valid time for measures.

In this case study, the resulting schema was shown to the users. As was indicated in the example of the analysis-driven approach, the users were interested only in information related to teaching activities and to administrative affiliation and payment for professors. Therefore, the Project participation and Research fact relationships were excluded from the schema. Finally, the modified schema and the corresponding mappings between the data in source systems and in the data warehouse were delivered.

Late Inclusion of Temporal Support

This approach is used when the source systems do not include temporal data or when this data exists but the users or designers prefer first to derive schemas without temporal support, leaving the temporal aspects to a later stage. The three steps in Fig. 7.5a and the first two steps in Fig. 7.5b are conducted in the same way as for a conventional data warehouse. Then, after the users have determined which elements should be left for analysis purposes, an additional analysis is conducted to specify the temporal support required (the step of adding temporal support in Fig. 7.5b). Similarly to the analysis-driven approach above, designers must check whether the temporal data required by the users is available in the source systems. If this is not the case, a decision about how to capture changes to data, as described previously for the analysis-driven approach, should be made.

Since this approach combines the source-driven approach to conventional-data-warehouse design with a step of adding temporal support as described for the analysis-driven approach above, we shall not include an example of its use.

Analysis/Source-Driven Approach

The analysis/source-driven approach combines the two previously described approaches. The steps of this approach are given in Fig. 7.8. As in the case of spatial data warehouses, we provide two different options for the designer. The early inclusion of temporal support considers temporal data starting from the first steps in both chains. Alternatively, in the case of late inclusion of temporal support, conceptual schemas are derived as for a conventional data warehouse. Then, after the matching of the schemas obtained from the analysis and source chains, a step of adding temporal support is executed. This is performed by analyzing the schema resulting from the matching process to determine which elements should include temporal support, as described for the analysis-driven approach.

7.5.2 Logical and Physical Design

Similarly to the case of spatial data warehouses, in this section we shall refer firstly to logical and physical schemas for temporal data warehouses and secondly to the ETL process.

Schema Definition

We presented in Sect. 5.12 a two-step approach to the logical-level design of temporal data warehouses. The first step translates MultiDim schemas into ER schemas. Then, depending on the needs of the developer, the corresponding relational or object-relational schema can easily be developed.

In Sect. 5.10, we discussed the advantages of implementing temporal data warehouses using object-relational databases. Further, in Sects. 5.11.1–5.11.4, we presented the mapping of the various elements of the MultiDim model into the object-relational model. In this section, we use the Program hierarchy in Fig. 7.9 to demonstrate a mapping into the object-relational model based on SQL:2003, and to discuss various aspects of physical-level design in the context of Oracle 10g.

We first give the definition of the temporality types. We define a set of intervals as a multiset, i.e., an unordered collection of values without maximum cardinality:

```
create type IntervalType as (FromDate date, ToDate date);
create type IntervalSetType as IntervalType multiset;
```

Then, we create the temporal attributes of the levels forming the Program hierarchy as follows:

```
create type ResponsibleValue as (
    Name character varying(25), VT IntervalSetType );
create type ResponsibleType as ResponsibleValue multiset;
create type NoStudentsValue as ( Number integer, VT IntervalSetType );
create type NoStudentsType as NoStudentsValue multiset;
create type NoEmployeesValue as ( Number integer, VT IntervalSetType );
create type NoEmployeesType as NoEmployeesValue multiset;
```

The above definitions are used to create levels and links between them as follows:

```
create type FacultyType as (
    LS IntervalSetType,
    FacultyName character varying(25),
    Dean character varying(25),
    NoEmployees NoEmployeesType )
    ref is system generated;
create table Faculty of FacultyType (
    constraint FacultyPK primary key (FacultyName),
    ref is FId system generated );
create type DepartmentType as (
    LS IntervalSetType,
    DepartName character varying(25),
    Chairman character varying(25),
    NoEmployees NoEmployeesType,
    FacRef ref(FacultyType) scope Faculty references are checked )
    ref is system generated;
```

```
create table Department of DepartmentType (
    constraint DepartmentPK primary key (DepartName),
    ref is DId system generated );
create type AcademicDegreeType as (
    LS IntervalSetType,
    DegreeName character varying(25),
    Description character varying(125),
    Responsible ResponsibleType,
    NoStudents NoStudentsType,
    DepartRef ref(DepartmentType) scope Department references are checked )
    ref is system generated;
create table AcademicDegree of AcademicDegreeType (
    constraint AcademicDegreePK primary key (DegreeName),
    ref is ADId system generated );
create type DegreeYearType as (
    LS IntervalSetType,
    YearNumber integer,
    Description character varying(125),
    NoStudents NoStudentsType,
    AcademicDegreeRef ref(AcademicDegreeType) scope AcademicDegree
    references are checked )
    ref is system generated;
create table DegreeYear of DegreeYearType (
    constraint DegreeYearPK primary key (YearNumber),
    ref is DYId system generated );
```

As already explained in Sects. 5.11.1 and 5.11.2, translating the above object-relational schema into the physical model using Oracle 10g is straightforward, since Oracle provides constructs similar to those of SQL:2003. However, it is important to consider the differences at the physical level when implementing collections.

Oracle provides two different types of collections, i.e., varying arrays, which store an ordered set of elements in a single row, and table types, which allow one to have unordered sets and to create nested tables. Varying arrays are in general stored "in line" in a row, if they are smaller than 4000 bytes or not explicitly specified as a large object (LOB). Even though varying arrays cannot be indexed to improve query performance, they do not require any join to retrieve the data.

On the other hand, rows in a nested table can have identifiers, can be indexed, and are not necessarily brought to memory when the main table is accessed, if the data from the nested table is not required. A nested table in Oracle must be stored in a separate table that includes a column called nested_table_id. This contains the keys to the associated parent table that are necessary in order to perform a join operation when related data from both tables is accessed. The nested_table_id column can be used for defining "index-organized tables", in which the records in the nested table are grouped according to the parent record, to which they belong. Another option is to index the nested_table_id column, if heap files (see Sect. 2.4) are used as a storage option for the nested table.

The implementation in Oracle of an excerpt from the above SQL:2003 schema is as follows:

```
create type IntervalType as object (FromDate date, ToDate date);
create type IntervalSetType as array(10) of IntervalType;
create type NoEmployeesValue as object (
     Value number(10), VT IntervalSetType);
create type NoEmployeesType as table of NoEmployeesValue;
create type DepartmentType as object (
     LS IntervalSetType,
     DepartName varchar2(25),
     Chairman varchar2(25),
     NoEmployees NoEmployeesType );
create table Department of DepartmentType (
     constraint DepartmentPK primary key (DepartName) )
     nested table NoEmployees store as NoEmpNT (
          primary key (nested_table_id)
          organization index compress )
     object identifier is system generated;
```

The clause nested table NoEmployees store as NoEmpNT creates a nested table called NoEmpNT to store the numbers of employees working in departments during various periods of time. This table is ordered according to the nested_table_id column and is also compressed to decrease the storage capacity required.

Further, recall from Sect. 5.11.3 that there are several different alternatives for implementing temporal hierarchies. Therefore, during the physical-design phase, developers should choose the alternative that corresponds best to query performance requirements.

Since physical-level design deals with the task of ensuring efficient execution of queries and facilitating the management of the temporal data warehouse, the aspects specified earlier for conventional-data-warehouse design should also be considered here, such as indexing, materialized views, and fragmentation.

Implementation of the ETL Process

Sections 6.7.2 and 6.8.2 refer to various issues that developers should consider during the implementation of the ETL process. These issues exist independently of whether a conventional or a temporal data warehouse is being developed. However, for temporal data warehouses, an additional difficulty arises when users' requirements for temporal data cannot be satisfied by the data available in the source systems. If the decision is made that the changes to the data will be tracked in the data staging area or in the data warehouse, suitable storage and processes should be implemented. As described in Sect. 6.8.2, in some situations the DBMS can provide facilities for capturing changes in source systems, such as the mechanism provided by Oracle 10g based on the publisher-subscriber paradigm.

7.6 Method Summary

This section summarizes the proposed methods for the design of spatial and temporal data warehouses, as illustrated in Figs. 7.12–7.14. Our methods are based on that described in Chap. 6 for designing conventional data warehouses. These methods include four phases, i.e., requirements specification, conceptual design, logical design, and physical design. Our methods propose three approaches for the requirements specification phase, which are based on the users' demands, the data in the source systems, or a combination of the two, leading to the analysis-driven, source-driven, and analysis/source-driven approaches, respectively. Further, the method for designing conventional data warehouses is extended by considering the users' knowledge about spatial/temporal data, and the presence of spatial/temporal data in the source systems. Similarly to the case of conventional data warehouses, the subsequent phase of conceptual design differs in some steps, depending on the approach chosen for requirements specification, while the logical- and physical-design phases remain the same in all approaches.

7.6.1 Analysis-Driven Approach

The analysis-driven approach, illustrated in Fig. 7.12, starts with the identification of the users who are able to express their analysis needs in terms of business goals. Since these users could have or not knowledge about spatial/temporal data, two different options exist (see Sects. 7.3.1 and 7.5.1).

The early inclusion of spatial/temporal support targets users who are familiar with spatial/temporal concepts, as described in Chaps. 4 and 5. Therefore, all the steps in the requirements specification and conceptual-design phases take the spatial/temporal elements of the application into account. On the other hand, the late inclusion of spatial/temporal support can be applied when users do not have knowledge about spatial/temporal data, or when they prefer to first express their analysis needs as for conventional data warehouses, and to include spatial/temporal support later on. This approach follows the same steps as for conventional-data-warehouse design (Sects. 6.4.1 and 6.5.1), with the addition of a step needed to extend the initial multidimensional schema with spatial/temporal elements (see Fig. 7.12).

During the two subsequent phases, the logical and physical schemas and the ETL processes are specified and implemented, taking the particularities of spatial/temporal data warehouses into account, as described in Sects. 7.3.2 and 7.5.2.

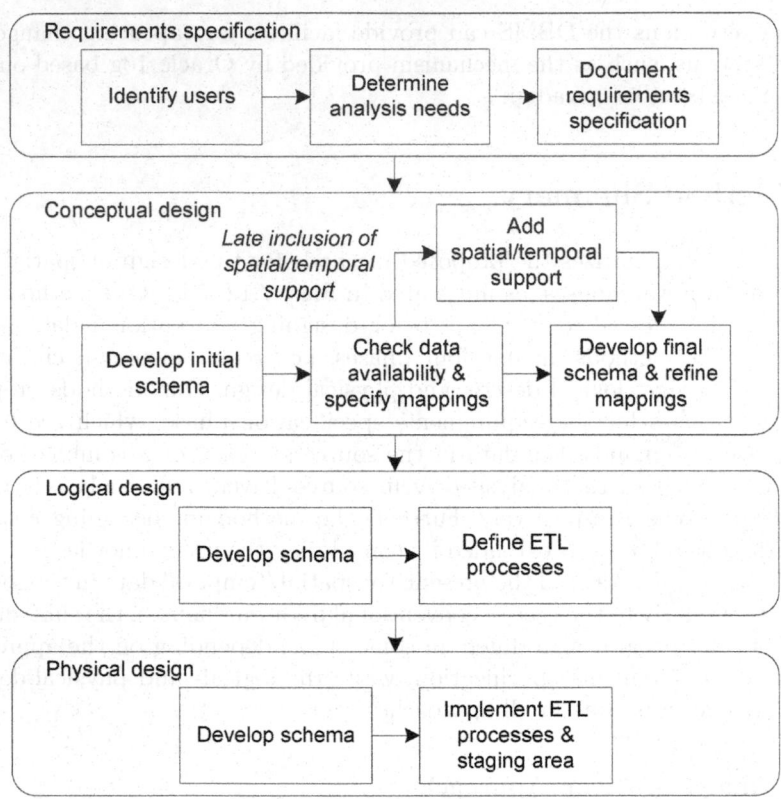

Fig. 7.12. Steps in the analysis-driven approach for spatial and temporal data
warehouses

7.6.2 Source-Driven Approach

The source-driven approach, illustrated in Fig. 7.13, first identifies the source
systems that may contain data required for analysis purposes. Similarly to the
analysis-driven approach, early or late inclusion of spatial/temporal support
may be considered (see Sects. 7.3.1 and 7.5.1).

The early inclusion of spatial/temporal support is used when the source
systems contain spatial/temporal data. In this case, all steps of the requirements specification and conceptual-design phases refer explicitly to the spa-
tial/temporal elements of the application, as described in Chaps. 4 and 5.
On the other hand, when the source systems do not provide spatial/temporal
data, or when the corresponding derivation process is complex, late inclusion
of spatial/temporal support can be used. In this case, the requirements specification and conceptual-design phases are conducted as for conventional data
warehouses (see Sects. 6.4.2 and 6.5.2); the only difference consists in the fact

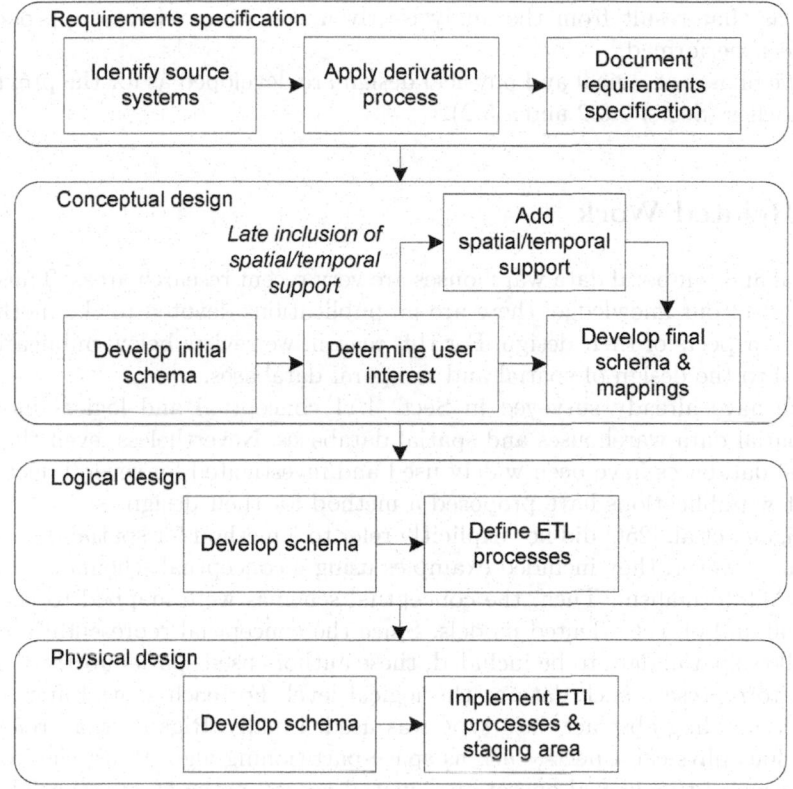

Fig. 7.13. Steps in the source-driven approach for spatial and temporal data warehouses

that after determining user interest, an additional step must be included to add the necessary spatial/temporal support.

The two subsequent phases of logical and physical design are the same as for the analysis-driven approach described above.

7.6.3 Analysis/Source-Driven Approach

Figure 7.14 summarizes the analysis/source-driven approach for designing spatial or temporal data warehouses. As mentioned in Sects. 7.3.1 and 7.5.1, this approach is the combination of the analysis-driven and source-driven approaches. The combined approach also allows the designer to choose between early or late inclusion of spatial/temporal support, as can be seen in Fig. 7.14. Therefore, spatial/temporal support may be considered from the beginning of the requirements specification phase or, later, after the matching of the schemas that result from the analysis-driven and source-driven approaches has been performed.

The phases of logical and physical design are developed as for the previous approaches (Sects. 7.3.2 and 7.5.2).

7.7 Related Work

Spatial and temporal data warehouses are very recent research areas. Thus, to the best of our knowledge, there are no publications devoted to the methodological aspects of their design. For this reason, we review below publications related to the design of spatial and temporal databases.

We have already surveyed in Sect. 4.11 conceptual and logical models for spatial data warehouses and spatial databases. Nevertheless, even though spatial databases have been widely used and investigated for several decades, very few publications have proposed a method for their design.

Rigaux et al. [256] did not explicitly refer to a method for spatial-database design; however, they included examples using a conceptual schema based on the OMT formalism. Then, the conceptual schemas were mapped to the relational and object-oriented models. Since the conceptual representation did not allow spatial data to be included, these authors used spatial abstract data types to represent such data at the logical level. For each spatial attribute, a corresponding abstract data type was used. Finally, Rigaux et al. referred to various physical aspects, such as space-partitioning algorithms, the implementation of topological operations, spatial access methods, and spatial join operations.

Shekar and Chawla [268] also referred to a three-step design method for spatial databases. First, they used the ER model to represent spatial and nonspatial data. For the former, they used multivalued attributes, which were later mapped to separate relational tables and handled as nonspatial data. Additionally, they extended the ER model with pictograms to represent spatial data, which allowed them to reduce the number of elements in the ER schema. This also simplified the relational schema resulting from the mapping process. Then, various physical aspects were discussed, such as spatial indexes, file structures, clustering, and spatial joins.

Worboys and Duckham [315] explicitly presented conceptual and logical design and a mapping between them. To represent conceptual schemas containing spatial data, they used either an extended ER model or UML notation. Then, the ER schemas were mapped to relational schemas and the UML schemas were mapped to object-oriented schemas. Similarly to other publications mentioned above, Worboys and Duckham referred to several physical features required for representing and handling spatial data in DBMSs.

Parent et al. [227] proposed the MADS conceptual model for designing spatiotemporal databases. They also referred to a logical representation, which was achieved by a two-step process. In the first step, the MADS constructs

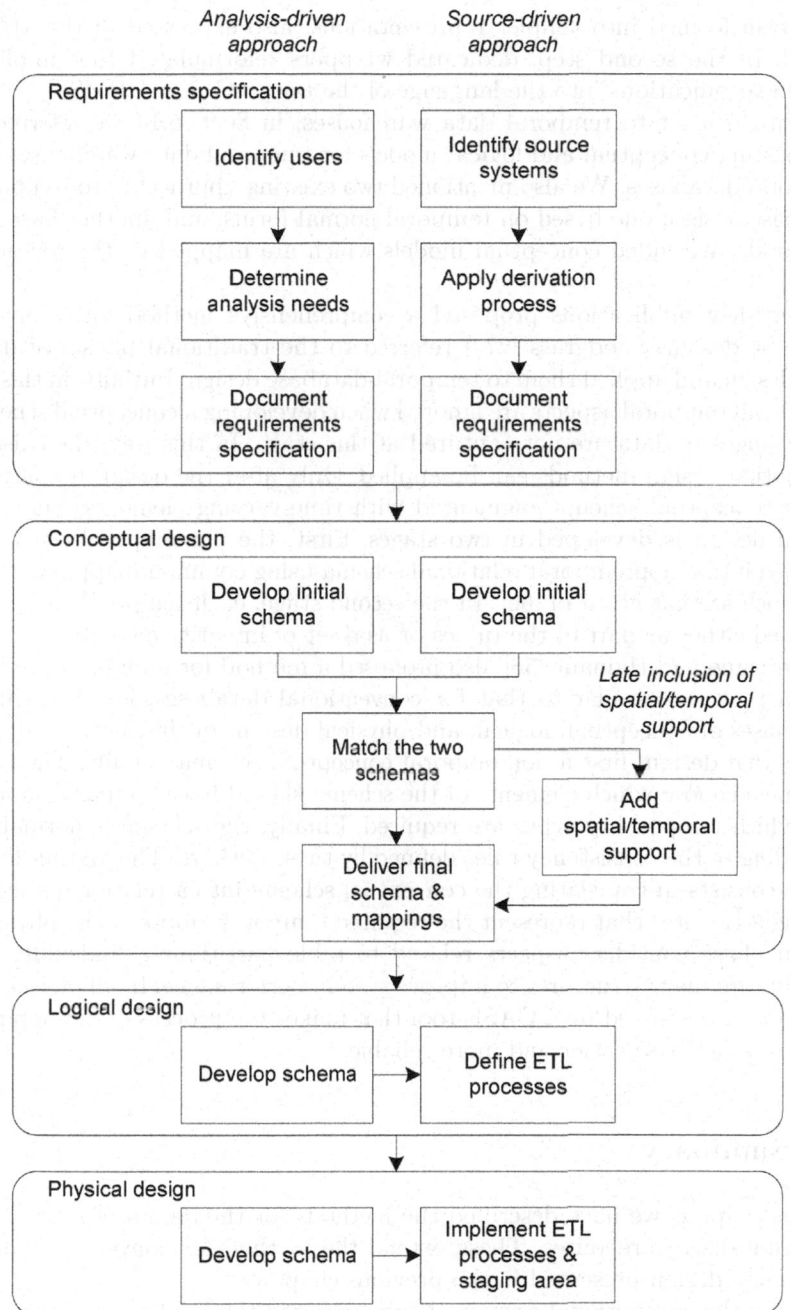

Fig. 7.14. Steps in the analysis/source-driven approach for spatial and temporal data warehouses

were transformed into simpler representations, also expressed in the MADS model. In the second step, dedicated wrappers reformulated the simplified MADS specifications into the language of the target GIS or DBMS.

With respect to temporal data warehouses, in Sect. 5.14 we referred to the existing conceptual and logical models for temporal data warehouses and temporal databases. We also mentioned two existing approaches to temporal-database design: one based on temporal normal forms, and another based on temporally extended conceptual models which are mapped to the relational model.

Very few publications proposed a comprehensive method for temporal-database design. Snodgrass [274] referred to the traditional phases of database design and applied them to temporal-database design. Initially, in this approach, all temporal aspects are ignored when developing a conceptual schema, i.e., changes in data are not captured at this stage. In this way, the existing conceptual-design methods can be applied. Only after the design is complete is the conceptual schema augmented with time-varying elements. Similarly, logical design is developed in two stages. First, the nontemporal schema is mapped into a nontemporal relational schema using common mapping strategies, such as that given in [66]. In the second stage, each temporal element is included either as part of the tables or as a set of integrity constraints.

Detienne and Hainaut [56] also proposed a method for temporal-database design that was similar to that for conventional databases, i.e., it included the phases of conceptual, logical, and physical design. In this method, during conceptual design, first a nontemporal conceptual schema is built. Then, the designers choose which elements of the schema should have temporal support and which temporality types are required. Finally, the schema is normalized according to the consistency rules defined by these authors. The logical-design phase consists in translating the conceptual schema into a relational one and adding attributes that represent the required temporal support. The physical-design phase considers aspects related to table partitioning, indexing, and creating auxiliary structures to improve system performance. In all phases, the designers are assisted by a CASE tool that makes the process of developing a temporal database easier and more reliable.

7.8 Summary

In this chapter, we have described the methods for the design of spatial and temporal data warehouses. These extend the method for conventional-data-warehouse design presented in the previous chapter.

As in the conventional case, we have presented three different approaches to requirements specification that rely the analysis needs of users (or the business), the data in source systems, or both. As was explained in the previous chapter, these three approaches differ in the involvement required of the users in the development process and in the expertise required of the developer team. Additionally, for each of the three approaches we considered two situations, where we allow the inclusion of spatial or temporal support either in an early stage of the requirements specification phase or after a conventional (i.e., nonspatial and nontemporal) conceptual schema has been developed. The choice between these two possibilities depend on the users' knowledge about spatial or temporal data and the availability of spatial or temporal data in the source systems.

The method presented in this chapter allows users to express their needs for including spatial and temporal data in analysis processes. It also indicates to designers and implementers how and when spatial or temporal support may be included during the data warehouse design process. Our method provides designers with several different design alternatives: they can first choose one of the three approaches to requirements specification, and then apply either early or late inclusion of spatial or temporal support to the chosen approach. As a consequence, designers can choose from the various design alternatives the one that best fits the specific skills of the development team and the particularities of the data warehouse project at hand.

8

Conclusions and Future Work

8.1 Conclusions

Today, many organizations use data warehouse and online analytical processing (OLAP) systems to support their decision-making processes. These systems use a multidimensional model to express users' analysis requirements. A multidimensional model includes measures that represent the focus of analysis, dimensions used to analyze measures according to various viewpoints, and hierarchies that provide the possibility to consider measures at different levels of detail.

Nevertheless, the current multidimensional models raise several important issues. Firstly, there is not a commonly agreed conceptual model for representing multidimensional data. Secondly, the existing conceptual multidimensional models do not provide associated logical representations, and thus implementers do not have guidance about mapping conceptual schemas to logical schemas. Thirdly, many kinds of complex hierarchies that arise in real-world situations are not addressed by current multidimensional models. Fourthly, these models do not allow spatial data, even though such data can improve the analysis process. Finally, the traditional time dimension only allows changes to measures to be represented, leaving to implementers the responsibility of managing changes to dimension data. As a response to these issues, in this book we have proposed the MultiDim model, a conceptual multidimensional model that is able to express data requirements for data warehouse and OLAP applications. The model covers conventional, spatial, and temporal data.

Chapter 3 defined the MultiDim model, which comprises fact relationships, measures, dimensions, and hierarchies. For the latter, we studied in a systematic way the various kinds of hierarchies that exist in real-world applications and in the scientific literature. On the basis of this study, we proposed a conceptual classification of such hierarchies and provided a graphical notation for them. This notation allows a clear distinction between the various kinds of hierarchies, taking into account their differences at the schema and instance levels. Further, we provided a mapping of the MultiDim model to the relational and object-relational models based on the SQL:2003 standard. Since the MultiDim model is based on the entity-relationship (ER) model, this mapping was performed using well-known rules. We discussed alternative representations of the various kinds of hierarchies at a logical level, comparing them and stating which situations they were more suitable in.

The proposed conceptual multidimensional model and its associated logical mapping should benefit users, designers, and implementers of data warehouse and OLAP systems and applications. Using the MultiDim model, designers are able to represent better the analysis requirements of decision-making users than when using conventional ER or UML models or logical models. Most of the existing conceptual multidimensional models do not distinguish between the different kinds of hierarchies proposed in this book, although some of these models can be extended to include such hierarchies. Further, a conceptual representation of hierarchies offers a common vision of these hierarchies for data warehouse and OLAP system implementers; it provides the requirements for extending the functionality of current OLAP tools. Moreover, the mapping of the various kinds of hierarchies to the relational and object-relational models can guide implementers in the physical design of a data warehouse. Since there is currently a lack of logical-level representations for the various kinds of hierarchies, designers must apply various strategies at the implementation level to transform some kinds of hierarchies into simpler ones.

Chapter 4 extended the MultiDim model with spatial data. We proposed the inclusion of spatial dimensions with spatial hierarchies, spatial fact relationships, and spatial measures. We referred to the classification of hierarchies given in Chap. 3 and showed its applicability to spatial hierarchies. We indicated that a summarizability problem may occur, owing to the various topological relationships that may exist between levels in a hierarchy. We classified these relationships according to the complexity of the process required to aggregate measures. Moreover, we showed that when a fact relationship relates more than one spatial dimension, a topological relationship between them may be required. We distinguished two possible situations that can arise when spatiality is applied to measures: when a spatial measure is represented by a geometry, and when a traditional measure is the result of a calculation that uses spatial or topological operators. A particular feature of our model is that it allows the inclusion of spatial measures without the presence of spatial dimensions. Moreover, we discussed the necessity of having spatial aggregation functions for spatial measures in order to perform roll-up operations along hierarchies. Finally, we proposed a mapping to the object-relational model and provided examples using Oracle 10g Spatial. The mapping was complemented with additional integrity constraints needed to ensure semantic equivalence between conceptual and logical schemas. These integrity constraints were mainly implemented using triggers.

The spatial extension of the MultiDim model aims at expanding the analysis capabilities of existing data warehouse and OLAP applications. The inclusion of spatial data allows a more global perspective for the decision-making activities, which can be put into their geographic context. In addition, spatial decision making is increasingly becoming a necessity owing to the recent technological advances in mobile and location-based technologies and the way these have influenced our lives. By integrating spatiality into a conceptual multidimensional model, decision-making users can represent their analysis needs in an abstract manner without considering complex implementation issues. Further, developers of spatial OLAP tools can have a common vision of the various features that comprise a spatial multidimensional model and of the roles they play. Mapping the spatial multidimensional model to an object-relational model shows the feasibility of implementing spatial data warehouses in current database management systems.

Chapter 5 was dedicated to the temporal extension of the MultiDim model. We proposed the inclusion of several different temporality types, namely valid time, transaction time, and lifespan, which are obtained from the source systems, in addition to the loading time generated in a temporal data warehouse. We included temporal support for levels, hierarchies, and measures. We used an orthogonal approach, where levels may include temporal support independently of whether they have time-varying attributes. For hierarchies, we discussed various cases that arise when changes to levels or to the links between them are required to be kept. By means of real-world examples, we showed the necessity of considering various temporality types for measures. We also referred to the problem of dealing with the presence of different time granularities in the context of measures. Finally, we included a two-step mapping of our conceptual model, first to the ER model and then to the object-relational model. We gave examples of this mapping using Oracle 10g.

The temporal extension of the MultiDim model allows temporal support to be included as an integral part of a data warehouse. It allows the analysis spectrum for decision-making users to be expanded, which may allow them to establish cause-effect relationships better by taking into account the evolution in time of all elements under analysis. Being a conceptual model that integrates temporal semantics, it allows many technical issues required when coping with time-varying information to be hidden from users. Indeed, we showed that schemas represented using our model are less complex and require less technical knowledge than schemas represented in the ER model. Further, we showed that the object-relational model represents the temporal semantics better than does the relational model, by allowing related data and its corresponding timestamps to be grouped in the same structures. Nevertheless, an object-relational representation is not suitable for all time-varying elements of temporal data warehouses, for example measures.

Chapter 6 described a design method for conventional data warehouses. Our method is composed of several phases, which correspond to those used for designing traditional databases, although these phases have been adapted to the data warehouse context. For the requirements specification phase, we proposed three different approaches that use as the driving force either the analysis requirements, the data available in the source systems, or a combination of both. We also briefly referred to logical-level design, including the ETL process. Finally, we proposed several recommendations for physical design and gave examples of their implementation using Oracle 10g.

Chapter 7 extended the proposed method for conventional-data-warehouse design by the inclusion of spatial and temporal data. We modified the requirements specification phase and considered two approaches to including spatial or temporal support: either from the beginning of the requirements-gathering process or later, after the creation of an initial conceptual schema that does not have any spatial or temporal support.

Methodological support for data warehouse development is essential owing to the intrinsic complexity of this task. This support is increasingly important when spatial and temporal information is included, owing in particular to the necessity of accurately capturing spatial and temporal semantics in a data warehouse. Since our method is model- and software-independent, it can be used as a general framework for data warehouse development in various application domains. The method provides the data warehouse developer team with a systematic specification of the various approaches that can be used in the requirements-gathering phase. The guidelines given allow one of these approaches to be chosen according to the knowledge and experience of the developer team and the particularities of the data warehouse project. Our method also helps users, since designers can choose an approach that fits better to the users' time constraints, their identification with business goals, their motivation for using the data warehouse, and their involvement in the project. Further, implementers can benefit from the specifications of the data warehouse structures and ETL processes developed during the conceptual and logical phases.

Appendix A presents the formalization of the MultiDim model. This is based on denotational semantics and includes specifications of the syntax, the semantics, and the associated constraints. This formalization of the model provides the necessary essentials for its future implementation.

Appendix B summarizes the notation used in this book to represent entity-relationship, relational, and object-relational models, and also to represent conventional, spatial, and temporal multidimensional models.

8.2 Future Work

The work reported in this book may be continued in several directions. We present next several areas in which extensions may be developed.

8.2.1 Conventional Data Warehouses

The conceptual multidimensional model presented in this book does not consider operations for querying and aggregating data contained in data warehouses. However, these are essential in order to integrate data warehouse and

OLAP systems into the decision-making process. Therefore, it is necessary to define such operations at a conceptual level, and to study how to implement them using the operations available in current relational and object-relational systems. In particular, an interesting is the definition of aggregation procedures for those kinds of hierarchies for which summarizability conditions do not hold, i.e., unbalanced, generalized, noncovering, and nonstrict hierarchies. Some solutions for unbalanced, noncovering, and nonstrict hierarchies have been proposed by Pedersen et al. [231]. The approach described there transforms these hierarchies into balanced, covering, and strict hierarchies, respectively. However, for the latter, the additional structures created may not correspond to users' needs. Some insights that are useful for developing customized aggregation procedures can be obtained from [106, 206, 242].

Another aspect is the inclusion of other constructs of the ER model. Currently, the MultiDim model includes entity types, relationship types, and simple monovalued attributes with their usual semantics. Additionally, our model offers explicit support for representing various kinds of hierarchies. However, we have not considered other ER constructs, such as weak entity types, multivalued attributes, composite attributes, and generalization/specialization relationships. The inclusion of these features is not straightforward and requires an analysis of their usefulness in multidimensional modeling. Some ideas could be taken from [8], where the problem of representing generalization/specialization hierarchies in a multidimensional model is discussed, or from [4], where a drill-across operation was proposed for dimensions represented by generalization/specialization relationships.

Further, even though the multidimensional storage in MOLAP systems is vendor-specific, it would be an interesting topic to investigate how the proposed hierarchies could be mapped to array-based physical structures. This would give insights that could be useful for their future implementation on current MOLAP platforms. Various models can be used as a starting point (e.g., [34, 306]). References to these and other models are included in [307].

8.2.2 Spatial Data Warehouses

The MultiDim model supports spatial dimensions comprising spatial hierarchies, as well as spatial fact relationships and spatial measures. In addition, the model supports topological relationships for spatial hierarchies and spatial fact relationships. We have already discussed the impact that topological relationships between hierarchy levels may have on aggregation procedures. However, it is still necessary to extend this analysis, taking into account the various topological relationships that may exist between spatial measures. For example, if a spatial measure represents the geometry of a building and the same building is used for different purposes, for instance for concerts and exhibitions, aggregation of the areas dedicated to entertainment will count the area of the building twice. The double-counting problem is well known in the field of conventional data warehouses. However, to our knowledge, only [232] has dealt with this aspect in the spatial context; that publication proposed a classification of topological relationships between spatial measures that avoids the problem of double counting. However, this is still an open research area that requires one to combine research related to topological relationships between measures and between hierarchy levels.

Another issue for spatial data warehouses is their extension to the management of three-dimensional (3D) objects. Many application domains, such as urban planning, telecommunications engineering, disaster management, and molecular biology, require 3D objects. Several conceptual models supporting 3D objects have been proposed in the literature (e.g., [2, 149, 158, 305]). These models consider both geometrical aspects and topological aspects [326]. Very few publications have addressed the combination of data warehouses and 3D objects. As an example, the BioMap data warehouse [171, 172] integrates biological data sources in order to provide integrated sequence/structure/function resources that support analysis, mining, and visualization of functional genomic data. Since this domain is at an exploratory stage, it is necessary to analyze various kinds of applications that manipulate 3D objects in order to determine their analytical requirements. Extending the MultiDim model in this direction requires first the definition of 3D spatial data types, 3D topological operators, and spatial operations and functions that can operate on 3D data types. After that, issues such as the aggregation of spatial measures and the classification of topological relationships between spatial levels with respect to measure aggregation should be addressed.

Further, as explained in Sect. 4.1, our model uses a discrete (or object) view of spatial data. However, a continuous (or field) view is more suitable for representing spatial phenomena such as temperature, altitude, soil cover, or pollution. Both kinds of views are important for spatial applications [227]. Therefore, the MultiDim model should be extended by the inclusion of field data. Various aspects should be considered, such as the representation of field data in the model, the spatial hierarchies formed by levels represented by field data, spatial measures representing continuous phenomena and their aggregation, and fact relationships that include spatial dimensions that contain field data. Several approaches to these problems have been proposed in the spatial-database community. For example, [227] integrated spatial object and field data in a conceptual spatiotemporal model called MADS, [117] presented algorithms that can be used for creating spatial hierarchies based on field data, and Tomlin's algebra [294] or advanced map operations [68] can be used to analyze how to manage spatial measures represented by field data.

Another important issue is to cope with multiple representations of spatial data, i.e., allowing the same real-world object to have several geometries. Dealing with multiple representations is a common requirement in spatial databases (e.g., [55, 227]), in particular as a consequence of dealing with multiple levels of spatial resolution. This is also an important aspect in the data warehouse context, since spatial data may be integrated from source systems containing data at different spatial resolutions. This integration process may either select one representation from those available (we have implicitly assumed this situation in our work) or include multiple representations in a multidimensional model. Our conceptual model could be extended to allow multiple representations of spatial data. This extension could be based on ideas proposed in [19, 227]. However, this raises some important issues. For example, if levels forming a hierarchy can have multiple representations, additional conditions may be necessary to establish meaningful roll-up and drill-down operations. Further, managing multiple representations of measures and levels requires one to establish conditions for aggregation procedures. Some interesting insights can be obtained from [299], where constraints for multiply represented spatial objects related by aggregation relationships are specified.

8.2.3 Temporal Data Warehouses

With respect to temporal data warehouses, we have discussed the issue of the presence of different time granularities for measures in the source systems and in the data warehouse; we have also briefly mentioned the problem of measure aggregation when the time granularities for measures and dimensions are different. Further analysis of the management of multiple granularities among the different levels forming a hierarchy is required. To the best of our knowledge, only [264] discusses issues related to differences in granularities between measures and dimensions in relation to obtaining precomputed aggregations of measures. On the other hand, several publications have dealt with multiple time granularities in the context of temporal databases (e.g., [21, 51, 196, 310, 313]). These publications could provide solutions for managing multiple time granularities in temporal data warehouses.

Even though operations in temporal data warehouses have been proposed in several publications (e.g., [194, 204]), the implementation of these operations in object-relational DBMSs has not been yet studied. As we mentioned in Sects. 5.11.2 and 5.11.3, several alternatives exist for representing temporality types and temporal links between hierarchy levels. Therefore, the performance of OLAP operations in the various implementation options should be evaluated.

Further, as briefly mentioned in Sects. 5.13.2 and 5.14, various techniques exist for aggregating measures. Some of them consider schema versioning, while others refer only to changes at the instance level. Several solutions propose a mapping of data from different versions of a schema in order to perform an analysis according to the current or a previous version of the schema. The variety of approaches to aggregating measures in the presence of time-varying dimensions is confusing, and it is difficult to determine which approach better corresponds to a particular analysis task. Therefore, one important contribution would be a systematic study and classification of the existing approaches, indicating the situations in which it would be most suitable to apply each of them.

Our temporal multidimensional model assumes a unique multidimensional schema that allows changes in data to be captured. Nevertheless, in many real-life situations it is necessary to consider changes in schema, which induces the necessity to manage several data warehouse schemas. Many publications have already addressed schema evolution and schema versioning in data warehouses (e.g., [25, 63, 79, 316]). Our conceptual multidimensional model could be extended to deal with multiple schemas. For example, this extension could be done in a way similar to that proposed in [16, 203, 204, 316], where, after performing temporal aggregation for each version of the schema/instances (corresponding to temporal aggregations based on instant grouping [278]), users must decide whether the results can be integrated.

8.2.4 Spatiotemporal Data Warehouses

An important extension to our model would be to combine spatial and temporal support into a conceptual multidimensional model for spatiotemporal applications.

The field of spatiotemporal data management results from the convergence of the GIS and temporal-database research communities and aims at providing an integrated approach to managing both spatial and temporal information. Spatiotemporal data management has become increasingly important owing to recent technological advances, such as GPS systems, wireless communication, mobile devices, and ubiquitous computing. This has led to the concept of moving objects, which are objects whose location in time is relevant for the purposes of an application. Examples of moving objects include people on the move, fleets of vehicles, migrating animals, and natural phenomena such as hurricanes and tsunamis.

Current spatiotemporal applications exploit mostly the current location of moving objects. This is typically the case for location-based services, which contextualize the services provided according to the current location of the object requesting them. Nevertheless, analyzing the whole history of the locations of moving objects, or their trajectories [283], would allow one to establish behavioral patterns that might provide new solutions in many applications with high societal and environmental impact, such as traffic management, transportation networks, and the protection of endangered species.

A spatiotemporal multidimensional model is essential for supporting such innovative applications. However, as was the case when spatiotemporal data models were developed, combining spatial and temporal support into a single multidimensional model must be done carefully. This implies revisiting the concepts proposed in this book, as well as introducing new ones.

Defining a spatiotemporal multidimensional model raises many issues. We can mention, for instance, temporality types for time-varying spatial dimensions and measures, aggregation of time-varying spatial measures, and conditions for traversing spatial hierarchies with time-varying links and levels. Many insights that could be useful in developing this extension could be obtained from research related either to spatiotemporal databases (e.g., [39, 65, 92, 153, 217, 227]) or to spatiotemporal data warehouses (e.g., [24, 185, 199, 221, 222, 223, 236, 293, 308]). A recent publication that focuses on trajectory data warehouses is [77].

8.2.5 Design Methods

Defining a data warehouse design method is still an ongoing research issue. Since the requirements specification phase is the most important phase in the overall data warehouse design process, the various approaches proposed should be evaluated. Although some insights may be taken from [165], the comparison there was empirical and was applied to data warehouse projects with various characteristics. Therefore, it is necessary to determine a set of evaluation criteria that allow one to compare the various approaches used in the requirements specification phase. Then, these criteria can be applied, considering data warehouses (or data marts) with similar analysis purposes and data availability.

Another research issue is related to the synthesis of current approaches to determining which aggregates should be precomputed. There are many publications that refer to this issue (e.g., [6, 224]). However, these publications are usually highly technical and cannot be used as a general recommendation for data warehouse developers. Currently, developers must rely on their intuition or on commercial tools, which automatically select the aggregates to be precomputed.

Since spatial and temporal data warehouses are very recent research areas, there is a lack of knowledge about and experience in their development. Therefore, the various solutions proposed for conventional data warehouses, in particular for the requirements specification phase, may be extended by addressing spatial and temporal data. For example, in the source-driven approach, a specification of how to derive in an automatic or semiautomatic way spatial (or temporal) multidimensional schemas from the underlying spatial (or temporal) source systems could be helpful to data warehouse designers. Moreover, a similar extension should be devised for the matching process required in the analysis/source-driven approach. Although several solutions have been proposed for conventional data warehouses, for example [28, 78, 238], to our knowledge there are no such proposals for spatial or temporal data warehouses.

An interesting research issue is the validation of the proposed method for the design of spatial and temporal data warehouses in real-world applications. Since several different approaches to requirements specification are possible, i.e., analysis-driven, source-driven, and analysis/source-driven, and for each of them two solutions have been proposed, i.e., early or late inclusion of spatial or temporal support, six different scenarios should be considered for the design of both spatial and temporal data warehouses. Evaluating them in real-life applications would help to refine the proposed method and to specify in more detail how the development of each phase should be conducted.

A

Formalization of the MultiDim Model

The following formalization was inspired by [86]. We first describe the notation, assumptions, and metavariables required for defining the abstract syntax and the semantics of the MultiDim model. Next, we give the abstract syntax of the model, which allows the graphical representation to be translated to the equivalent textual representation. We show this syntax by three examples, of a conventional, a spatial, and a temporal data warehouse. Finally, after describing the auxiliary functions, we define the semantics of the MultiDim model.

A.1 Notation

We use *SET*, *FSET*, and *TF* to denote the class of sets, the class of finite sets, and the class of total functions, respectively. Given $S_1, S_2, \ldots, S_n \in SET$, $S_i \uplus S_j$ indicates a disjoint union of sets, $S_i \cup S_j$ denotes a union of sets, and $S_1 \times S_2 \times \ldots \times S_n$ represents the Cartesian product over the sets $S_1, S_2, \ldots, S_n$.

We write finite sets as $\{c_1, c_2, \ldots, c_n\}$, lists as $\langle c_1, c_2, \ldots, c_n \rangle$, and elements of a Cartesian product as $(c_1, c_2, \ldots, c_n)$. For any set, we use $\perp$ to denote an undefined value of the set.

A.2 Predefined Data Types

A data signature describes the predefined data types, operations, and predicates. The MultiDim model includes the basic data types *int*, *real*, and *string*; the inclusion of other data types is straightforward. These predefined data types and the operations and predicates on them have the usual semantics, and this interpretation is fixed, that is, it is defined once and for all.

The syntax of a data signature *DS* may be given as follows:

- the sets *DATA*, *OPNS*, *PRED* $\in$ *FSET*,

- a function $input \in TF$ such that $input : OPNS \to DATA^*$,
- a function $output \in TF$ such that $output : OPNS \to DATA$, and
- a function $args \in TF$ such that $args : PRED \to DATA^+$.

If $\sigma \in OPNS$, $input(\sigma) = \langle d_1, \ldots, d_n \rangle$, and $output(\sigma) = d$, this is denoted as $\sigma : \langle d_1, \ldots, d_n \rangle \to d$. If $\pi \in PRED$, with $args(\pi) = \langle d_1, \ldots, d_n \rangle$, this is denoted as $\pi : \langle d_1, \ldots, d_n \rangle$.

The predefined data types and some operators and predicates on them are as follows.

$$
\begin{aligned}
DATA &\supseteq \{int, real, string\} \\
OPNS &\supseteq \{ +_i, -_i, *_i : & int \times int & \to int \\
& +_r, -_r, *_r : & real \times real & \to real \\
& /_i : & int \times int & \to real \\
& /_r : & real \times real & \to real \\
& cat : & string \times string & \to string \\
& \ldots \} \\
PRED &\supseteq \{ <_i, >_i, \leq_i, \geq_i, \neq_i : & int \times int & \\
& <_r, >_r, \leq_r, \geq_r, \neq_r : & real \times real & \\
& <_s, >_s, \leq_s, \geq_s, \neq_s : & string \times string & \\
& \ldots \}
\end{aligned}
$$

A.3 Metavariables

$S_D \in SchemaDecl$ – MultiDim schema declarations
$D_D \in DimDecl$ – dimension declarations
$L_D \in LevDecl$ – level declarations
$PC_D \in PCRelDecl$ – parent-child relationship declarations
$F_D \in FactRelDecl$ – fact relationship declarations
$IC_D \in ICDecl$ – integrity constraint declarations
$H_D \in HierDecl$ – hierarchy declarations
$A_D \in AttDecl$ – attribute declarations
$M_D \in MeasDecl$ – measure declarations
$I_S \in InvSpec$ – the set of level involvement specifications
$T_S \in TempSpec$ – the set of specifications for temporal support
$D \in Dimensions$ – the set of dimension names
$F \in FactRels$ – the set of fact relationship names
$L \in Levels$ – the set of level names
$PC \in PCRels$ – the set of parent-child relationship names
$Q \in 2^{PCRels}$ – the set of subsets of parent-child relationship names
$H \in Hier$ – the set of hierarchy names
$A \in Attributes$ – the set of attribute names
$K \in 2^{Attributes}$ – the set of subsets of attribute names

$M \in Measures$ – the set of measure names

$d \in DATA$ – the set of basic data types supported by the MultiDim model

$min, max \in Integer\ constants$ – the set of integer constants

$add \in \{additive, semiadditive, nonadditive\}$ – the set of additivity types for measures

$sp \in \{Geo, SimpleGeo, ComplexGeo, Point, Line, OrientedLine, Surface, SimpleSurface, PointSet, LineSet, OrientedLineSet, SurfaceSet, SimpleSurfaceSet\}$ – the set of spatial data types

$topo \in \{meets, overlaps, intersects, contains, inside, covers, coveredBy, equals, disjoint, crosses\}$ – the set of topological relationships

$temp \in \{LS, VT, TT, LT\}$ – the set of temporality types

$t \in \{Time, SimpleTime, ComplexTime, Instant, InstantSet, Interval, IntervalSet\}$ – the set of temporal data types

$g \in \{sec, min, hour, day, week, month, year\}$ – the set of granules for temporality

$sync \in \{meets, overlaps, intersects, contains, inside, covers, coveredBy, equals, disjoint, starts, finishes, precedes, succeeds\}$ – the set of synchronization relationships

A.4 Abstract Syntax

$$S_D ::= D_D; L_D; PC_D; F_D; IC_D;$$
$$D_D ::= D_{D_1}; D_{D_2}$$
$$\quad\quad | \text{ \bf Dimension } D \text{ \bf includes level } L$$
$$\quad\quad | \text{ \bf Dimension } D \text{ \bf includes } H_D$$
$$L_D ::= L_{D_1}; L_{D_2}$$
$$\quad\quad | \text{ \bf Level } L \text{ \bf has } A_D$$
$$\quad\quad | \text{ \bf Level } L \text{ \bf with spatiality } sp \text{ \bf has } A_D$$
$$\quad\quad | \text{ \bf Level } L \text{ \bf with temporality } T_S \text{ \bf has } A_D$$
$$PC_D ::= PC_{D_1}; PC_{D_2}$$
$$\quad\quad | \text{ \bf P-C relationship } PC \text{ \bf involves } L_1, L_2$$
$$\quad\quad | \text{ \bf P-C relationship } PC \text{ \bf involves } L_1, L_2$$
$$\quad\quad\quad \text{ \bf has distributing factor}$$
$$\quad\quad | \text{ \bf P-C relationship } PC \text{ \bf involves } L_1, L_2$$
$$\quad\quad\quad \text{ \bf with temporality } T_S$$
$$\quad\quad | \text{ \bf P-C relationship } PC \text{ \bf involves } L_1, L_2$$
$$\quad\quad\quad \text{ \bf with temporality } T_S \text{ \bf has distributing factor}$$
$$F_D ::= F_{D_1}; F_{D_2}$$
$$\quad\quad | \text{ \bf Fact relationship } F \text{ \bf involves } I_S$$

$\qquad$ | **Fact relationship** F **involves** I_S **has** M_D

$IC_D ::= IC_{D_1} ; IC_{D_2}$

$\qquad$ | K **is primary key of** L

$\qquad$ | **Participation of** L **in** PC **is** (min, max)

$\qquad$ | **Instant participation of** L **in** PC **is** (min, max)

$\qquad$ | **Lifespan participation of** L **in** PC **is** (min, max)

$\qquad$ | **Exclusive participation of** L **in** Q

$\qquad$ | **Topological relationship** L_1 *topo* L_2 **in** PC

$\qquad$ | **Synchronization relationship** L_1 *sync* L_2 **in** PC

$\qquad$ | **Topological relationship** *topo* **between** I_S **in** F

$\qquad$ | **Synchronization relationship** *sync* **between** I_S **in** F

$H_D ::= H_{D_1}, H_{D_2}$

$\qquad$ | **hierarchy** H **composed of** Q

$A_D ::= A_{D_1}, A_{D_2}$

$\qquad$ | **Attribute** A **of** A'_D

$\qquad$ | **Derived attribute** A **of** A'_D

$A'_D ::= \mathbf{type}\ \ d$

$\qquad$ | **spatial type** sp

$\qquad$ | **type** d **with temporality** T_S

$M_D ::= M_{D_1}, M_{D_2}$

$\qquad$ | **Measure** M **of** A'_D **is** *add*

$\qquad$ | **Derived measure** M **of** A'_D **is** *add*

$I_S ::= I_{S_1}, I_{S_2}$

$\qquad$ | L

$\qquad$ | L **as** *role*

$T_S ::= T_{S_1}, T_{S_2}$

$\qquad$ | $(temp, t, g)$

$d ::= int \mid real \mid string$

$add ::= additive \mid semiadditive \mid nonadditive$

$sp ::= Geo \mid SimpleGeo \mid ComplexGeo \mid Point \mid Line \mid OrientedLine$

$\qquad$ | $Surface \mid SimpleSurface \mid PointSet \mid LineSet \mid OrientedLineSet$

$\qquad$ | $SurfaceSet \mid SimpleSurfaceSet$

$temp ::= LS \mid VT \mid TT \mid LT$

$t ::= Time \mid SimpleTime \mid ComplexTime \mid Instant \mid Interval$

$\qquad$ | $InstantSet \mid IntervalSet$

$g ::= sec \mid min \mid hour \mid day \mid month \mid year$

$topo ::= meets \mid overlaps \mid intersects \mid contains \mid inside \mid covers \mid$

$\qquad coveredBy \mid equals \mid disjoint \mid crosses$

$sync ::= meets \mid overlaps \mid intersects \mid contains \mid inside \mid covers \mid$
$\qquad coveredBy \mid equals \mid disjoint \mid starts \mid finishes \mid precedes \mid succeeds$

A.5 Examples Using the Abstract Syntax

In this section we show examples of the textual representation of schemas for a conventional, a spatial, and a temporal data warehouse. For brevity, only part of the textual representation is given.

A.5.1 Conventional Data Warehouse

We give next the textual representation of the schema shown in Fig. A.1.

Level Definitions

Level *Customer* **has**
    Attribute *Customer id* **of type** *integer*,
    Attribute *Customer name* **of type** *string*,
    ... ;
Level *Product* **has**
    Attribute *Product number* **of type** *integer*,
    Attribute *Product name* **of type** *string*,
    ... ;
Level *Store* **has**
    Attribute *Store number* **of type** *integer*,
    Attribute *Store name* **of type** *string*,
    ... ;
...

Parent-Child Relationship Definitions

P-C relationship *CustSect* **involves** *Customer, Sector*;
P-C relationship *CustProf* **involves** *Customer, Profession*;
P-C relationship *SectBra* **involves** *Sector, Branch*;
P-C relationship *ProfBra* **involves** *Profession, Branch*;
P-C relationship *DateMonth* **involves** *Date, Month*;
P-C relationship *MonthYear* **involves** *Month, Year*;
P-C relationship *EmpSect1* **involves** *Employee, Section*
    has distributing factor;
P-C relationship *EmpSect2* **involves** *Employee, Section*;

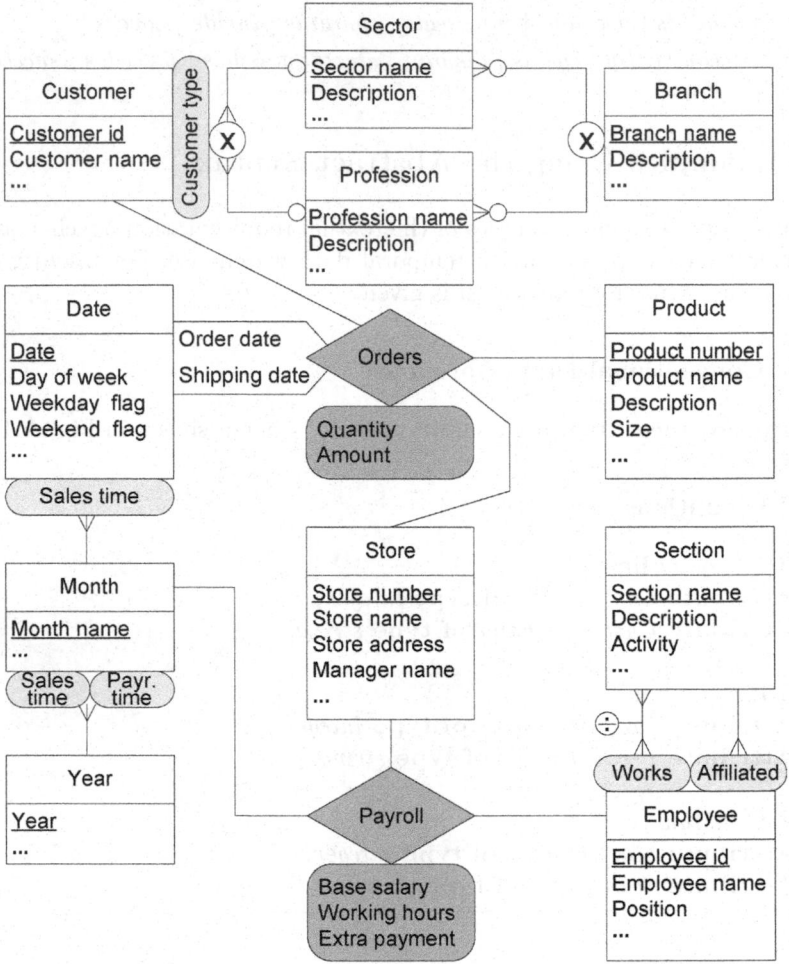

Fig. A.1. Example of a multidimensional schema containing several hierarchies

Dimension Definitions

Dimension *Customer* **includes**
  hierarchy *Customer type* **composed of** *CustSect, CustProf, SectBra, ProfBra*;
Dimension *Date* **includes**
  hierarchy *Sales time* **composed of** *DateMonth, MonthYear*;
Dimension *Month* **includes**
  hierarchy *Payr. time* **composed of** *MonthYear*;
Dimension *Product* **includes level** *Product*;
Dimension *Store* **includes level** *Store*;

Dimension *Employee* **includes**
> **hierarchy** *Works* **composed of** *EmpSect1*,
> **hierarchy** *Affiliated* **composed of** *EmpSect2*;

Fact Relationship Definitions

Fact relationship *Orders* **involves**
> *Time* **as** *Order date*, *Time* **as** *Shipping date*, *Product*, *Customer*, *Store*

has
> **Measure** *Quantity* **of type** *int* **is** *additive*,
> **Measure** *Amount* **of type** *real* **is** *additive*;

Fact relationship *Payroll* **involves**
> *Store*, *Month*, *Employee*

has
> **Measure** *Base salary* **of type** *real* **is** *additive*,
> **Measure** *Working hours* **of type** *real* **is** *additive*,
> **Measure** *Extra payment* **of type** *real* **is** *additive*;

Constraint Definitions

Customer id **is primary key of** *Customer*;
Sector name **is primary key of** *Sector*;
. . .

Exclusive participation of *Customer* **in** *CustSect*, *CustProf*;
Exclusive participation of *Branch* **in** *SectBra*, *ProfBra*;
Participation of *Customer* **in** *CustSect* **is** $(0, 1)$;
Participation of *Sector* **in** *CustSect* **is** $(1, n)$;
Participation of *Customer* **in** *CustProf* **is** $(0, 1)$;
Participation of *Profession* **in** *CustProf* **is** $(1, n)$;
. . .

A.5.2 Spatial Data Warehouse

The textual representation of the schema shown in Fig. A.2 is given next.

Level Definitions

Level *State* **with spatiality** *SurfaceSet* **has**
> **Attribute** *State name* **of type** *string*,
> **Attribute** *State population* **of type** *integer*,
> **Attribute** *State area* **of type** *real*,
> **Attribute** *State major activity* **of type** *string*,
> **Attribute** *Capital name* **of type** *string*;
> **Attribute** *Capital location* **of spatial type** *Point*;
> . . . ;

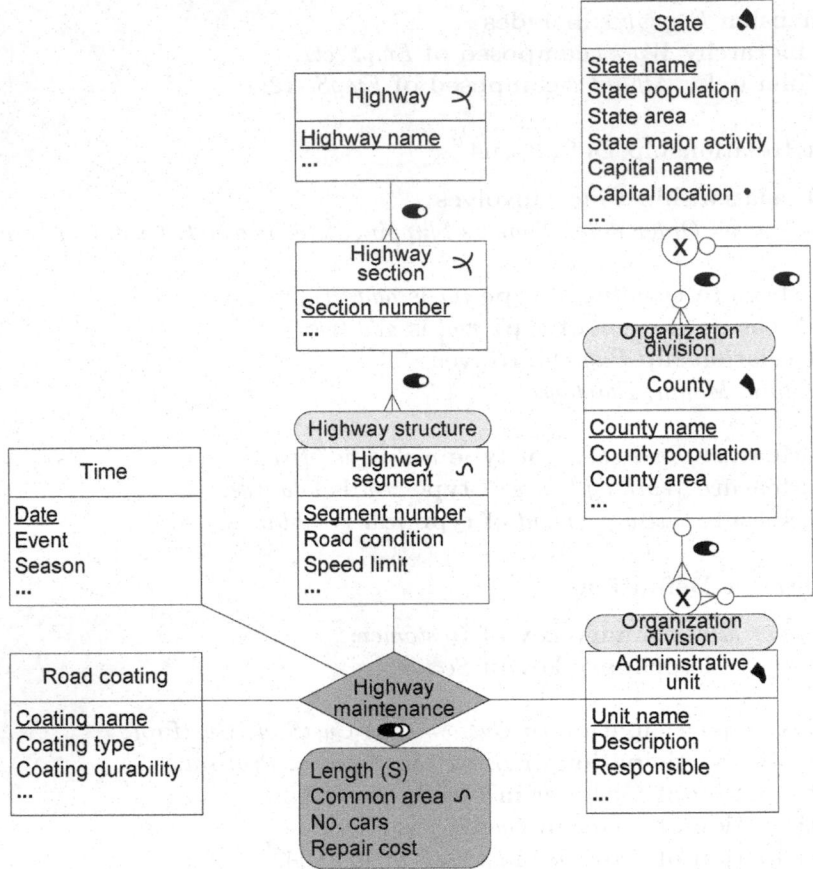

Fig. A.2. Example of a multidimensional schema including spatial and nonspatial
elements

Level *County* **with spatiality** *Surface* **has**
    Attribute *County name* **of type** *string*,
    Attribute *County population* **of type** *integer*,
    Attribute *County area* **of type** *real*;
    ... ;
Level *Highway* **with spatiality** *LineSet* **has**
    Attribute *Highway name* **of type** *string*,
    ... ;
Level *Highway section* **with spatiality** *LineSet* **has**
    Attribute *Section number* **of type** *string*,
    ... ;

Level *Highway segment* **with spatiality** *Line* **has**
 Attribute *Segment number* **of type** *string*,
 Attribute *Road condition* **of type** *string*,
 Attribute *Speed limit* **of type** *int*;
 ... ;
...

Parent-Child Relationship Definitions

P-C relationship *HSegHSec* **involves** *Highway segment*,
 Highway section;
P-C relationship *HSecH* **involves** *Highway section*, *Highway*;
P-C relationship *AdmCoun* **involves** *Administrative unit*, *County*;
P-C relationship *AdmSta* **involves** *Administrative unit*, *State*;
P-C relationship *CounSta* **involves** *County*, *State*;

Dimension Definitions

Dimension *Road coating* **includes level** *Road coating*;
Dimension *Time* **includes level** *Time*;
Dimension *Highway segment* **includes**
 hierarchy *Highway structure* **composed of** *HSegHSec*, *HSecH*;
Dimension *Administrative unit* **includes**
 hierarchy *Organization division* **composed of** *AdmCoun*, *AdmSta*,
 CounSta;

Fact Relationship Definitions

Fact relationship *Highway maintenance* **involves**
 Road coating, *Time*, *Highway segment*, *Administrative unit*
has
 Measure *Length* **of type** *real* **is** *additive*,
 Measure *Common area* **of spatial type** *Line* **is** *nonadditive*,
 Measure *No. cars* **of type** *integer* **is** *additive*,
 Measure *Repair cost* **of type** *real* **is** *additive*;

Constraint Definitions

Segment number **is primary key of** *Highway segment*;
County name **is primary key of** *County*;
...

Participation of *Highway segment* **in** *HSegHSec* **is** $(1, 1)$;
Participation of *Highway section* **in** *HSegHSec* **is** $(1, n)$;
Participation of *Highway section* **in** *HSecH* **is** $(1, 1)$;

Participation of *Highway* **in** *HSecH* **is** $(1, n)$;
. . .

Exclusive participation of *Administrative unit* **in** *AdmCoun, AdmState*;
Exclusive participation of *State* **in** *CounSta, AdmState*;
Topological relationship *Highway segment coveredBy Highway section*
     in *HSegHSec*;
Topological relationship *Highway section coveredBy Highway* **in** *HSecH*;
Topological relationship *County coveredBy State* **in** *CounSta*;
Topological relationship *Administrative unit coveredBy State* **in** *AdmSta*;
Topological relationship *overlaps* **between** *Highway segment,*
     Administrative unit **in** *Highway maintenance*;

A.5.3 Temporal Data Warehouse

The textual representation of the schema shown in Fig. A.3 is given next.

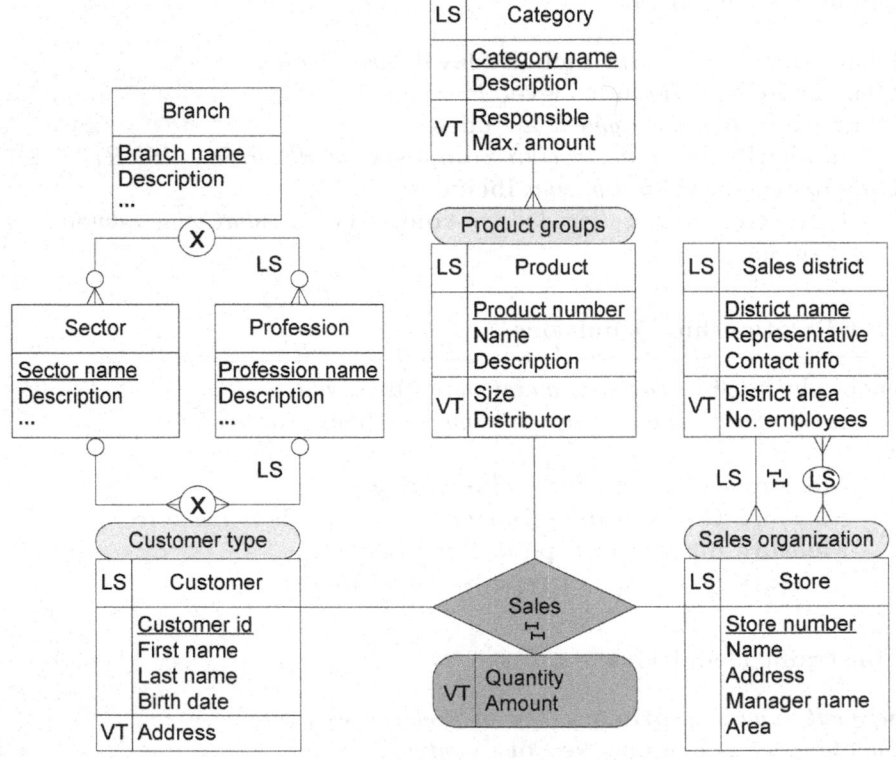

Fig. A.3. Example of a multidimensional schema including temporal and
nontemporal elements

Level Definitions

Level *Branch* **has**
    Attribute *Branch name* **of type** *string*,
    Attribute *Description* **of type** *string*,
    ... ;
Level *Sector* **has**
    Attribute *Sector name* **of type** *string*,
    Attribute *Description* **of type** *string*,
    ... ;
Level *Profession* **has**
    Attribute *Profession name* **of type** *string*,
    Attribute *Description* **of type** *string*,
    ... ;
Level *Customer* **with temporality** ($LS, IntervalSet, month$) **has**
    Attribute *Customer id* **of type** *integer*,
    Attribute *First name* **of type** *string*,
    Attribute *Last name* **of type** *string*,
    Attribute *Birth date* **of type** *string*,
    Attribute *Address* **of type** *string*
      with temporality ($VT, IntervalSet, month$),
Level *Product* **with temporality** ($LS, IntervalSet, month$) **has**
    Attribute *Product number* **of type** *integer*,
    Attribute *Name* **of type** *string*,
    Attribute *Description* **of type** *string*,
    Attribute *Size* **of type** *real*
      with temporality ($VT, IntervalSet, month$),
    Attribute *Distributor* **of type** *string*
      with temporality ($VT, IntervalSet, month$);
Level *Store* **with temporality** ($LS, IntervalSet, year$) **has**
    Attribute *Store number* **of type** *integer*,
    Attribute *Name* **of type** *string*,
    Attribute *Address* **of type** *string*,
    Attribute *Manager name* **of type** *string*,
    Attribute *Area* **of type** *string*;
... ;

Parent-Child Relationship Definitions

P-C relationship *SectBra* **involves** *Sector*, *Branch*;
P-C relationship *ProfBra* **involves** *Profession*, *Branch*
    with temporality ($LS, IntervalSet, month$);
P-C relationship *CustSect* **involves** *Customer*, *Sector*;
P-C relationship *CustProf* **involves** *Customer*, *Profession*
    with temporality ($LS, IntervalSet, month$);

P-C relationship *ProdCat* **involves** *Product, Category*;
P-C relationship *StoSD* **involves** *Store, Sales district*
  with temporality (*LS, IntervalSet, month*);

Dimension Definitions

Dimension *Product* **includes**
  hierarchy *Product groups* **composed of** *ProdCat*;
Dimension *Customer* **includes**
  hierarchy *Customer type* **composed of** *CustSect, CustProf,*
    SectBra, ProfBra;
Dimension *Store* **includes**
  hierarchy *Sales organization* **composed of** *StoSD*;

Fact Relationship Definitions

Fact relationship *Sales* **involves**
  Customer, Product, Store
has
  Measure *Quantity* **of type** *int*
    with temporality (*VT, Instant, month*) **is** *additive,*
  Measure *Amount* **of type** *real*
    with temporality (*VT, Instant, month*) **is** *additive*;

Constraint Definitions

Customer id **is primary key of** *Customer*;
Product number **is primary key of** *Product*;
Store number **is primary key of** *Store*;
  . . .
Instant participation of *Product* **in** *ProdCat* **is** $(1,1)$;
Lifespan participation of *Category* **in** *ProdCat* **is** $(1,n)$;
Instant participation of *Category* **in** *ProdCat* **is** $(1,1)$;
Lifespan participation of *Category* **in** *ProdCat* **is** $(1,n)$;
  . . .
Exclusive participation of *Customer* **in** *CustSect, CustProf*;
Exclusive participation of *Branch* **in** *SectBra, ProfBra*;
Synchronization relationship *Store overlaps Sales district* **in** *StoreSD*;
Synchronization relationship *overlaps* **between** *Customer, Product, Store*
  in *Sales*;

A.6 Semantics

In this section, we define the semantics of the textual representation of the
MultiDim model. We begin by defining the semantics of the predefined data

types, and the models of space and time. Then, after presenting some auxiliary functions, we give functions defining the semantics of the various components of the model.

A.6.1 Semantics of the Predefined Data Types

The semantics of the predefined data types is given by three functions:

- A function $\mathcal{D}[\![DATA]\!] \in TF$ such that $\mathcal{D}[\![DATA]\!] : DATA \to SET$. We assume $\forall d \in DATA \ (\perp \in \mathcal{D}[\![DATA]\!](d))$, where $\perp$ represents an undefined value indicating an incorrect use of a function or an error.
- A function $\mathcal{D}[\![OPNS]\!] \in TF$ such that $\mathcal{D}[\![OPNS]\!] : OPNS \to TF$ and $\sigma : d_1 \times \ldots \times d_n \to d$ implies $\mathcal{D}[\![OPNS]\!](\sigma) : \mathcal{D}[\![DATA]\!](d_1) \times \ldots \times \mathcal{D}[\![DATA]\!](d_n) \to \mathcal{D}[\![DATA]\!](d) \in DATA$ for every $d \in DATA$.
- A function $\mathcal{D}[\![PRED]\!] \in TF$ such that $\mathcal{D}[\![PRED]\!] : PRED \to REL$ and $\pi : d_1 \times \ldots \times d_n \to d$ implies $\mathcal{D}[\![PRED]\!](\pi) \subseteq \mathcal{D}[\![DATA]\!](d_1) \times \ldots \times \mathcal{D}[\![DATA]\!](d_n) \to \mathcal{D}[\![DATA]\!](d) \in DATA$ for every $d \in DATA$.

For example, the semantics of the predefined data types and one of their operators is defined as follows:

$$\mathcal{D}[\![DATA]\!](int) = \mathbb{Z} \cup \{\perp\}$$
$$\mathcal{D}[\![DATA]\!](real) = \mathbb{R} \cup \{\perp\}$$
$$\mathcal{D}[\![DATA]\!](string) = \mathbb{A}^* \cup \{\perp\}$$
$$\mathcal{D}[\![+_i]\!] : \mathcal{D}[\![DATA]\!](int) \times \mathcal{D}[\![DATA]\!](int) \to \mathcal{D}[\![DATA]\!](int)$$
$$= \begin{cases} i_1 \times i_2 \to i_1 + i_2 & \text{if } i_1, i_2 \in \mathbb{Z} \\ \perp & \text{otherwise} \end{cases}$$

A.6.2 The Space Model

We use a Euclidean plane $\mathbb{R}^2$ as a basis for modeling spatial objects. The formal framework employs point set theory and point set topology to define the point sets that are admissible for the various spatial data types. The following formalization is based on [267], which defines three types for complex points, complex lines, and complex regions. A complex point may include several points; a complex line may be composed of several lines composing a network; and a complex region may be a multipart region, possibly with holes.

A value of type *point* is defined as a finite set of isolated points in the plane, i.e.,

$$point = \{P \subset \mathbb{R}^2 \mid P \text{ is finite}\}.$$

A value of this type is called a *complex point*. If $P \in point$ is a singleton, i.e., $\|P\| = 1$, P is called a *single point*.

To define simple lines, we need a few definitions. We assume a Euclidean distance function $d : \mathbb{R}^2 \times \mathbb{R}^2 \to \mathbb{R}$, where $d(p, q) = d((x_1, x_2), (y_1, y_2)) =$

$\sqrt{(x_1 - x_2)^2 + (y_1 - y_2)^2}$. The Euclidean distance allows us to define a neighborhood of a point in $\mathbb{R}^2$. Let $q \in \mathbb{R}^2$ and $\epsilon \in \mathbb{R}^+$. The set $N_\epsilon(q) = \{p \in \mathbb{R}^2 \mid d(p,q) < \epsilon\}$ is called the *open neighborhood of radius ϵ and center q*. Any open neighborhood with center q is denoted by $N(q)$.

We define next the notion of a *continuous mapping* which preserves neighborhood relations between mapped points in two spaces of the plane. These mappings are also called *topological transformations* and include translation, rotation, and scaling. Let $X \subset \mathbb{R}$ and $f : X \to \mathbb{R}^2$. Then f is said to be *continuous at a point* $x_0 \in X$ if, given an arbitrary number $\epsilon > 0$, there exists a number $\delta > 0$ (usually depending on ϵ) such that for every $x \in N_\delta(x_0) \cap X$ we obtain the result that $f(x) \in N_\epsilon(f(x_0))$. The mapping f is said to be *continuous on X* if it is continuous at every point of X. For a function $f : X \to Y$ and a set $A \subseteq X$, we introduce the notation $f(A) = \{f(x) \mid x \in A\}$. Further, let $\mathbb{N}_0 = \mathbb{N} \cup \{0\}$.

Complex lines are defined as the union of the images of a finite number of continuous mappings. The spatial data type *line* is defined as

$line = \{L \in \mathbb{R}^2 \mid$ (i) $L = \bigcup_{i=1}^{n} f_i([0,1])$ with $n \in \mathbb{N}_0$
                  (ii) $\forall 1 \leq i \leq n : f_i : [0,1] \to \mathbb{R}^2$ is a continuous mapping
                  (ii) $\forall 1 \leq i \leq n : |f_i([0,1])| > 1\}$.

A value of this type is called a *complex line*. Each f_i defines a component of the complex line, where the values $f_i(0)$ and $f_i(1)$ are called the *end points* of f_i. The first condition allows a line geometry to be the empty set (if $n = 0$). The third condition avoids degenerate line geometries consisting of only a single point.

Particular cases of lines can be distinguished. A *single-component line* models all curves that can be drawn on a sheet of paper from a starting point to an end point without lifting the pen. Figure A.4 shows some examples. A *simple line* is obtained if, during the drawing process, the pen does not meet an already occupied point of the line. In a *self-touching line*, one of its end points touches its interior, while in a *closed line*, the end points coincide. Finally, a *non-self-intersecting line* is either a simple line or a closed line.

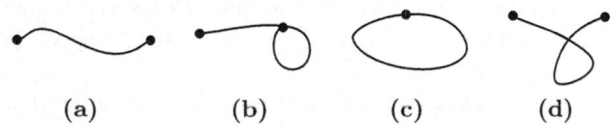

(a)   (b)   (c)   (d)

Fig. A.4. Examples of single-component lines, where the thick points indicate the start and end points: (a) a simple line; (b) a self-touching line; (c) a closed line; and (d) a self-intersecting line

The above definition of lines determines the point set of a line. However, this definition is too general, since it allows multiple representations of the same geometry. For example, the self-intersecting line in Fig. A.4d can be represented by a single continuous mapping or by three continuous mappings, one for the closed part and two others for the remaining parts. Reference [267] provides another structured definition, based on non-self-intersecting lines, that gives a unique representation of lines. We refer to [267] for the precise definition. This canonical representation of lines is then used to define the boundary of a complex line, which is the set of end points of all non-self-intersecting lines.

In order to define complex regions, a few concepts are needed. Let $X \subseteq \mathbb{R}^2$ and $q \in \mathbb{R}^2$. q is an *interior point of* X if there exists a neighborhood N such that $N(q) \subseteq X$. q is a *exterior point of* X if there exists a neighborhood N such that $N(q) \cap X = \emptyset$. q is a *boundary point of* X if a is neither an interior nor an exterior point of X. q is a *closure point of* X if q is either an interior or a boundary point of X. The sets of all interior, exterior, and boundary points of X are called, respectively, the *interior*, the *exterior*, and the *boundary* of X and are denoted by $I(X)$, $E(X)$, and $B(X)$. Finally, the set of all closure points of X is called the *closure of* X, denoted by $C(X)$, and is given by $C(X) = I(X) \cup B(X)$.

A point q is a *limit point* of X if, for every neighborhood $N(q)$, it holds that $(N - \{q\}) \cap X \neq \emptyset$. X is called an *open set* in $\mathbb{R}^2$ if $X = I(X)$. X is called a *closed set* in $\mathbb{R}^2$ if every limit point of X is a point of X. Arbitrary point sets do not necessarily form a region. However, open and closed point sets in $\mathbb{R}^2$ are also inadequate models for complex regions, since they can suffer from undesired geometric anomalies. An open point set may have missing lines and points in the form of cuts and punctures, i.e., its boundary is missing. A complex region defined as a closed point set admits isolated or dangling points and lines. Regular closed point sets avoid these anomalies. Let $X \subseteq \mathbb{R}^2$. X is called a *regular closed point set* if $X = C(I(X))$. The effect of the interior operation is to eliminate dangling points, dangling lines, and boundary parts. The effect of the closure operation is to eliminate cuts and punctures by appropriately supplementing the set with points and adding the boundary. For example, closed neighborhoods, as defined above, are regular closed sets.

For the specification of the *region* data type, it is necessary to define bounded and connected sets. Two sets $X, Y \subseteq \mathbb{R}^2$ are said to be *separated* if $X \cap C(X) = \emptyset = C(X) \cap Y$. A set $X \subseteq \mathbb{R}^2$ is *connected* if it is not the union of two nonempty separated sets. Let $q = (x, y) \in \mathbb{R}^2$. The *norm* of q is then defined as $||q|| = \sqrt{x^2 + y^2}$. A set $X \subseteq \mathbb{R}^2$ is said to be *bounded* if there exists a number $r \in \mathbb{R}^+$ such that $||q|| < r$ for every $q \in X$.

The definition of the spatial data type *region* is as follows:

$region = \{$ (i) R is regular closed

    (ii) R is bounded

    (iii) The number of connected sets of R is finite $\}$.

A value of this type is called a *complex region*. Note that the geometry of a region can also be empty (an empty set).

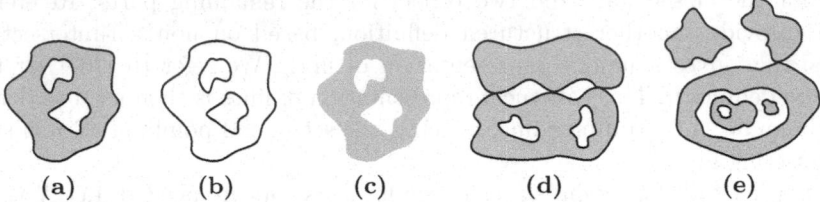

(a)   (b)   (c)   (d)   (e)

Fig. A.5. Examples of complex regions: (a) a face with two holes; (b) its boundary; (c) its interior; (d) a complex region with two faces; and (e) a complex region with five faces

Particular cases of regions can be distinguished. A *simple region* is a bounded, regular closed set homeomorphic (i.e., topologically equivalent) to a closed neighborhood in $\mathbb{R}^2$. Simple regions do not consist of several components and do not contain holes. A *face* is simple region with holes. A face with two holes is shown in Fig. A.5a, and its boundary and its interior are shown in Figs. A.5b,c, respectively. Finally, a *complex region* is a finite set of faces that are disjoint, or that meet in one or several single boundary points, or lie inside a hole in another face and possibly share one or several single boundary points with the boundary of the hole. Faces that have connected boundary parts in common with other faces or holes are disallowed. Figures A.5d,e show examples of a region with two and five faces, respectively.

As was the case for complex lines, the above definition of complex regions allows multiple representations of the same geometry. For example, the geometry of Fig. A.5d can be represented by two component regions, the lower one having two holes, or by a single-component region with four holes. Reference [267] provides a more structured definition based on simple regions with holes that allows a unique representation of complex regions. Further, this structured definition allows one to define the interior, the boundary, and the exterior of a complex region.

On the basis of the definitions of the boundary, interior, and exterior of a complex point, a complex line, and a complex surface, the topological predicates can be defined as follows.

$a \; meets \; b \Leftrightarrow (I(a) \cap I(b) = \emptyset) \wedge (a \cap b \neq \emptyset)$

$a \; crosses \; b \Leftrightarrow (Dim(I(a) \cap I(b)) < \max(Dim(I(a)), Dim(I(b))) \wedge$
$\qquad (a \cap b \neq a) \wedge (a \cap b \neq b) \wedge (a \cap b \neq \emptyset)$

$a \; overlaps \; b \Leftrightarrow Dim(I(a)) = Dim(I(b)) = Dim(I(a) \cap I(b)) \wedge$
$\qquad (a \cap b \neq a) \wedge (a \cap b \neq b)$

$a \; contains \; b \Leftrightarrow (I(a) \cap I(b) \neq \emptyset) \wedge (a \cap b = b) \Leftrightarrow b \; inside \; a$

$a \; covers \; b \Leftrightarrow (a \cap b = b) \Leftrightarrow (b - a) = \emptyset \Leftrightarrow a \; coveredBy \; b$

$a \; disjoint \; b \Leftrightarrow (a \cap b = \emptyset) \Leftrightarrow \neg a \; intersects \; b$

$a \; equals \; b \Leftrightarrow (a \cap b = a) \wedge (a \cap b = b) \Leftrightarrow (a - b) \cup (b = a) = \emptyset$

We can now define the semantics of our hierarchy of spatial data types given in Sect. 4.1.2. We define the following value domains:

D_{point} = the domain of simple points
D_{line} = the domain of single-component lines
D_{region_h} = the domain of simple regions with holes
D_{region} = the domain of simple regions

We can now define the semantics of the spatial data types as follows:

$\mathcal{D}[\![SPATIAL]\!](Geo) =$
$\quad \mathcal{D}[\![SPATIAL]\!](SimpleGeo) \cup \mathcal{D}[\![SPATIAL]\!](ComplexGeo)$
$\mathcal{D}[\![SPATIAL]\!](SimpleGeo) = \mathcal{D}[\![SPATIAL]\!](Point) \cup$
$\quad \mathcal{D}[\![SPATIAL]\!](Line) \cup \mathcal{D}[\![SPATIAL]\!](OrientedLine) \cup$
$\quad \mathcal{D}[\![SPATIAL]\!](Surface) \cup \mathcal{D}[\![SPATIAL]\!](SimpleSurface)$
$\mathcal{D}[\![SPATIAL]\!](ComplexGeo) = \mathcal{D}[\![SPATIAL]\!](PointSet) \cup$
$\quad \mathcal{D}[\![SPATIAL]\!](LineSet) \cup \mathcal{D}[\![SPATIAL]\!](OrientedLineSet) \cup$
$\quad \mathcal{D}[\![SPATIAL]\!](SurfaceSet) \cup \mathcal{D}[\![SPATIAL]\!](SimpleSurfaceSet)$
$\mathcal{D}[\![SPATIAL]\!](Point) = D_{point}$
$\mathcal{D}[\![SPATIAL]\!](Line) = D_{line}$
$\mathcal{D}[\![SPATIAL]\!](OrientedLine) = D_{line}$
$\mathcal{D}[\![SPATIAL]\!](Surface) = D_{region\_h}$
$\mathcal{D}[\![SPATIAL]\!](SimpleSurface) = D_{region}$
$\mathcal{D}[\![SPATIAL]\!](PointSet) = 2^{D_{point}}$
$\mathcal{D}[\![SPATIAL]\!](LineSet) = 2^{D_{line}}$
$\mathcal{D}[\![SPATIAL]\!](OrientedLineSet) = 2^{D_{line}}$
$\mathcal{D}[\![SPATIAL]\!](SurfaceSet) = 2^{D_{region\_h}}$
$\mathcal{D}[\![SPATIAL]\!](SimpleSurfaceSet) = 2^{D_{region}}$

A.6.3 The Time Model

We assume that the real time line is represented in the database by a baseline clock that is discrete and bounded at both ends [71, 86, 135]. Time domains are then ordered, finite sets of elements isomorphic to finite subsets of the integer numbers. The nondecomposable elements of a time domain are called *chronons*. Depending on the requirements of the application, consecutive chronons can be grouped into larger units called *granules*, such as seconds, minutes, or days. We denote a granule by g. The granule g_{now} denotes the granule representing the current time. The *granularity* represents the size of a granule, i.e., it is the time unit used for specifying the duration of a granule.

Following Gregersen and Jensen [86], we include a domain for each combination of the temporality types $temp \in \{LS, VT, TT, LT\}$ and the granularities g. These domains are denoted by $D_{temp}^g = \{g_1^{temp}, g_2^{temp}, \ldots, g_n^{temp}\}$, for example $D_{VT}^{month} = \{Jan, Feb, Mar, \ldots, Dec\}$. The domain of each temporal type is the union of the domains represented by different granularities, i.e., $D_{temp} = \bigcup_g (D_{temp}^g)$, for example $D_{TT} = \bigcup_g (D_{TT}^g)$ for TT.

Real-world instants are represented by granules according to the chosen granularity; for example, we may have a granule $g^{day} = 02/10/2006$ using a granularity of a day. A time interval is defined as the time between two instants called the *begin* and *end* instants, i.e., $[g_{begin}, g_{end}]^g$, for example $[25/09/2006, 02/10/2006]^{day}$. Thus, a time interval is a sequence of consecutive granules between the starting (g_{begin}) and ending (g_{end}) granules with a granularity g, for example all days between $25/09/2006$ and $02/10/2006$. We also use sets of instants and sets of intervals. A set of instants over time domains is a finite union of instants, i.e., $IS^g = g_1^g \cup \ldots \cup g_n^g$. Furthermore, an interval set, or a temporal element, over time domains is a finite union of intervals, i.e., $TE^g = [g_{begin_1}, g_{end_1}]^g \cup \ldots \cup [g_{begin_n}, g_{end_n}]^g$.

A.6.4 Semantic Domains

The MultiDim model includes the following value domains:

$D_S \cup \{\bot\}$ – the set of surrogates
$D_S^L \subseteq D_S$ – the set of surrogates assigned to $L \in Levels$
$D_S^{PC} \subseteq D_S$ – the set of surrogates assigned to $PC \in PCRels$
$D_{LS} = \bigcup_g (D_{LS}^g) \cup \{\bot\}$ – the lifespan domain
$D_{VT} = \bigcup_g (D_{VT}^g) \cup \{\bot\}$ – the valid time domain
$D_{TT} = \bigcup_g (D_{TT}^g) \cup \{\bot\}$ – the transaction time domain
$D_{LT} = \bigcup_g (D_{LT}^g) \cup \{\bot\}$ – the loading time domain
$\mathcal{D}[\![DATA]\!]$ – the set of basic domains
$\mathcal{D}[\![SPATIAL]\!]$ – the set of spatial domains

A.6.5 Auxiliary Functions

This section presents some auxiliary functions required for defining the semantic functions.

The function *attOf* takes as its argument a level declaration or an attribute declaration and returns the attribute names:

$attOf(\textbf{Level } L \textbf{ has } A_D) =$
$\quad attOf(\textbf{Level } L \textbf{ with spatiality } sp \textbf{ has } A_D) =$
$\quad\quad attOf(\textbf{Level } L \textbf{ with temporality } T_S \textbf{ has } A_D) = attOf(A_D)$
$attOf(A_{D_1}, A_{D_2}) = attOf(A_{D_1}) \cup attOf(A_{D_2})$
$attOf(\textbf{Attribute } A \textbf{ of } A'_D) = A$
$attOf(\textbf{Derived attribute } A \textbf{ of } A'_D) = A$

The function *measOf* takes as its argument a measure declaration and returns the measure names:

$measOf(M_{D_1}, M_{D_2}) = measOf(M_{D_1}) \cup measOf(M_{D_2})$
$measOf(\textbf{Measure } M \textbf{ of } A'_D \textbf{ is } add) =$
$\quad measOf(\textbf{Derived measure } M \textbf{ of } A'_D \textbf{ is } add) = M$

The function *tempOf* takes as its argument a temporal specification and returns the temporality types of the specification, i.e., a subset of $\{LS, VT, TT, LT\}$:

$tempOf(T_{S_1}, T_{S_2}) = tempOf(T_{S_1}) \cup tempOf(T_{S_2})$
$tempOf((temp, t, g)) = \{temp\}$

The function *instants* takes as its argument a temporal specification and returns the set of instant tuples of a temporal specification. More precisely, the function returns a set of functions t. The domain of each function is the set of temporality types of the specification. The value domain of the temporality types is their underlying temporal domain:

$instants(T_S) = \{t \mid t \in TF \wedge dom(t) = \{tempOf(T_S)\} \wedge$
$\quad \forall T_i \in tempOf(T_S)((T_i, t, g) \in T_S \wedge t[T_i] \in D^g_{T_i})\}$

The function *contains* takes as its arguments an instant tuple c of a temporal specification and a set of instances of a parent-child relationship and returns the subset of the instances that contain c in its temporal support:

$contains(c, \{t_1, \ldots, t_n\}) =$
$$\begin{cases} \emptyset & \text{if } n = 0 \\ contains(c, \{t_1, \ldots, t_{n-1}\}) \cup \{t_n\} & \text{if } \forall T_i \in dom(c)(c[T_i] \in t_n[T_i]) \\ contains(c, \{t_1, \ldots, t_{n-1}\}) & \text{otherwise} \end{cases}$$

The function *cnt* takes as its arguments a level member m, a level L, and a set of instances of a parent-child relationship and returns the number of tuples in the parent-child set in which the member m participates:

$cnt(e, L, \{t_1, \ldots, t_n\}) =$
$$\begin{cases} 0 & \text{if } n = 0 \\ cnt(e, L, \{t_1, \ldots, t_{n-1}\}) & \text{if } n \geq 1 \wedge t_n[s_L] \neq e \\ cnt(e, L, \{t_1, \ldots, t_{n-1}\}) + 1 & \text{if } n \geq 1 \wedge t_n[s_L] = e \end{cases}$$

The function *lifespan* takes as its arguments an identifier of a level member m and a level L and returns the lifespan of the member, if any, or the empty set otherwise:

$$lifespan(m, L) = \begin{cases} t[LS] & \text{if } LS \in tempOf(L) \wedge \exists t \in \mathcal{L}(L)(t[s] = m) \\ \emptyset & \text{otherwise} \end{cases}$$

The predicate *inSch* takes as its first argument the name of a level, of a parent-child relationship, or of a fact relationship, and a schema declaration as its second argument. It returns *true* if the element mentioned is declared in the schema and *false* otherwise:

$$inSch(L, S_D) = inSch(L, D_D; L_D; PC_D; F_D; IC_D;) = inSch(L, L_D) =$$
$$\begin{cases} true & \text{if \textbf{Level} } L \textbf{ has } A_D \in L_D \\ true & \text{if \textbf{Level} } L \textbf{ with spatiality } sp \textbf{ has } A_D \in L_D \\ true & \text{if \textbf{Level} } L \textbf{ with temporality } T_S \textbf{ has } A_D \in L_D \\ false & \text{otherwise} \end{cases}$$

$$inSch(PC, S_D) = inSch(PC, D_D; L_D; PC_D; F_D; IC_D;) = inSch(PC, PC_D) =$$
$$\begin{cases} true & \text{if \textbf{P-C relationship} } PC \textbf{ involves } L_1, L_2 \in L_D \\ true & \text{if \textbf{P-C relationship} } PC \textbf{ involves } L_1, L_2 \\ & \textbf{has distributing factor } \in L_D \\ true & \text{if \textbf{P-C relationship} } PC \textbf{ involves } L_1, L_2 \\ & \textbf{with temporality } T_S \in L_D \\ true & \text{if \textbf{P-C relationship} } PC \textbf{ involves } L_1, L_2 \\ & \textbf{with temporality } T_S \textbf{ has distributing factor } \in L_D \\ false & \text{otherwise} \end{cases}$$

$$inSch(F, S_D) = inSch(F, D_D; L_D; PC_D; F_D; IC_D;) = inSch(F, F_D) =$$
$$\begin{cases} true & \text{if \textbf{Fact relationship} } F \textbf{ involves } I_S \in L_D \\ true & \text{if \textbf{Fact relationship} } F \textbf{ involves } I_S \textbf{ has } M_D \in L_D \\ false & \text{otherwise} \end{cases}$$

The function *parOf* takes as its argument the name of a parent-child relationship and returns the levels that participate in that relationship:

$$parOf(PC) =$$
$$\begin{cases} parOf(\textbf{P-C relationship } PC\ldots) & \text{if \textbf{P-C relationship} } PC\ldots \in S_D \\ \bot & \text{otherwise} \end{cases}$$
$$parOf(\textbf{P-C relationship } PC \textbf{ involves } L_1, L_2) =$$
$$parOf(\textbf{P-C relationship } PC \textbf{ involves } L_1, L_2 \textbf{ has } \ldots) =$$
$$parOf(\textbf{P-C relationship } PC \textbf{ involves } L_1, L_2 \textbf{ with } \ldots) = \{L_1, L_2\}$$

The function *tempSpec* takes as its argument the name of a parent-child relationship and returns the specification of its temporal support if the relationship is temporal, and the empty set otherwise:

$$tempSpec(PC) =$$
$$\begin{cases} \emptyset & \text{if \textbf{P-C} relationship } PC \text{ \textbf{involves} } L_1, L_2 \in E_D \\ \emptyset & \text{if \textbf{P-C} relationship } PC \text{ \textbf{involves} } L_1, L_2 \\ & \text{\textbf{has distributing factor}} \in E_D \\ T_S & \text{if \textbf{P-C} relationship } PC \text{ \textbf{involves} } L_1, L_2 \\ & \text{\textbf{with temporality} } T_S \in E_D \\ T_S & \text{if \textbf{P-C} relationship } PC \text{ \textbf{involves} } L_1, L_2 \\ & \text{\textbf{with temporality} } T_S \text{ \textbf{has distributing factor}} \in E_D \\ \bot & \text{otherwise} \end{cases}$$

Recall that a level may participate several times in a fact relationship, using different roles. The functions *role* and *level* take the involvement of a level in a fact relationship and provide the role name and the level name, respectively:

$$role(L) = L$$
$$role(L \text{ \textbf{as} } role) = role$$
$$level(L) = level(L \text{ \textbf{as} } role) = L$$

A.6.6 Semantic Functions

We give next the signatures and definitions of the semantic functions.

The semantic function $\mathcal{I}$ determines the surrogate sets of the levels that are involved in a fact relationship or in a parent-child relationship:

$$\mathcal{I} : Levels \rightarrow D_S^L$$
$$\mathcal{I}[\![I_{S_1}, I_{S_2}]\!] = \mathcal{I}[\![I_{S_1}]\!] \times \mathcal{I}[\![I_{S_2}]\!]$$
$$\mathcal{I}[\![L]\!] = \mathcal{I}[\![L \text{ \textbf{as} } role]\!] = \begin{cases} D_S^L & \text{if } L \in Levels \\ \bot & \text{otherwise} \end{cases}$$

The semantic function $\mathcal{T}$ determines the time domains of the temporal support specified for a given level, parent-child relationship, or attribute:

$$\mathcal{T} : TempSpec \rightarrow D_{VT} \cup D_{TT} \cup D_{LS} \cup D_{LT}$$
$$\mathcal{T}[\![T_{S_1}, T_{S_2}]\!] = \mathcal{T}[\![T_{S_1}]\!] \times \mathcal{T}[\![T_{S_2}]\!]$$
$$\mathcal{T}[\![(temp, Instant, g)]\!] = D_{temp}^g$$
$$\mathcal{T}[\![(temp, InstantSet, g)]\!] = 2^{D_{temp}^g}$$
$$\mathcal{T}[\![(temp, Interval, g)]\!] = 2^{D_{temp}^g}$$
$$\mathcal{T}[\![(temp, IntervalSet, g)]\!] = 2^{D_{temp}^g}$$

The semantic function $\mathcal{A}$ defines the value domains of attribute declarations. If an attribute is of a predefined data type or a spatial data type, then the value domain is that of the specified data type. If the attribute is of a type that includes temporal support, this indicates that the value of this attribute changes over time. Therefore, the value domain of this attribute is a function from a time domain to a value domain:

$$\mathcal{A} : \textit{Attributes} \times \textit{DATA} \times \textit{SPATIAL} \times \textit{TempSpec} \rightarrow$$
$$\mathcal{D}[\![DATA]\!] \cup \mathcal{D}[\![SPATIAL]\!] \cup (\mathcal{T}[\![T_S]\!] \rightarrow \mathcal{D}[\![DATA]\!])$$
$$\mathcal{A}[\![A_{D_1}, A_{D_2}]\!] = \mathcal{A}[\![A_{D_1}]\!] \times \mathcal{A}[\![A_{D_2}]\!]$$
$$\mathcal{A}[\![\textbf{Attribute } A \textbf{ of } A'_D]\!] = \mathcal{A}[\![A'_D]\!]$$
$$\mathcal{A}[\![\textbf{Derived attribute } A \textbf{ of } A'_D]\!] = \mathcal{A}[\![A'_D]\!]$$
$$\mathcal{A}[\![\textbf{type } d]\!] = \begin{cases} \mathcal{D}[\![DATA]\!](d) & \text{if } d \in DATA \\ \bot & \text{otherwise} \end{cases}$$
$$\mathcal{A}[\![\textbf{spatial type } sp]\!] = \begin{cases} \mathcal{D}[\![SPATIAL]\!](s) & \text{if } sp \in SPATIAL \\ \bot & \text{otherwise} \end{cases}$$
$$\mathcal{A}[\![\textbf{type } d \textbf{ with temporality } T_S]\!] =$$
$$\begin{cases} \mathcal{T}[\![T_S]\!] \rightarrow \mathcal{D}[\![DATA]\!](d) & \text{if } d \in DATA \wedge T_S \in TempSpec \\ \bot & \text{otherwise} \end{cases}$$

The semantic function $\mathcal{M}$ defines the value domains of measure declarations. Since measures are attributes of fact relationships, the function $\mathcal{M}$ is based on the semantic function $\mathcal{A}$:

$$\mathcal{M} : \textit{Measures} \times \textit{DATA} \times \textit{SPATIAL} \times \textit{TempSpec} \rightarrow$$
$$\mathcal{D}[\![DATA]\!] \cup \mathcal{D}[\![SPATIAL]\!] \cup (\mathcal{T}[\![T_S]\!] \rightarrow \mathcal{D}[\![DATA]\!])$$
$$\mathcal{M}[\![M_{D_1}, M_{D_2}]\!] = \mathcal{M}[\![M_{D_1}]\!] \times \mathcal{M}[\![M_{D_2}]\!]$$
$$\mathcal{M}[\![\textbf{Measure } M \textbf{ of } A'_D \textbf{ is } add]\!] = \mathcal{A}[\![A'_D]\!]$$
$$\mathcal{M}[\![\textbf{Derived measure } M \textbf{ of } A'_D \textbf{ is } add]\!] = \mathcal{A}[\![A'_D]\!]$$

The function $\mathcal{S}$ defines the semantics of a MultiDim schema composed of definitions of levels, parent-child relationships, fact relationships, and integrity constraints. It defines each component of the underlying database, and predicates that ensure validity and consistency of the database:

$$\mathcal{S} : S_D \rightarrow \mathcal{S}[\![S_D]\!]$$
$$\mathcal{S}[\![S_D]\!] = \mathcal{S}[\![L_D; PC_D; F_D; IC_D]\!]$$
$$\mathcal{S}[\![L_D; PC_D; F_D; IC_D]\!] = \mathcal{L}[\![L_D]\!] \uplus \mathcal{PC}[\![PC_D]\!] \uplus \mathcal{F}[\![F_D]\!] \uplus \mathcal{IC}[\![IC_D]\!]$$

The function $\mathcal{L}$ defines the semantics of levels. The attributes of a level have an associated value domain. The association between a set of attributes $A = \{A_1, A_2, \ldots, A_n\}$ and a set of value domains D is given by a function $dom : A \rightarrow D$. A member of a level together with its attributes can be seen as a tuple. A tuple t over a set of attributes A is actually a function that associates each attribute $A_i \in A$ with a value from the value domain $dom(A_i)$. For an attribute A, we denote this value by $t[A]$.

The semantics of a level is thus a set of functions t (tuples). The domain of each function t is the surrogate attribute s and the set of attribute names belonging to the level L. If this level has spatial support, this means that the members of the level have an associated geometry; in this case the domain of each function t has, in addition, an attribute *geom*. The value domain of the surrogate attribute s is the set D_S^L of surrogate values assigned to the level L, and the value domain of the geometry and of the attributes of the level L is determined by the semantics of the attribute declarations.

If the level has temporal support, this means that the database stores the lifespan, transaction time, and/or loading time for the members of the level. Recall that the lifespan indicates the time during which the corresponding real-world member exists, the transaction time refers to the time during which the member is current in the database, and the loading time refers to the time when the member was introduced into the data warehouse. Therefore, the timestamps recording these temporality types must be associated with the member.

The function $\mathcal{L}$, when applied to a composition of level definitions, returns the disjoint union of the functions applied to each component. This is because each level defines a unique set of tuples, which is stored separately in the database:

$\mathcal{L} : Levels \times SPATIAL \times TempSpec \times AttDecl \rightarrow \mathcal{I}[\![L]\!] \times \mathcal{A}[\![A_D]\!] \cup$
$\quad \mathcal{I}[\![L]\!] \times \mathcal{D}[\![Geo]\!] \times \mathcal{A}[\![A_D]\!] \cup \mathcal{I}[\![L]\!] \times \mathcal{T}[\![T_S]\!] \times \mathcal{A}[\![A_D]\!]$

$\mathcal{L}[\![L_{D_1}; L_{D_2}]\!] = \mathcal{L}[\![L_{D_1}]\!] \uplus \mathcal{L}[\![L_{D_2}]\!]$

$\mathcal{L}[\![\textbf{Level } L \textbf{ has } A_D]\!] =$
$\quad \{t \mid t \in TF \wedge dom(t) = \{s, attOf(A_D)\} \wedge t[s] \in D_S^L \wedge$
$\quad \forall A_i \in attOf(A_D) \ (t[A_i] \in \mathcal{A}[\![\textbf{Attribute } A_i \textbf{ of } A_D']\!] \vee$
$\quad t[A_i] \in \mathcal{A}[\![\textbf{Derived attribute } A_i \textbf{ of } A_D']\!])\}$

$\mathcal{L}[\![\textbf{Level } L \textbf{ with spatiality } sp \textbf{ has } A_D]\!] =$
$\quad \{t \mid t \in TF \wedge dom(t) = \{s, geom, attOf(A_D)\} \wedge t[s] \in D_S^L \wedge$
$\quad t[geom] = \mathcal{D}[\![SPATIAL]\!](sp) \wedge$
$\quad \forall A_i \in attOf(A_D) \ (t[A_i] \in \mathcal{A}[\![\textbf{Attribute } A_i \textbf{ of } A_D']\!] \vee$
$\quad t[A_i] \in \mathcal{A}[\![\textbf{Derived attribute } A_i \textbf{ of } A_D']\!])\}$

$\mathcal{L}[\![\textbf{Level } L \textbf{ with temporality } T_S \textbf{ has } A_D]\!] =$
$\quad \{t \mid t \in TF \wedge dom(t) = \{s, tempOf(T_S), attOf(A_D)\} \wedge t[s] \in D_S^L \wedge$
$\quad \forall T_i \in tempOf(T_S)(t[T_i] \in \mathcal{T}[\![(T_i, t, g)]\!]) \wedge$
$\quad \forall A_i \in attOf(A_D)(t[A_i] \in \mathcal{A}[\![\textbf{Attribute } A_i \textbf{ of } A_D']\!] \vee$
$\quad t[A_i] \in \mathcal{A}[\![\textbf{Derived attribute } A_i \textbf{ of } A_D']\!])\}$

The function $\mathcal{PC}$ define the semantics of parent-child relationships. A parent-child relationship relates a child and a parent level and may have, in addition, temporal support and/or a distributing factor. The semantics of the relationship is thus a set of tuples t relating a child and a parent member. Members are identified through their surrogates, with the value domain defined by $\mathcal{I}$. If the relationship includes a distributing factor, the domain of the function t includes additionally an attribute d; its value domain is the set of real numbers. If the relationship includes temporal support, the timestamps for the various temporality types must be kept.

Since each parent-child relationship defines a unique set of tuples, if the function $\mathcal{PC}$ is applied to a composition of parent-child relationship definitions, it returns the disjoint union of the functions applied to each component:

$\mathcal{PC} : PCRels \times Levels \times Levels \times TempSpec \rightarrow$
$\quad \mathcal{I}[\![L]\!] \times \mathcal{I}[\![L]\!] \cup \mathcal{I}[\![L]\!] \times \mathcal{I}[\![L]\!] \times \mathcal{D}[\![DATA]\!] \cup$
$\quad \mathcal{I}[\![L]\!] \times \mathcal{I}[\![L]\!] \times \mathcal{T}[\![T_S]\!] \cup \mathcal{I}[\![L]\!] \times \mathcal{I}[\![L]\!] \times \mathcal{T}[\![T_S]\!] \times \mathcal{D}[\![DATA]\!]$
$\mathcal{PC}[\![PC_{D_1}; PC_{D_2}]\!] = \mathcal{PC}[\![PC_{D_1}]\!] \uplus \mathcal{PC}[\![PC_{D_2}]\!]$
$\mathcal{PC}[\![\textbf{P-C relationship } PC \textbf{ involves } L_1, L_2]\!] =$
$\quad \{t \mid t \in TF \wedge dom(t) = \{s_{L_1}, s_{L_2}\} \wedge t[s_{L_1}] \in \mathcal{I}[\![L_1]\!] \wedge t[s_{L_2}] \in \mathcal{I}[\![L_2]\!]\}$
$\mathcal{PC}[\![\textbf{P-C relationship } PC \textbf{ involves } L_1, L_2$
$\quad \textbf{has distributing factor }]\!] =$
$\quad \{t \mid t \in TF \wedge dom(t) = \{s_{L_1}, s_{L_2}, d\} \wedge t[s_{L_1}] \in \mathcal{I}[\![L_1]\!] \wedge$
$\quad t[s_{L_2}] \in \mathcal{I}[\![L_2]\!] \wedge t[d] \in \mathcal{D}[\![DATA]\!](real)\}$
$\mathcal{PC}[\![\textbf{P-C relationship } PC \textbf{ involves } L_1, L_2 \textbf{ with temporality } T_S]\!] =$
$\quad \{t \mid t \in TF \wedge dom(t) = \{s_{L_1}, s_{L_2}, tempOf(T_S)\} \wedge t[s_{L_1}] \in \mathcal{I}[\![L_1]\!] \wedge$
$\quad t[s_{L_2}] \in \mathcal{I}[\![L_2]\!] \wedge \forall T_i \in tempOf(T_S)(t[T_i] \in \mathcal{T}[\![(T_i, t, g)]\!]) \wedge$
$\quad (LS \in tempOf(T_S) \Rightarrow t[LS] \subseteq lifespan(s_{L_1}, L_1) \cap lifespan(s_{L_2}, L_2))\}$
$\mathcal{PC}[\![\textbf{P-C relationship } PC \textbf{ involves } L_1, L_2 \textbf{ with temporality } T_S$
$\quad \textbf{has distributing factor }]\!] =$
$\quad \{t \mid t \in TF \wedge dom(t) = \{s_{L_1}, s_{L_2}, tempOf(T_S), d\} \wedge t[s_{L_1}] \in \mathcal{I}[\![L_1]\!] \wedge$
$\quad t[s_{L_2}] \in \mathcal{I}[\![L_2]\!] \wedge \forall T_i \in tempOf(T_S)(t[T_i] \in \mathcal{T}[\![(T_i, t, g)]\!]) \wedge$
$\quad (LS \in tempOf(T_S) \Rightarrow t[LS] \subseteq lifespan(s_{L_1}, L_1) \cap lifespan(s_{L_2}, L_2)) \wedge$
$\quad t[d] \in \mathcal{D}[\![DATA]\!](real)\}$

Note that the last two definitions above enforce the constraint that the lifespan of an instance of a temporal parent-child relationship must be included in the intersection of the lifespans of its participating members.

The function $\mathcal{F}$ defines the semantics of fact relationships. A fact relationship relates several levels and may have attributes. Its semantics is thus a set of tuples t defining a member from each of its levels, and also values for its attributes. Recall that a level may participate several times in a fact relationship, using different roles. If this is the case, the role name is used instead of the level name in the domain of the function t. Members are identified through their surrogates, with the value domain defined by $\mathcal{I}$. If the fact relationship has measures, the domain of the function t includes additionally the set of measure names. The value domains of these measures are determined by the semantics of the measure declarations. As in the case of levels and parent-child relationships, a fact relationship defines a unique set of tuples that are stored separately in the database:

$\mathcal{F} : FactRels \times InvSpec \times MeasDecl \rightarrow \mathcal{I}[\![I_S]\!] \cup \mathcal{I}[\![I_S]\!] \times \mathcal{M}[\![M_D]\!]$
$\mathcal{F}[\![F_{D_1}; F_{D_2}]\!] = \mathcal{F}[\![F_{D_1}]\!] \uplus \mathcal{F}[\![F_{D_2}]\!]$
$\mathcal{F}[\![\textbf{Fact relationship } F \textbf{ involves } I_S]\!] =$
$\quad \{t \mid t \in TF \wedge dom(t) = \{\bigcup_{L_i \in I_S} s_{role(L_i)}\} \wedge$
$\quad \forall L_i \in I_S (t[s_{role(L_i)}] \in \mathcal{I}[\![level(L_i)]\!])\}$
$\mathcal{F}[\![\textbf{Fact relationship } F \textbf{ involves } I_S \textbf{ has } M_D]\!] =$

$\{t \mid t \in TF \wedge dom(t) = \{\bigcup_{L_i \in I_S} s_{role(L_i)}, measOf(M_D)\} \wedge$
$\forall M_i \in measOf(M_D) \, (t[M_i] \in \mathcal{M}[\![\mathbf{Measure} \ M_i \ \mathbf{of} \ A'_D \ \mathbf{is} \ add]\!] \vee$
$t[M_i] \in \mathcal{M}[\![\mathbf{Derived \ measure} \ M_i \ \mathbf{of} \ A'_D \ \mathbf{is} \ add]\!]) \wedge$
$\forall L_i \in I_S \, (t[s_{role(L_i)}] \in \mathcal{I}[\![level(L_i)]\!])\}$

The function $\mathcal{IC}$ defines the semantics of integrity constraints. The semantics of a constraint is a set of predicates that a database must satisfy. In the textual representation, all constraints are separate constructs, and so the predicates must first verify that the constructs (e.g, levels and relationships) mentioned in the constraints belong to the schema using the function $inSch$:

$\mathcal{IC} : IC_D \rightarrow PRED$
$\mathcal{IC}[\![IC_{D_1}; IC_{D_2}]\!] = \mathcal{IC}[\![IC_{D_1}]\!] \wedge \mathcal{IC}[\![IC_{D_2}]\!]$

The primary-key constraint ensures that the values of the key attributes are unique for all members of the level:

$\mathcal{IC}[\![K \ \mathbf{is \ primary \ key \ of} \ L]\!] = inSch(L, S_D) \wedge$
$\quad ((K \subseteq attOf(\mathbf{Level} \ L \ \mathbf{has} \ A_D) \wedge \forall t_i, t_j \in \mathcal{L}[\![\mathbf{Level} \ L \ \mathbf{has} \ A_D]\!]$
$\quad (t_i[K] = t_j[K] \Rightarrow t_i[s] = t_j[s])) \vee$
$\quad (K \subseteq attOf(\mathbf{Level} \ L \ \mathbf{with \ spatiality} \ sp \ \mathbf{has} \ A_D) \wedge$
$\quad \forall t_i, t_j \in \mathcal{L}[\![\mathbf{Level} \ L \ \mathbf{with \ spatiality} \ sp \ \mathbf{has} \ A_D]\!] \wedge$
$\quad (t_i[K] = t_j[K] \Rightarrow t_i[s] = t_j[s])) \vee$
$\quad (K \subseteq attOf(\mathbf{Level} \ L \ \mathbf{with \ temporality} \ T_S \ \mathbf{has} \ A_D) \wedge$
$\quad \forall t_i, t_j \in \mathcal{L}[\![\mathbf{Level} \ L \ \mathbf{with \ temporality} \ T_S \ \mathbf{has} \ A_D]\!] \wedge$
$\quad (\mathcal{T}[\![T_S]\!] \rightarrow t_i[K] - \mathcal{T}[\![T_S]\!] \rightarrow t_j[K] \rightarrow \mathcal{T}[\![T_S]\!] \rightarrow t_i[s] = \mathcal{T}[\![T_S]\!] \rightarrow t_j[s]))$

The cardinality constraints ensure that a child member can be related to a minimum of min and a maximum of max parent members. Three cases must be considered: the usual cardinality constraints for nontemporal relationships, and the instant and lifespan cardinality constraints for temporal relationships. In the case of cardinality constraints of the usual kind, for every member m of a level L we use the function cnt to determine the number of tuples belonging to the semantics of the parent-child relationship in which m participates. Instant cardinality constraints must be satisfied for each instant tuple c of the temporal domain of the relationship. Therefore, we use the function $contains$ to obtain the subset of the tuples belonging to the semantics of the parent-child relationship that contain the instant tuple c. The function cnt is then applied to this subset. In addition, if a level participating in a relationship is temporal, then for each instant belonging to the lifespan of a member m, that member must be related to a valid member of the other level through an instance of the relationship. Finally, lifespan cardinality constraints must be satisfied during the whole temporal domain of the relationship:

$\mathcal{IC}[\![\mathbf{Participation \ of} \ L \ \mathbf{in} \ PC \ \mathbf{is} \ (min, max)]\!] =$
$\quad inSch(L, S_D) \wedge inSch(PC, S_D) \wedge L \in parOf(PC) \wedge \forall m \in D_S^L$
$\quad (min \leq cnt(m, L, \mathcal{PC}[\![\mathbf{P\text{-}C \ relationship} \ PC \ldots]\!]) \leq max)$
$\mathcal{IC}[\![\mathbf{Instant \ participation \ of} \ L \ \mathbf{in} \ PC \ \mathbf{is} \ (min, max)]\!] =$

$inSch(L, S_D) \wedge inSch(PC, S_D) \wedge L \in parOf(PC) \wedge$
$\forall c \in instants(tempSpec(PC)) \; \forall m \in D_S^L(min \leq$
$cnt(m, L, contains(c, \mathcal{PC}[\![\textbf{P-C relationship } PC \ldots]\!])) \leq max) \wedge$
$\forall m \in D_S^L \; \forall c \in lifespan(m, L) \; \exists t \in \mathcal{PC}[\![\textbf{P-C relationship } PC \ldots]\!]$
$(t[s_L] = m \wedge c \in t[LS])$

$\mathcal{IC}[\![\textbf{Lifespan participation of } L \textbf{ in } PC \textbf{ is } (min, max)]\!] =$
$inSch(L, S_D) \wedge inSch(PC, S_D) \wedge L \in parOf(PC) \wedge \forall m \in D_S^L$
$(min \leq cnt(m, L, \mathcal{PC}[\![\textbf{P-C relationship } PC \ldots]\!]) \leq max)$

A member of a level that participates exclusively in a set of parent-child relationships (i.e., a member of a splitting or joining level) cannot be involved in more than one of these relationships:

$\mathcal{IC}[\![\textbf{Exclusive participation of } L \textbf{ in } Q]\!] = inSch(L, S_D) \wedge$
$\forall PC_i \in Q(inSch(PC_i, S_D)) \wedge \neg(\exists PC_i, PC_j \in Q$
$\exists t_1 \in \mathcal{PC}[\![\textbf{P-C relationship } PC_i \textbf{ involves } \ldots]\!]$
$\exists t_2 \in \mathcal{PC}[\![\textbf{P-C relationship } PC_j \textbf{ involves } \ldots]\!]$
$(i \neq j \wedge t_1[s_L] = t_2[s_L]))$

Topological constraints in parent-child relationships ensure that if a parent and a child member participate in the relationship, their geometries satisfy the given topological relationship. We give next a parameterized semantics of a constraint, where $TOPO$ represents any topological relationship defined in the MultiDim model:

$\mathcal{IC}[\![\textbf{Topological relationship } L_1 \; TOPO \; L_2 \textbf{ in } PC]\!] =$
$inSch(L_1, S_D) \wedge inSch(L_2, S_D) \wedge inSch(PC, S_D) \wedge$
$\forall t \in \mathcal{PC}[\![\textbf{P-C Relationship } PC \textbf{ involves } L_1, L_2 \ldots]\!]$
$\forall t_1 \in \mathcal{L}[\![\textbf{Level } L_1 \textbf{ with spatiality } sp_1 \ldots]\!]$
$\forall t_2 \in \mathcal{L}[\![\textbf{Level } L_2 \textbf{ with spatiality } sp_2 \ldots]\!]$
$(t[s_{L_1}] = t_1[s_{L_1}] \wedge t[s_{L_2}] = t_2[s_{L_2}] \Rightarrow t_1[geom] \; TOPO \; t_2[geom])$

Topological constraints in fact relationships ensure that for each tuple in the semantics of the fact relationship and for every pair of levels that are restricted in the relationship, the geometries of the corresponding members must satisfy the given topological relationship. Recall that a level may participate several times in a fact relationship, using different roles. The functions *role* and *level* are used in the definition below to obtain the role name and the level name. We give next a parameterized semantics of a constraint, where $TOPO$ represents any topological relationship defined in the MultiDim model:

$\mathcal{IC}[\![\textbf{Topological relationship } TOPO \textbf{ between } I_S \textbf{ in } F]\!] =$
$\forall L_i \in I_S(inSch(L_i, S_D)) \wedge inSch(F, S_D) \wedge$
$\forall t \in \mathcal{F}[\![\textbf{Fact Relationship } F \textbf{ involves } \ldots]\!]$
$\forall L_1 \in I_S \forall L_2 \in I_S$
$\forall t_1 \in \mathcal{L}[\![\textbf{Level } level(L_1) \textbf{ with spatiality } sp_1 \ldots]\!]$
$\forall t_2 \in \mathcal{L}[\![\textbf{Level } level(L_2) \textbf{ with spatiality } sp_2 \ldots]\!]$
$(t[s_{role(L_1)}] = t_1[s_{L_1}] \wedge t[s_{role(L_2)}] = t_2[s_{L_2}] \Rightarrow$
$t_1[geom] \; TOPO \; t_2[geom])$

Synchronization constraints in parent-child relationships ensure that if a parent and a child member participate in the relationship, their lifespans satisfy the given synchronization relationship. We give next a parameterized semantics of a constraint, where $SYNC$ represents any synchronization relationship defined in the MultiDim model:

$\mathcal{IC}[\![$**Synchronization relationship** L_1 $SYNC$ L_2 **in** $PC]\!] =$
  $inSch(L_1, S_D) \wedge inSch(L_2, S_D) \wedge inSch(PC, S_D) \wedge$
  $\forall t \in \mathcal{PC}[\![$**P-C Relationship** PC **involves** $L_1, L_2 \ldots]\!]$
  $\forall t_1 \in \mathcal{L}[\![$**Level** L_1 **with temporality** $T_{S_1} \ldots]\!]$
  $\forall t_2 \in \mathcal{L}[\![$**Level** L_2 **with temporality** $T_{S_2} \ldots]\!]$
  $(t[s_{L_1}] = t_1[s_{L_1}] \wedge t[s_{L_2}] = t_2[s_{L_2}] \Rightarrow$
  $lifespan(t_1, L_1)$ $SYNC$ $lifespan(t_2, L_2)$

Synchronization constraints in fact relationships ensure that for each tuple in the semantics of the fact relationship and for every pair of levels that are restricted in the relationship, the lifespans of the corresponding members must satisfy the given synchronization relationship. As in the case of topological relationships above, the functions *role* and *level* are used in the definition below to obtain the role name and the level name. We give next a parameterized semantics of a constraint, where $SYNC$ represents any synchronization relationship defined in the MultiDim model:

$\mathcal{IC}[\![$**Synchronization relationship** $SYNC$ **between** I_S **in** $F]\!] =$
  $\forall L_i \in I_S(inSch(L_i, S_D)) \wedge inSch(F, S_D) \wedge$
  $\forall t \in \mathcal{F}[\![$**Fact Relationship** F **involves** $\ldots]\!]$
  $\forall L_1 \in I_S \forall L_2 \in I_S$
  $\forall t_1 \in \mathcal{L}[\![$**Level** $level(L_1)$ **with temporality** $T_{S_1} \ldots]\!]$
  $\forall t_2 \in \mathcal{L}[\![$**Level** $level(L_2)$ **with temporality** $T_{S_2} \ldots]\!]$
  $(t[s_{role(L_1)}] = t_1[s_{L_1}] \wedge t[s_{role(L_2)}] = t_2[s_{L_2}] \Rightarrow$
  $lifespan(t_1, L_1)$ $SYNC$ $lifespan(t_2, L_2)$

Note that in the above formalization, dimensions, hierarchies, and additivity types of measures do not have semantic interpretations. However, they are needed for defining meaningful OLAP operations. Dimensions are required for the drill-across operation that allows measures from different fact relationships to be compared. Hierarchies are needed for defining aggregations for the roll-up and drill-down operations. Additivity types of measures are used to define meaningful aggregations for the roll-up operation. Such OLAP operations are beyond the scope of this book.

B

Graphical Notation

In the following, we summarize the graphical notation used in this book.

B.1 Entity-Relationship Model

We show below the notation used to represent the ER constructs that we refer to in this book.

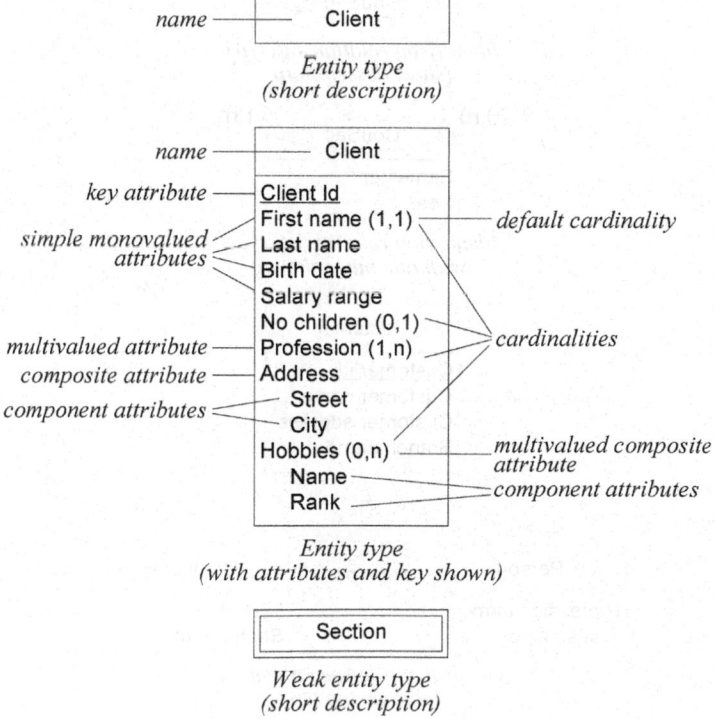

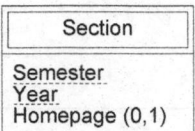

Weak entity type
(with attributes and partial key shown)

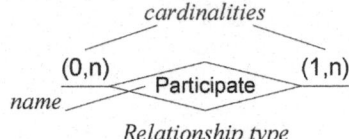

Relationship type
(short description)

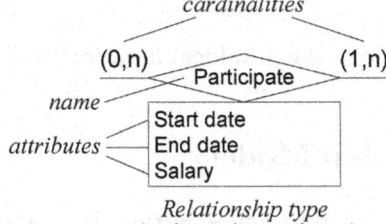

Relationship type
(with attributes shown)

Identifying relationship type
(short description)

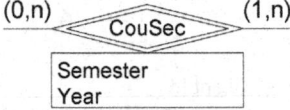

Identifying relationship type
(with attributes shown)

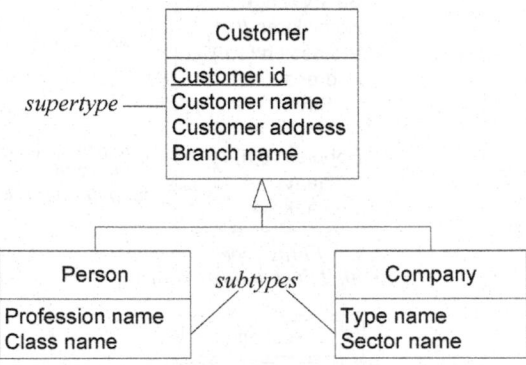

Generalization/specialization
relationship type

B.2 Relational and Object-Relational Models

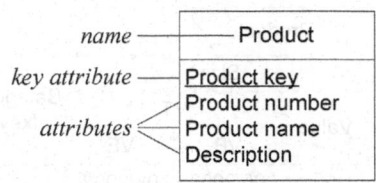

Relational table
(with attributes and primary key shown)

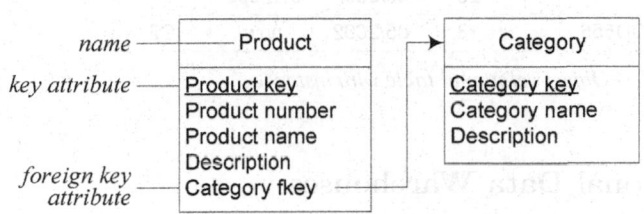

Referential integrity

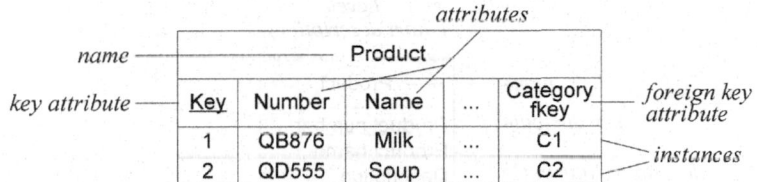

Relational table with instances

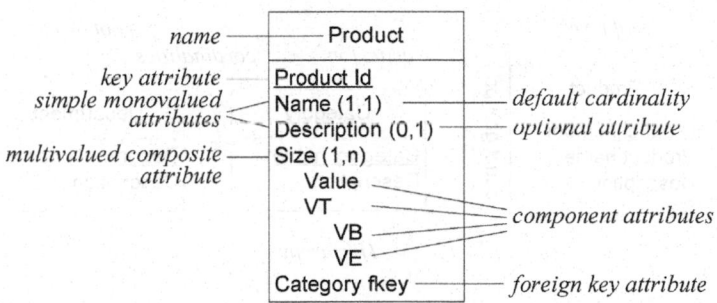

Object-relational table
(with attributes and primary key shown)

| | | surrogate attribute | | name | | multivalued composite attribute | | component attributes | | foreign key attribute |

Product

			Size*			
Sid	Product Id	...	Value	VT		Category fkey
				VB	VE	
1	QB876	...	10	05/2002	08/2002	C1
			10	08/2003	now	
			20	09/2002	07/2003	
2	QD555	...	18	05/2002	now	C2

instances

Object-relational table with instances

B.3 Conventional Data Warehouses

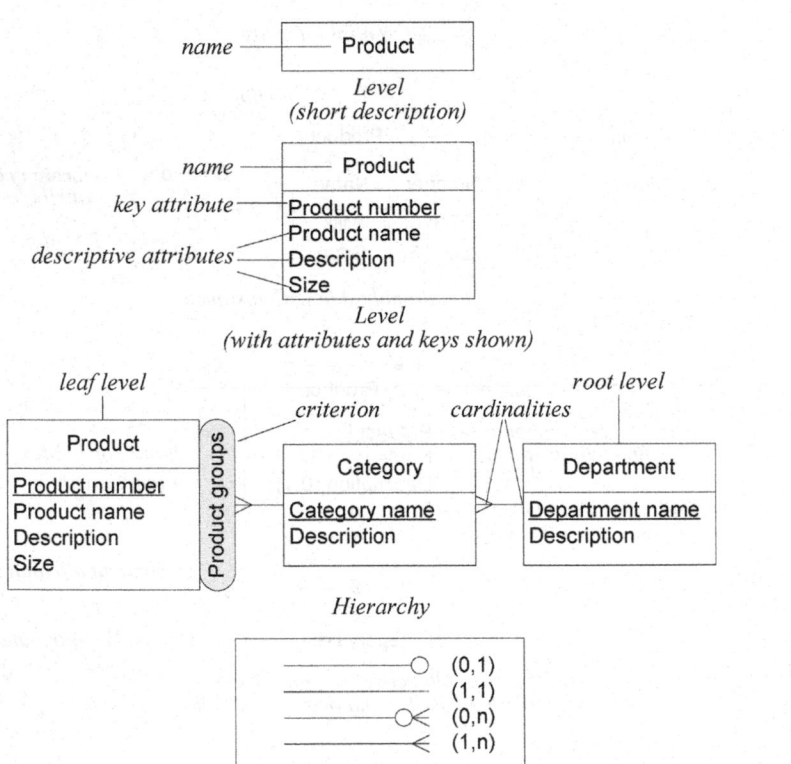

name ——— Product

*Level
(short description)*

name ——— Product
key attribute ——— Product number
Product name
descriptive attributes ——— Description
Size

*Level
(with attributes and keys shown)*

leaf level   criterion   cardinalities   root level

Product | Product groups | Category | Department

Product number
Product name
Description
Size

Category name
Description

Department name
Description

Hierarchy

——————○	(0,1)
——————	(1,1)
————○<	(0,n)
————<	(1,n)

Cardinalities

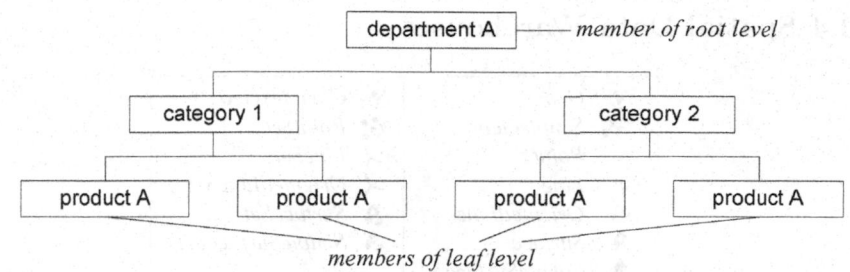

Members of different hierarchy levels

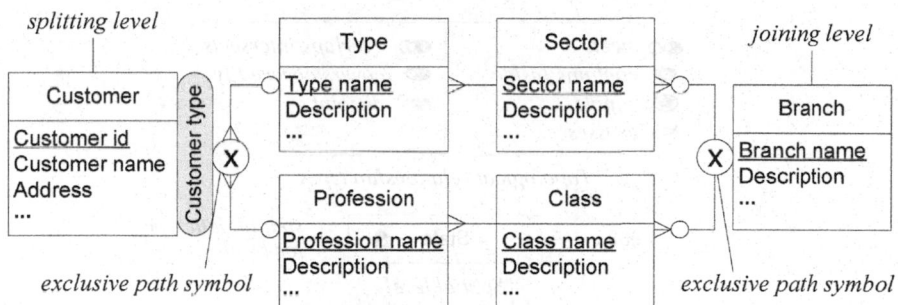

Generalized hierarchy

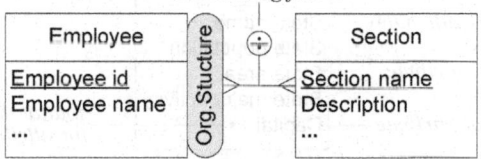

Nonstrict hierarchy

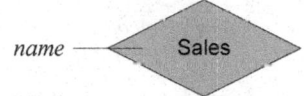

Fact relationship
(short description)

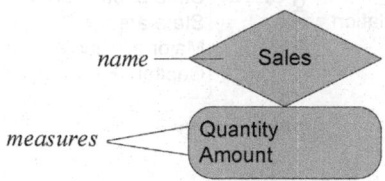

Fact relationship
(with measures shown)

B.4 Spatial Data Warehouses

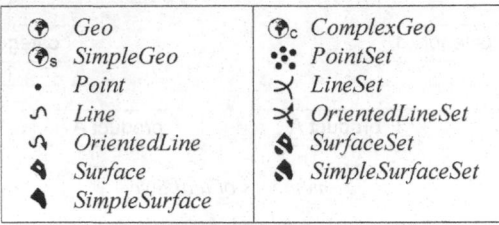

Spatial data types

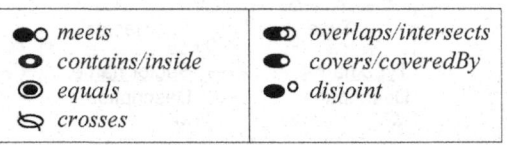

Topological relationship types

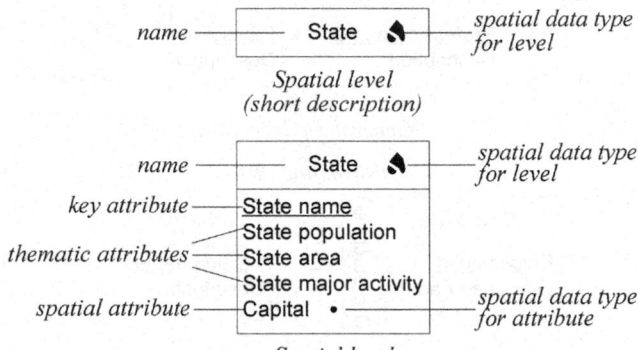

Spatial level
(short description)

Spatial level
(with attributes and key shown)

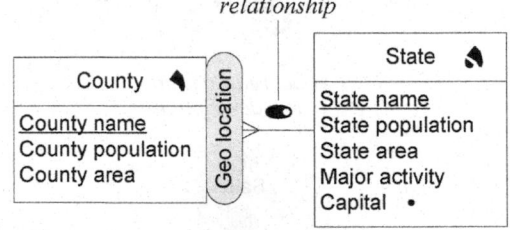

Spatial hierarchy

Spatial fact relationship
(short description)

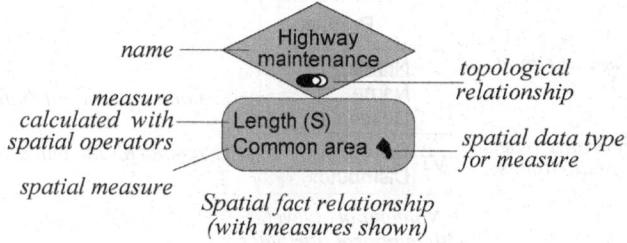

name — *topological relationship*

measure calculated with spatial operators

spatial data type for measure

spatial measure

Spatial fact relationship
(with measures shown)

B.5 Temporal Data Warehouses

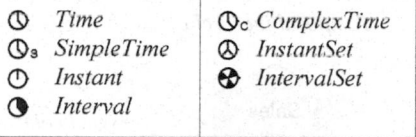

VT	Valid time
TT	Transaction time
BT	Bitemporal time
LS	Lifespan
LT	Loading time

Temporality types

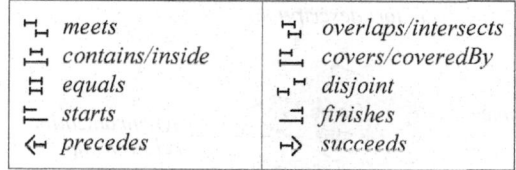

Temporal data types

Synchronization predicates

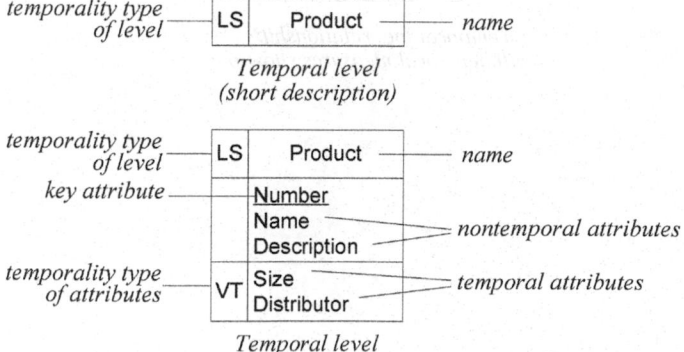

temporality type of level — *name*

Temporal level
(short description)

temporality type of level — *name*

key attribute

nontemporal attributes

temporality type of attributes

temporal attributes

Temporal level
(with attributes and key shown)

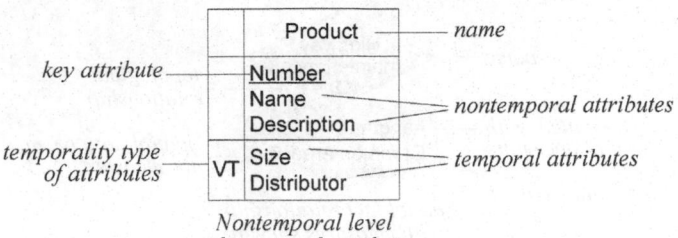

*Nontemporal level
with temporal attributes*

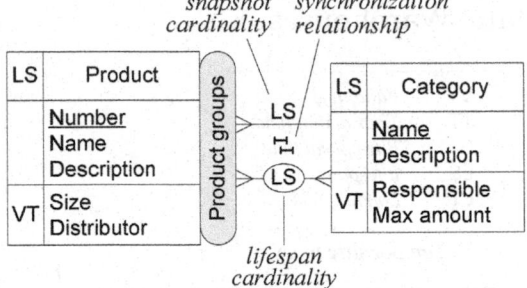

Temporal hierarchy

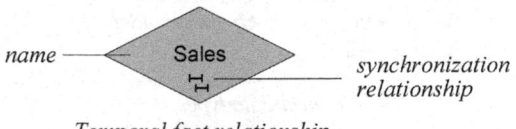

*Temporal fact relationship
(short description)*

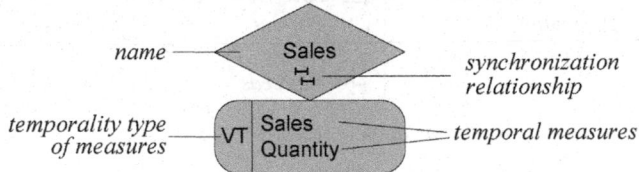

*Temporal fact relationship
(with temporal measures shown)*

References

1. A. El Abbadi, M.L. Brodie, S. Chakravarthy, U. Dayal, N. Kamel, G. Schlageter, and K.-Y. Whang, editors. *Proceedings of the 26th International Conference on Very Large Data Bases, VLDB'00.* Morgan Kaufmann, 2000.
2. A. Abdul-Rahman and M. Pilouk. *Spatial Data Modelling for 3D GIS.* Springer, 2008.
3. A. Abelló and C. Martín. A bitemporal storage structure for a corporate data warehouse. In [40], pages 177–183.
4. A. Abelló, J. Samos, and F. Saltor. On relationships offering new drill-across possibilities. In [289], pages 7–13.
5. A. Abelló, J. Samos, and F. Saltor. YAM^2 (yet another multidimensional model): An extension of UML. *Information Systems,* 32(6):541–567, 2006.
6. S. Agarwal, R. Agrawal, P. Deshpande, A. Gupta, J. Naughton, R. Ramakrishnan, and S. Sarawagi. On the computation of multidimensional aggregates. In [12], pages 506–521.
7. S. Agrawal, S. Chaudhuri, and V. Narasayya. Automated selection of materialized views and indexes for SQL databases. In [1], pages 496–505.
8. J. Akoka, I. Comyn-Wattiau, and N. Prat. Dimension hierarchies design from UML generalizations and aggregations. In [155], pages 442 445.
9. J. Allen. Maintaining knowledge about temporal intervals. *Communications of the ACM,* 26(11):832–843, 1983.
10. A. Arctur and M. Zeiler. *Designing Geodatabases: Case Studies in GIS Data Modeling.* ESRI Press, 2004.
11. H. Arisawa, K. Moriya, and T. Miura. Operations and the properties on non-first-normal-form relational databases. In M. Schkolnick and C. Thanos, editors, *Proceedings of the 9th International Conference on Very Large Data Bases, VLDB'83,* pages 197–204. Morgan Kaufmann, 1983.
12. M.P. Atkinson, M.E. Orlowska, P. Valduriez, S.B. Zdonik, and M.L. Brodie, editors. *Proceedings of the 25th International Conference on Very Large Data Bases, VLDB'99.* Morgan Kaufmann, 1999.
13. C. Ballard, A. Beaton, D. Chiou, J. Chodagam, M. Lowry, A. Perkins, R. Phillips, and J. Rollins. *Leveraging DB2 Data Warehouse Edition for Business Intelligence.* IBM Redbooks SG24-7274-00, 2006.

14. C. Ballard, D. Herreman, D. Schau, R. Bell, E. Kim, and A. Valencic. *Data Modeling Techniques for Data Warehousing*. IBM Redbooks SG24-2238-00, 1998.

15. A. Bauer, W. Hümmer, and W. Lehner. An alternative relational OLAP modeling approach. In [139], pages 189–198.

16. B. Bebel, J. Eder, C. Koncilia, T. Morzy, and R. Wrembel. Creation and management of versions in multiversion data warehouse. In H. Haddad, A. Omicini, R. Wainwright, et al., editors, *Proceedings of the ACM Symposium on Applied Computing, SAC'04*, pages 717–723. ACM Press, 2004.

17. Y. Bédard. Visual modeling of spatial databases towards spatial PVL and UML. *Geomatica*, 53(2):169–186, 1999.

18. Y. Bédard, T. Merrett, and J. Han. Fundaments of spatial data warehousing for geographic knowledge discovery. In H. Miller and J. Han, editors, *Geographic Data Mining and Knowledge Discovery*, pages 53–73. Taylor & Francis, 2001.

19. Y. Bédard, M. Proulx, S. Larrivée, and E Bernier. Modeling multiple representations into spatial data warehouses: A UML-based approach. In *Proceedings of the Joint International Symposium of the International Society for Photogrammetry and Remote Sensing, ISPRS 2002, Commission IV*, 2002.

20. A. Belussi, M. Negri, and G. Pelagatti. GeoUML: A geographic conceptual model defined through specialization of ISO TC211 standards. In *Proceedings of the 10th EC-GI&GIS Workshop: ESDI State of the Art*, 2004.

21. C. Bettini, S. Jajodia, and X. Wang. *Time Granularities in Databases, Data Mining, and Temporal Reasoning*. Springer, 2000.

22. M. Blaschka, C. Sapia, and G. Höfling. On schema evolution in multidimensional databases. In [200], pages 153–164.

23. R. Bliujute, S. Slatenis, G. Slivinskas, and C.S. Jensen. Systematic change mangement in dimensional data warehousing. Technical Report TR-23, Time Center, 1998.

24. M. Body, M. Miquel, Y. Bédard, and A. Tchounikine. A multidimensional and multiversion structure for OLAP applications. In [289], pages 1–6.

25. M. Body, M. Miquel, Y. Bédard, and A. Tchounikine. Handling evolution in multidimensional structures. In U. Dayal, K. Ramamritham, and T. Vijayaraman, editors, *Proceedings of the 19th International Conference on Data Engineering, ICDE'03*, pages 581–592. IEEE Computer Society Press, 2003.

26. M. Böhnlein and A. Ulbrich-vom Ende. Deriving initial data warehouses structures from the conceptual data models of the underlying operational information systems. In [282], pages 15–21.

27. M. Böhnlein and A. Ulbrich-vom Ende. Business process oriented development of data warehouse structures. In R. Jung and R. Winter, editors, *Proceedings of Data Warehousing 2000*, pages 3–21. Physica-Verlag, 2000.

28. A. Bonifati, F. Cattaneo, S. Ceri, A. Fuggetta, and S. Paraboschi. Designing data marts for data warehouses. *ACM Transactions on Software Engineering and Methodology*, 10(4):452–483, 2001.

29. G. Booch, I. Jacobson, and J. Rumbaugh. *The Unified Modeling Language: User Guide*. Addison-Wesley, second edition, 2005.

30. K. Borges, A. Laender, and C. Davis. Spatial data integrity constraints in object-oriented geographic data modeling. In [190], pages 1–6.

31. R. Bruckner, B. List, and J. Schiefer. Developing requirements for data warehouse systems with use cases. In *Proceedings of the 7th Americas' Conference on Information Systems, AMCIS'01*, pages 329–335, 2001.

32. R. Bruckner and A. Min Tjoa. Capturing delays and valid times in data warehouses – towards timely consistent analyses. *Journal of Intelligent Information Systems*, 19(2):169–190, 2002.

33. L. Cabibbo, I. Panella, and R. Torlone. DaWaII: a tool for the integration of autonomous data marts. In L. Liu, A. Reuter, K.Y. Whang, and J. Zhang, editors, *Proceedings of the 22nd International Conference on Data Engineering, ICDE'06*, pages 158–114. IEEE Computer Society Press, 2006.

34. L. Cabibbo and R. Torlone. A logical approach to multidimensional databases. In [265], pages 183–197.

35. L. Cabibbo and R. Torlone. The design and development of a logical system for OLAP. In [139], pages 1–10.

36. L. Cabibbo and R. Torlone. On the integration of autonomous data marts. In *Proceedings of the 16th International Conference on Scientific and Statistical Database Management, SSDBM'04*, pages 223–234, 2004.

37. L. Cabibbo and R. Torlone. Integrating heterogeneous multidimensional databases. In J. Frew, editor, *Proceedings of the 17th International Conference on Scientific and Statistical Database Management, SSDBM'05*, pages 205–214, 2005.

38. J. Cabot, A. Olivé, and E. Teniente. Representing temporal information in UML. In P. Stevens, J. Whittle, and G. Booch, editors, *Proceedings of the 6th International Conference on the Unified Modeling Language, UML'2003*, LNCS 2863, pages 44–59. Springer, 2003.

39. E. Camossi, M. Bertolotto, E. Bertino, and G. Guerrini. A multigranular spatiotemporal data model. In E. Hoel and P. Rigaux, editors, *Proceedings of the 11th ACM Symposium on Advances in Geographic Information Systems, ACM-GIS'03*, pages 94–101. ACM Press, 2003.

40. O. Camp, J. Filipe, S. Hammoudi, and M. Piattini, editors. *Proceedings of the 5th International Conference on Enterprise Information Systems, ICEIS'03*, 2003.

41. L. Carneiro and A. Brayner. X-META: A methodology for data warehouse design with metadata management. In [157], pages 13–22.

42. P. Chamoni and S. Stock. Temporal structure in data warehousing. In [200], pages 353–358.

43. C. Chan and Y. Ioannidis. Bitmap index design and evaluation. In L.M. Haas and A. Tiwary, editors, *Proceedings of the ACM SIGMOD International Conference on Management of Data, SIGMOD'98*, pages 355–366. ACM Press, 1998.

44. S. Chaudhuri, M. Datar, and V. Narasayya. Index selection for databases: A hardness study and a principled heuristic solution. *IEEE Transactions on Knowledge and Data Engineering*, 16(11):1313–1323, 2004.

45. P. Chen. The Entity-Relationship model - towards a unified view of data. *ACM Transaction on Database Systems*, 1(1):9–36, 1976.

46. T. Chenoweth, D. Schuff, and R. Louis. A method for developing dimensional data marts. *Communications of the ACM*, 46(12):93–98, 2003.

47. G. Chin. *Agile Project Management: How to Succeed in the Face of Changing Project Requirements*. American Management Association, AMACOM, 2003.

48. J. Clifford, A. Croker, and A. Tuzhilin. On completeness of historical relational query languages. *ACM Transactions on Database Systems*, 19(1):64–116, 1994.

49. J. Clifford, C.E. Dyreson, T. Isakowitz, C.S. Jensen, and R.T. Snodgrass. On the semantics of "now" in databases. *ACM Transactions on Database Systems*, 22(2):171–214, 1997.

50. E.F. Codd, S.B. Codd, and C.T. Salley. Providing OLAP (On-Line Analytical Processing) to user-analysts: An IT mandate. Technical report, E.F. Codd and Associates, 1993.

51. C. Combi, M. Franceschet, and A. Peron. Representing and reasoning about temporal granularities. *Journal of Logic and Computation*, 4(1):52–77, 2004.

52. H. Darwen. Valid time and transaction time proposals: Language design aspects. In [71], pages 195–210.

53. C.J. Date. *An Introduction to Database Systems*. Addison-Wesley, eighth edition, 2003.

54. C.J. Date, H. Darwen, and N. Lorentzos. *Temporal Data and the Relational Model*. Morgan Kaufmann, 2003.

55. C. Davis and A. Laender. Multiple representations in GIS: Materialization through map generalization, geometric, and spatial analysis operations. In [190], pages 60–65.

56. V. Detienne and J.L. Hainaut. CASE tool support for temporal database design. In [155], pages 208–224.

57. A. Di Pasquale, L. Forlizzi, C.S. Jensen, Y. Manolopoulos, E. Nardelli, D. Pfoser, G. Proietti, S. Saltenis, Y. Theodoridis, T. Tzouramanis, and M. Vassilakopoulos. Access methods and query processing techniques. In [153], pages 203–261.

58. S.W. Dietrich and S.D. Urban. *An Advanced Course in Database Systems: Beyond Relational Databases*. Pearson Prentice Hall, 2005.

59. C.E. Dyreson, W.S. Evans, H. Lin, and R.T. Snodgrass. Efficiently supported temporal granularities. *IEEE Transactions on Knowledge and Data Engineering*, 12(4):568–587, 2000.

60. J. Eder and S. Kanzian. Logical design of generalizations in object-relational databases. In G. Gottlob, A. Benczúr, and J. Demetrovic, editors, *Proceedings of the 8th East European Conference on Advances in Databases and Information Systems, ADBIS'04*, 2004.

61. J. Eder and C. Koncilia. Changes of dimension data in temporal data warehouses. In Y. Kambayashi, W. Winiwater, and M. Arikawa, editors, *Proceedings of the 3rd International Conference on Data Warehousing and Knowledge Discovery, DaWaK'01*, LNCS 2114, pages 284–293. Springer, 2001.

62. J. Eder, C. Koncilia, and T. Morzy. A model for a temporal data warehouse. In *Proceedings of the International Workshop on Open Enterprise Solutions: Systems, Experiences, and Organization, OESSEO'01*, pages 48–54, 2001.

63. J. Eder, C. Koncilia, and T. Morzy. The COMET metamodel for temporal data warehouses. In A. Banks Pidduck, J. Mylopoulos, C.C. Woo, and M. Özsu, editors, *Proceedings of the 14th International Conference on Advanced Information Systems Engineering, CAiSE'02*, LNCS 2348, pages 83–99. Springer, 2002.

64. M.J. Egenhofer, A. Frank, and J. Jackson. A topological data model for spatial databases. In A. Buchman, O. Günter, T. Smith, and Y. Wang, editors, *Proceedings of the 1st Symposium on Design and Implementation of Large Spatial Databases, SSD'89*, LNCS 409, pages 271–286. Springer, 1989.

65. M.J. Egenhofer and R. Golledge, editors. *Spatial and Temporal Reasoning in Geographic Information Systems*. Oxford University Press, 1998.

66. R. Elmasri and S. Navathe. *Fundamentals of Database Systems.* Addison-Wesley, fourth edition, 2003.
67. R. Elmasri and G. Wuu. A temporal model and query language for ER databases. In *Proceedings of the 6th International Conference on Data Engineering, ICDE'90,* pages 76–83. IEEE Computer Society Press, 1990.
68. M. Erwig and M. Schneider. Formalization of advanced map operations. In *Proceedings of the 9th International Symposium on Spatial Data Handling, SDH'00,* pages 3–17, 2000.
69. M. Erwig, M. Schneider, and R.H. Güting. Temporal objects for spatio-temporal data models and a comparison of their representations. In Y. Kambayashi, D. Lee, E. Lim, M. Mohania, and Y. Masunaga, editors, *Proceedings of the ER'98 International Workshops: Advances in Database Technologies,* LNCS 1552, pages 454–465. Springer, 1998.
70. ESRI, Inc. ArcGIS data models. `http://www.esri.com/software/arcgisdatamodels/index.html`, 2004.
71. O. Etzion, S. Jajodia, and S. Sripada, editors. *Temporal Databases: Research and Practice.* LNCS 1399. Springer, 1998.
72. F. Ferri, E. Pourabbas, M. Rafanelli, and F. Ricci. Extending geographic databases for a query language to support queries involving statistical data. In O. Günter and H. Lenz, editors, *Proceedings of the 12th International Conference on Scientific and Statistical Database Management, SSDBM'00,* pages 220–230. IEEE Computer Society Press, 2000.
73. R. Fidalgo, V. Times, J. Silva, and F. Souza. GeoDWFrame: A framework for guiding the design of geographical dimensional schemes. In Y. Kambayashi, M. Mohania, and W. Wös, editors, *Proceedings of the 6th International Conference on Data Warehousing and Knowledge Discovery, DaWaK'04,* LNCS 3181, pages 26–37. Springer, 2004.
74. W. Filho, L. Figueiredo, M. Gattas, and P. Carvalho. A topological data structure for hierarchical planar subdivisions. Technical Report CS-95-53, University of Waterloo, 1995.
75. M. Franklin, B. Moon, and A. Ailamaki, editors. *Proceedings of the ACM SIGMOD International Conference on Management of Data, SIGMOD'02.* ACM Press, 2002.
76. G.M. Freitas, A.H.F. Laender, and M.L. Campos. MD2: Getting users involved in the development of data warehouse application. In [157], pages 3–12.
77. F. Giannotti and D. Pedreschi, editors. *Mobility, Data Mining and Privacy: Geographic Knowledge Discovery.* Springer, 2008.
78. P/ Giorgini, S. Rizzi, and M. Garzetti. Goal-oriented requirements analysis for data warehouse design. In [281], pages 47–56.
79. M. Golfarelli, J. Lechtenbörger, S. Rizzi, and G. Vossen. Schema versioning in data warehouses. In S. Wang, Y. Dongqing, K. Tanaka, et al., editors, *Proceedings of the ER'04 International Workshops: Conceptual Modeling for Advanced Application Domains,* LNCS 3289, pages 415–428. Springer, 2004.
80. M. Golfarelli, J. Lechtenbörger, S. Rizzi, and G. Vossen. Schema versioning in data warehouses: Enabling cross-version querying via schema augmentation. *Data & Knowledge Engineering,* 59:435–459, 2006.
81. M. Golfarelli, D. Maio, and S. Rizzi. Conceptual design of data warehouses from E/R schemes. In *Proceedings of the 31st Hawaii International Conference on System Sciences, HICSS-31,* pages 334–343, 1998.

82. M. Golfarelli and S. Rizzi. A methodological framework for data warehouse design. In I.-Y. Song and T.J. Teorey, editors, *Proceedings of the 1st ACM International Workshop on Data Warehousing and OLAP, DOLAP'98*, pages 3–9. ACM Press, 1998.

83. M. Golfarelli and S. Rizzi. Designing the data warehouse: Key steps and crucial issues. *Journal of Computer Science and Information Management*, 2(3):1–14, 1999.

84. E.S. Grant, R. Chennamaneni, and H. Reza. Towards analyzing UML class diagram models to object-relational database systems transformations. In *Proceedings of the 24th IASTED International Conference on Databases and Applications, DBA 2006*, pages 129–134. IASTED/ACTA Press, 2006.

85. J. Gray, S. Chaudhuri, A. Basworth, A. Layman, D. Reichart, M. Venkatrao, F. Pellow, and H. Pirahesh. Data cube: A relational aggregation operator generalizing group-by, cross-tab, and sub-totals. *Data Mining and Knowledge Discovery*, 1(1):29–53, 1997.

86. H. Gregersen and C.S. Jensen. Conceptual modeling of time-varying information. Technical Report TR-35, Time Center, 1998.

87. H. Gregersen and C.S. Jensen. Temporal entity-relationship models: A survey. *IEEE Transactions on Knowledge and Data Engineering*, 11(3):464–497, 1999.

88. H. Gregersen and C.S. Jensen. On the ontological expressiveness of temporal extensions to the entity-relationship model. Technical Report TR-69, Time Center, 2002.

89. H. Gregersen, L. Mark, and C.S. Jensen. Mapping temporal ER diagrams to relational schemas. Technical Report TR-39, Time Center, 1998.

90. A. Gupta and I.S. Mumick. *Materialized Views: Techniques, Implementations, and Applications*. MIT Press, 1999.

91. H. Gupta and I.S. Mumick. Selection of views to materialize in a data warehouse. *IEEE Transactions on Data and Knowledge Engineering*, 17(1):24–43, 2005.

92. R.H. Güting and M. Schneider. *Moving Objects Databases*. Morgan Kaufmann, 2005.

93. A. Guttman. R-trees: A dynamic index structure for spatial searching. In B. Yormark, editor, *Proceedings of the ACM SIGMOD International Conference on Management of Data, SIGMOD'84*, pages 46–57. ACM Press, 1984.

94. K. Hahn, C. Sapia, and M. Blaschka. Automatically generating OLAP schemata from conceptual graphical models. In R. Missaoui and I.-Y. Song, editors, *Proceedings of the 3rd ACM International Workshop on Data Warehousing and OLAP, DOLAP'00*, pages 9–16. ACM Press, 2000.

95. J.L. Hainaut, J. Henrard, J.M. Hick, D. Roland, and V. Englebert. CASE tools for database engineering. In L. Rivero, J. Doorn, and V. Ferraggine, editors, *Encyclopedia of Database Technologies and Applications*, pages 59–65. Idea Group, 2005.

96. J.L. Hainaut, J.M. Hick, J. Henrard, D. Roland, and V. Englebert. Knowledge transfer in database reverse engineering: A supporting case study. In D. Baxter, A. Quilici, and C. Verhoef, editors, *Proceedings of the 4th Working Conference on Reverse Engineering, WCRE'97*, pages 194–205. IEEE Computer Society Press, 1997.

97. M. Hammer and J. Champy. *Reengineering the Corporation: A Manifesto for Business Revolution*. Harper Collins, 1993.

98. J. Han and M. Kamber. *Data Mining: Concepts and Techniques*. Morgan Kaufmann, second edition, 2005.

99. J. Han, K. Koperski, and N. Stefanovic. GeoMiner: A system prototype for spatial data mining. In J. Peckham, editor, *Proceedings of the ACM SIGMOD International Conference on Management of Data, SIGMOD'97*, pages 553–556. ACM Press, 1997.

100. S. Harinath and S. Quinn. *SQL Server Analysis Services 2005 with MDX*. Wiley, 2006.

101. J. Henrard, J.L. Hainaut, J.M. Hick, D. Roland, and V. Englebert. Data structure extraction in database reverse engineering. In P. Chen, D. Embley, J. Kouloumdjian, S.W. Liddle, and J.F. Roddick, editors, *Proceedings of the ER'99 Workshops: Advances in Conceptual Modeling*, LNCS 1727, pages 149–160. Springer, 1999.

102. J. Henrard, J.M. Hick, P. Thiran, and J.L. Hainaut. Strategies for data reengineering. In A. van Deursen and E. Burd, editors, *Proceedings of the 9th Working Conference on Reverse Engineering, WCRE'02*, pages 211–220. IEEE Computer Society Press, 2002.

103. W. Hümmer, W. Lehner, A. Bauer, and L. Schlesinger. A decathlon in multidimensional modeling: Open issues and some solutions. In Y. Kambayashi, W. Winiwater, and M. Arikawa, editors, *Proceedings of the 4th International Conference on Data Warehousing and Knowledge Discovery, DaWaK'02*, LNCS 2454, pages 275–285. Springer, 2002.

104. C.A. Hurtado and C. Gutierrez. Handling structural heterogeneity in OLAP. In [317], chapter 2, pages 27–57.

105. C. Hurtado, C. Gutierrez, and A. Mendelzon. Capturing summarizability with integrity constraints in OLAP. *ACM Transactions on Database Systems*, 30(3):854–886, 2005.

106. C. Hurtado and A. Mendelzon. Reasoning about summarizabiliy in heterogeneous multidimensional schemas. In J. Van den Bussche and V. Vianu, editors, *Proceedings of the 8th International Conference on Database Theory, ICDT'01*, LNCS 1973, pages 375–389. Springer, 2001.

107. C. Hurtado, A. Mendelzon, and A. Vaisman. Maintaining data cubes under dimension updates. In *Proceedings of the 15th International Conference on Data Engineering, ICDE'99*, pages 346–355. IEEE Computer Society Press, 1999.

108. C. Hurtado, A. Mendelzon, and A. Vaisman. Updating OLAP dimensions. In [282], pages 60–66.

109. B. Hüsemann, J. Lechtenbörger, and G. Vossen. Conceptual data warehouse design. In [137], page 6.

110. IBM. DB2 Spatial Extender and Geodetic Data Management Feature: User's Guide and Reference. ftp://ftp.software.ibm.com/ps/products/db2/info/vr9/pdf/letter/en_US/db2sbe90.pdf, 2006.

111. IBM. DB2 Alphablox cube server administrator's guide. http://publib.boulder.ibm.com/infocenter/ablxhelp/v8r4m0/index.jsp, 2007.

112. M. Ibrahim, J. Küng, and N. Revell, editors. *Proceedings of the 11th International Conference on Database and Expert Systems Applications, DEXA'00*, LNCS 1873. Springer, 2000.

113. C. Imhoff, N. Galemmo, and J. Geiger. *Mastering Data Warehouse Design*. Wiley, 2003.

114. M. Ince. Elements that paint a portrait of global power. The Times Higher, Education Supplement, 2004. World University Rankings Methodology.

115. M. Ince. Fine-tuning puts picture in much sharper focus. The Times Higher, Education Supplement, 2005. World University Rankings Methodology.

116. M. Ince. Insiders and outsiders lend a balanced view. The Times Higher, Education Supplement, 2006. World University Rankings Methodology.

117. M. Indulska and M. Orlowska. On aggregation issues in spatial data management. In X. Zhou, editor, *Proceedings of the 13th Australasian Database Conference, ADC'02*, pages 75–84. CRPIT Australian Computer Society, 2002.

118. W. Inmon. *Building the Data Warehouse*. Wiley, 2002.

119. Institute of Higher Education. Academic ranking of world universites. `http://ed.sjtu.edu.cn/rank/2003/2003main.htm`, 2003. Shanghai Jiao Tong University.

120. Institute of Higher Education. Ranking methodology. `http://ed.sjtu.edu.cn/rank/2003/methodology.htm`, 2003. Shanghai Jiao Tong University.

121. Institute of Higher Education. Academic ranking of world universites. `http://ed.sjtu.edu.cn/rank/2004/2004Main.htm`, 2004. Shanghai Jiao Tong University.

122. Institute of Higher Education. Ranking methodology. `http://ed.sjtu.edu.cn/rank/2004/Methodology.htm`, 2004. Shanghai Jiao Tong University.

123. Institute of Higher Education. Academic ranking of world universites. `http://ed.sjtu.edu.cn/rank/2005/ARWU2005TOP500list.htm`, 2005. Shanghai Jiao Tong University.

124. Institute of Higher Education. Ranking methodology. `http://ed.sjtu.edu.cn/rank/2005/ARWU2005Methodology.htm`, 2005. Shanghai Jiao Tong University.

125. Institute of Higher Education. Academic ranking of world universites. `http://ed.sjtu.edu.cn/rank/2006/ARWU2006TOP500list.htm`, 2006. Shanghai Jiao Tong University.

126. Institute of Higher Education. Ranking methodology. `http://ed.sjtu.edu.cn/rank/2006/ARWU2006Methodology.htm`, 2006. Shanghai Jiao Tong University.

127. ISO TC 211. Geographic information – Spatial referencing by coordinates: ISO 19111:2003, 2003.

128. ISO/IEC JTC 1/SC 32. Information Technology – Database languages – SQL Multimedia and Application Packages – Part 3: Spatial: ISO/IEC 13249-3:2006, 2006. Third edition.

129. H. Jagadish, L. Lakshmanan, and D. Srivastava. What can hierarchies do for data warehouses. In [12], pages 530–541.

130. M. Jarke, M. Lanzerini, C. Quix, T. Sellis, and P. Vassiliadis. Quality-driven data warehouse design. In [131], pages 165–179.

131. M. Jarke, M. Lenzerini, Y. Vassiluiou, and P. Vassiliadis. *Fundamentals of Data Warehouses*. Springer, second edition, 2003.

132. C.S. Jensen. *Temporal Database Management*. PhD thesis, University of Aalborg, 2003.

133. C.S. Jensen, A. Klygis, T. Pedersen, and I. Timko. Multidimensional data modeling for location-based services. *VLDB Journal*, 13(1):1–21, 2004.

134. C.S. Jensen, M. Schneider, B. Seeger, and V. Tsotras, editors. *Proceedings of the 7th International Symposium on Advances in Spatial and Temporal Databases, SSTD'01*, LNCS 2121. Springer, 2001.

135. C.S. Jensen and R.T. Snodgrass. Temporally enhanced database design. In M. Papazoglou, S. Spaccapietra, and Z. Tari, editors, *Advances in Object-Oriented Data Modeling*, pages 163–193. MIT Press, 2000.

136. C.S. Jensen, R.T. Snodgrass, and M. Soo. Extending normal forms to temporal relations. Technical Report TR-17, Time Center, 1992.

137. M. Jeusfeld, H. Shu, M. Staudt, and G. Vossen, editors. *Proceedings of the 2nd International Workshop on Design and Management of Data Warehouses, DMDW'00*. CEUR Workshop Proceedings, 2000.

138. H. Jullens. Spatial customer intelligence: The map shows the future. *GeoInformatics*, 3(12), 2000.

139. Y. Kambayashi, M. Mohania, and A. Min Tjoa, editors. *Proceedings of the 2nd International Conference on Data Warehousing and Knowledge Discovery, DaWaK'00*, LNCS 1874. Springer, 2000.

140. M. Kang, F. Pinet, M. Schneider, J. Chanet, and F. Vigier. How to design geographic databases? Specific UML profile and spatial OCL applied to wireless ad hoc networks. In *Proceedings of the 7th AGILE Conference on Geographic Information Science*, pages 197–207, 2004.

141. N. Karayannidis, A. Tsois, T. Sellis, R. Pieringer, V. Markl, F. Ramsak, R. Fenk, K. Elhardt, and R. Bayer. Processing star queries on hierarchically-clustered fact tables. In P.A. Bernstein, Y.E. Ioannidis, R. Ramakrishnan, and D. Papadias, editors, *Proceedings of the 28th International Conference on Very Large Data Bases, VLDB'02*, pages 730–741. Morgan Kaufmann, 2002.

142. V. Khatri, S. Ram, and R.T. Snodgrass. Augmenting a conceptual model with geospatiotemporal annotations. *IEEE Transactions on Knowledge and Data Engineering*, 16(11):1324–1338, 2004.

143. V. Khatri, S. Ram, and R.T. Snodgrass. On augmenting database design-support environments to capture the geo-spatio-temporal data semantics. *Information Systems*, 31(2):98–133, 2006.

144. KHEOPS Technologies. JMap spatial OLAP. http://www.kheops-tech.com/en/jmap/solap.jsp, 2004.

145. R. Kimball, L. Reeves, M. Ross, and W. Thornthwaite. *The Data Warehouse Lifecycle Toolkit: Expert Methods for Designing, Developing, and Deploying Data Warehouses*. Wiley, 1998.

146. R. Kimball and M. Ross. *The Data Warehouse Toolkit: The Complete Guide to Dimensional Modeling*. Wiley, second edition, 2002.

147. P. King. The database design process. In S. Spaccapietra, editor, *Proceedings of the 5th International Conference on the Entity-Relationship Approach, ER'86*, pages 475–488. North-Holland, 1986.

148. N. Kline and R.T. Snodgrass. Computing temporal aggregates. In P. Yu and A. Chen, editors, *Proceedings of the 11th International Conference on Data Engineering, ICDE'95*, pages 222–231. IEEE Computer Society Press, 1995.

149. T. Kolbe and R. Gröger. Towards unified 3D city models. In J. Schiewe, M. Hahn, M. Madden, and M. Sester, editors, *Proceedings of ISPRS Commission IV Joint Workshop on Challenges in Geospatial Analysis, Integration and Visualization II*, 2003.

150. C. Koncilia. A bi-temporal data warehouse model. In J. Eder and M. Missikoff, editors, *Proceedings of the 15th International Conference on Advanced Information Systems Engineering, CAiSE'03*, LNCS 2681, pages 77–80. Springer, 2003.

151. R.V. Kothuri, A. Godfrind, and E. Beinat. *Pro Oracle Spatial*. Apress, 2004.
152. Z. Kouba, K. Matoušek, and P. Mikšovský. On data warehouse and GIS integration. In [112], pages 604–613.
153. M. Koubarakis, T. Sellis, A. Frank, R.H. Güting, C.S. Jensen, A. Lorentzos, Y. Manolopoulos, E. Nardelli, B. Pernici, H.-J. Schek, M. Scholl, B. Theodoulidis, and N. Tryfona. *Spatio-Temporal Databases: The Chorochronos Approach*. Springer, 2003.
154. H. Kriegel, P. Kunath, M. Pfeifle, and M. Renz. Object-relational management of complex geographical objects. In [237], pages 109–117.
155. H. Kunii, S. Jajodia, and S. Solvberg, editors. *Proceedings of the 20th International Conference on Conceptual Modeling, ER'01*, LNCS 2224. Springer, 2001.
156. T. Lachev. *Applied Microsoft Analysis Services 2005*. Prologica Press, 2005.
157. L. Lakshmanan, editor. *Proceedings of the 4th International Workshop on Design and Management of Data Warehouses, DMDW'02*. CEUR Workshop Proceedings, 2002.
158. S. Larrivée, Y. Bédard, and J. Pouliot. How to enrich the semantics of geospatial databases by properly expressing 3D objects in a conceptual model. In [191], pages 999–1008.
159. J. Lechtenbörger. My favorite issues in data warehouse modeling. In [281], pages 87–88.
160. J. Lechtenbörger and G. Vossen. Multidimensional normal forms for data warehouse design. *Information Systems*, 28(5):415–434, 2003.
161. D. Leffingwell and D. Widrig. *Managing Software Requirements: A Unified Approach*. Addison-Wesley, 2000.
162. W. Lehner, J. Albrecht, and H. Wedekind. Normal forms for multidimensional databases. In [250], pages 63–72.
163. H. Lenz and A. Shoshani. Summarizability in OLAP and statistical databases. In Y. Ioannidis and D. Hansen, editors, *Proceedings of the 9th International Conference on Scientific and Statistical Database Management, SSDBM'97*, pages 132–143. IEEE Computer Society Press, 1997.
164. H. Lenz, P. Vassiliadis, M. Jeusfeld, and M. Staudt, editors. *Proceedings of the 5th International Workshop on Design and Management of Data Warehouses, DMDW'03*. CEUR Workshop Proceedings, 2003.
165. B. List, R. Bruckner, K. Machaczek, and J. Schiefer. Comparison of data warehouse development methodologies: Case study of the process warehouse. In A. Hameurlain, R. Cicchetti, and R. Traunmüller, editors, *Proceedings of the 13th International Conference on Database and Expert Systems Applications, DEXA'02*, LNCS 2453, pages 203–215. Springer, 2002.
166. B. List, J. Schiefer, and A. Min Tjoa. Process-oriented requirement analysis supporting the data warehouse design process: A use case driven approach. In [112], pages 593–603.
167. N.C. Liu and Y. Cheng. Academic ranking of world universities: Methodologies and problems. *Education in Europe*, 30(2):127–136, 2005.
168. S. Luján-Mora and J. Trujillo. A comprehensive method for data warehouse design. In [164].
169. S. Luján-Mora, J. Trujillo, and I. Song. A UML profile for multidimensional modeling in data warehouses. *Data & Knowledge Engineering*, 59(3):725–769, 2006.

170. S. Luján-Mora, P. Vassiliadis, and J. Trujillo. Data mapping diagrams for data warehouse design with UML. In P. Atzeni, W. Chu, H. Lu, S. Zhou, and T. Ling, editors, *Proceedings of the 23rd International Conference on Conceptual Modeling, ER'04*, LNCS 3288, pages 191–204. Springer, 2004.

171. M. Maibaum, G. Rimon, C. Orengo, N. Martin, and A. Poulovasillis. BioMap: Gene family based integration of heteregeneous biological databases using AutoMed metadata. In A. Min Tjoa and R. Wagner, editors, *Proceedings of the 15th International Workshop on Database and Expert Systems Applications, DEXA'04*, pages 384–388. IEEE Computer Society Press, 2004.

172. M. Maibaum, L. Zamboulis, G. Rimon, C. Orengo, N. Martin, and A. Poulovasillis. Cluster based integration of heterogeneous biological databases using the AutoMed toolkit. In *Proceedings of the 2nd International Workshop on Data Integration in the Life Sciences, DILS 2005*, LNCS 3615, pages 191–207. Springer, 2005.

173. E. Malinowski and E. Zimányi. OLAP hierarchies: A conceptual perspective. In A. Persson and J. Stirna, editors, *Proceedings of the 16th International Conference on Advanced Information Systems Engineering, CAiSE'04*, LNCS 3084, pages 477–491. Springer, 2004.

174. E. Malinowski and E. Zimányi. Representing spatiality in a conceptual multidimensional model. In [237], pages 12–21.

175. E. Malinowski and E. Zimányi. Spatial hierarchies and topological relationships in the Spatial MultiDimER model. In M. Jackson, D. Nelson, and S. Strik, editors, *Proceedings of the 22nd British National Conference on Databases, BNCOD'05*, LNCS 3567, pages 17–28. Springer, 2005.

176. E. Malinowski and E. Zimányi. A conceptual solution for representing time in data warehouse dimensions. In M. Stumptner, S. Hartmann, and Y. Kiyoki, editors, *Proceedings of the 3rd Asia-Pacific Conference on Conceptual Modelling, APCCM'06*, pages 45–54. CRPIT Australian Computer Society, 2006.

177. E. Malinowski and E. Zimányi. Hierarchies in a multidimensional model: From conceptual modeling to logical representation. *Data & Knowledge Engineering*, 59(2):348–377, 2006.

178. E. Malinowski and E. Zimányi. Inclusion of time-varying measures in temporal data warehouses. In Y. Manolopoulos, J. Filipe, P. Constantopoulos, and J. Cordeiro, editors, *Proceedings of the 8th International Conference on Enterprise Information Systems, ICEIS'06*, pages 181–186, 2006.

179. E. Malinowski and E. Zimányi. Object-relational representation of a conceptual model for temporal data warehouses. In E. Dubois and K. Pohl, editors, *Proceedings of the 18th International Conference on Advanced Information Systems Engineering, CAiSE'06*, LNCS 4001, pages 96–110. Springer, 2006.

180. E. Malinowski and E. Zimányi. Requirements specification and conceptual modeling for spatial data warehouses. In R. Meersman, Z. Tari, P. Herrero, et al., editors, *Proceedings of the OTM 2006 Workshops: On the Move to Meaningful Internet Systems*, LNCS 4277, pages 1616–1625. Springer, 2006.

181. E. Malinowski and E. Zimányi. A conceptual model for temporal data warehouses and its transformation to the ER and object-relational models. In *Data & Knowledge Engineering*, 2007. To appear.

182. E. Malinowski and E. Zimányi. Implementing spatial data warehouse hierarchies in object-relational DBMSs. In J. Cardoso, J. Cordeiro, and J. Felipe, editors, *Proceedings of the 9th International Conference on Enterprise Information Systems, ICEIS'07*, pages 186–191, 2007.

183. E. Malinowski and E. Zimányi. Logical representation of a conceptual model for spatial data warehouses. *GeoInformatica*, 11(4):431–457, 2007.
184. E. Malinowski and E. Zimányi. Spatial data warehouses: Some solutions and unresolved problems. In *Proceedings of the 3rd International Workshop on Databases for Next-Generation Researchers, SWOD'07*, 2007. To appear.
185. P. Marchand. *The Spatio-Temporal Topological Operator Dimension, a Hyperstructure for Multidimensional Spatio-temporal Explorations and Analysis.* PhD thesis, Université Laval Sainte-Foy, 2003.
186. E. Marcos, B. Vela, and J.M. Cavero. A methodological approach for object-relational database design. *Journal on Software and System Modelling*, 2(1):59–72, 2003.
187. C. Martín and A. Abelló. A temporal study of data sources to load a corporate data warehouse. In Y. Kambayashi, M. Mohania, and W. Wös, editors, *Proceedings of the 5th International Conference on Data Warehousing and Knowledge Discovery, DaWaK'03*, LNCS 2737, pages 109–118. Springer, 2003.
188. T. Martyn. Reconsidering multi-dimensional schemas. *SIGMOD Record*, 33(1):83–88, 2004.
189. J. Mazon, J. Trujillo, M. Serrano, and M. Piattini. Designing data warehouses: From business requirement analysis to multidimensional modeling. In *Proceedings of the 1st International Workshop on Requirements Engineering for Business Need and IT Alignment, REBN'05*, pages 44–53, 2005.
190. C. Medeiros, editor. *Proceedings of the 7th ACM Symposium on Advances in Geographic Information Systems, ACM-GIS'99.* ACM Press, 1999.
191. R. Meersman, Z. Tari, P. Herrero, et al., editors. *Proceedings of the OTM 2005 Workshops: On the Move to Meaningful Internet Systems*, LNCS 3762. Springer, 2005.
192. J. Melton. *Advanced SQL:1999. Understanding Object-Relational and Other Advanced Features.* Morgan Kaufmann, 2003.
193. J. Melton. SQL:2003 has been published. *SIGMOD Record*, 33(1):119–125, 2003.
194. A. Mendelzon and A. Vaisman. Temporal queries in OLAP. In [1], pages 243–253.
195. A. Mendelzon and A. Vaisman. Time in multidimensional databases. In [249], pages 166–199.
196. I. Merlo, E. Bertino, E. Ferrari, and G. Guerrini. A temporal object-oriented data model with multiple granularities. In *Proceedings of the 6th International Workshop on Temporal Representation and Reasoning, TIME'99*, pages 73–81. IEEE Computer Society Press, 1999.
197. P. Mikšovský and Z. Kouba. GOLAP: Geographical on-line analytical processing. In H. Mayr, F. Lazanský, G. Quirchmayr, and P. Vogel, editors, *Proceedings of the 12th International Conference on Database and Expert Systems Applications, DEXA'01*, LNCS 2113, pages 201–205. Springer, 2001.
198. A. Min Tjoa and J. Trujillo, editors. *Proceedings of the 8th International Conference on Data Warehousing and Knowledge Discovery, DaWaK'06*, LNCS 4081. Springer, 2006.
199. M. Miquel, A. Brisebois, Y. Bédard, and G. Edwards. Implementation and evaluation of hypercube-based method for spatio-temporal exploration and analysis. *ISPRS Journal of Photogrammetry & Remote Sensing*, 59(1–2):6–20, 2004.

200. M. Mohania and A. Min Tjoa, editors. *Proceedings of the 1st International Conference on Data Warehousing and Knowledge Discovery, DaWaK'99*, LNCS 1676. Springer, 1999.

201. D. Moody and M. Kortink. From enterprise models to dimensional models: A methodology for data warehouse and data mart design. In [137], page 5.

202. B. Moon, F. Vega, and V. Immanuel. Efficient algorithms for large-scale temporal aggregation. *IEEE Transactions on Knowledge and Data Engineering*, 15(3):744–759, 2003.

203. T. Morzy and R. Wrembel. Modeling a multiversion data warehouse: A formal approach. In [40], pages 120–127.

204. T. Morzy and R. Wrembel. On querying versions of multiversion data warehouse. In I.Y. Song and K. Davis, editors, *Proceedings of the 7th ACM International Workshop on Data Warehousing and OLAP, DOLAP'04*, pages 92–101. ACM Press, 2004.

205. H. Nemati, D. Steiger, L. Iyer, and R. Herschel. Knowledge warehouse: An architectural integration of knowledge management, decision support, artificial intelligence and data warehousing. *Decision Support Systems*, 33:143–161, 2002.

206. T. Niemi, J. Nummenmaa, and P. Thanisch. Logical multidimensional database design for ragged and unbalanced aggregation hierarchies. In [290], page 7.

207. Object Management Group. Common Warehouse Metamodel. http://www.omg.org/docs/formal/03-03-02.pdf, 2002.

208. J. O'Leary. Top performers on the global stage take a bow. The Times Higher, Education Supplement, 2004. World University Rankings Editorial.

209. E. O'Neil and G. Graefe. Multi-table joins through bitmapped join indices. *SIGMOD Record*, 24(3):8–11, 1995.

210. Open Geospatial Consortium Inc. OpenGIS Implementation Specification for Geographic information - Simple feature access - Part 2: SQL option. OGC 06-104r3, Version 1.2.0, 2006.

211. Open GIS Consortium. OpenGIS simple features specification for SQL. Technical Report Revision 1.1, 1999.

212. Oracle Corporation. OLAP option to Oracle database 10g. White paper, 2004.

213. Oracle Corporation. Oracle 10g database release 2 documentation. http://www.oracle.com/technology/documentation/database10gR2.html, 2006.

214. Oracle Corporation. Oracle OLAP application developer's guide 10g release 2. http://www.oracle.com/technology/products/bi/olap/index.html, 2006.

215. Oracle Corporation. Oracle Spatial: Topology and network data models, 10g release 2. http://www.oracle.com/technology/products/spatial/spatial_10g_doc_index.html, 2006.

216. Oracle Corporation. Oracle Spatial: User's guide and references, 10g release 2. http://www.oracle.com/technology/products/spatial/spatial_10g_doc_index.html, 2006.

217. T. Ott and F. Swiaczny. *Time-Integrative Geographic Information Systems: Management and Analysis of Spatio-temporal Data*. Springer, 2001.

218. M. Özsu and P. Valduriez. *Principles of Distributed Database Systems*. Prentice Hall, second edition, 1999.

219. F. Paim, A. Carvalho, and J. Castro. Towards a methodology for requirements analysis of data warehouse systems. In *Proceedings of the 16th Brazilian Symposium on Software Engineering, SBES'02*, pages 1–16, 2002.

220. F. Paim and J. Castro. DWARF: An approach for requirements definition and management of data warehouse systems. In *Proceedings of the 11th IEEE International Requirements Engineering Conference, RE'03*, pages 75–84. IEEE Computer Society Press, 2003.

221. D. Papadias and M.J. Egenhofer. Algorithms for hierarchical spatial reasoning. *Geomatica*, 1(3):251–273, 1997.

222. D. Papadias, P. Kalnis, J. Zhang, and Y. Tao. Efficient OLAP operations in spatial data warehouses. In [134], pages 443–459.

223. D. Papadias, Y. Tao, P. Kalnis, and J. Zhang. Indexing spatio-temporal data warehouses. *IEEE Data Engineering Bulletin*, 25(1):10–17, 2002.

224. S. Paraboschi, G. Sindoni, E. Baralis, and E Teniente. Materialized views in multidimensional databases. In [249], pages 222–251.

225. C. Parent, S. Spaccapietra, and E. Zimányi. Modeling time from a conceptual perspective. In G. Gardarin, J. French, N. Pissinou, and K. Makki, editors, *Proceedings of the 7th International Conference on Information and Knowledge Management, CIKM'98*, pages 432–440. ACM Press, 1998.

226. C. Parent, S. Spaccapietra, and E. Zimányi. Spatio-temporal conceptual models: Data structures + Space + Time. In [190], pages 26–33.

227. C. Parent, S. Spaccapietra, and E. Zimányi. *Conceptual Modeling for Traditional and Spatio-Temporal Applications: The MADS Approach*. Springer, 2006.

228. C.-S. Park, M.-H. Kim, and Y.-J. Lee. Finding an efficient rewriting of OLAP queries using materialized views in data warehouses. *Decision Support Systems*, 32(4):379–399, 2002.

229. PCI Geomatics. Solutions for Oracle 10g. GeoRaster ETL for Oracle. http://www.pcigeomatics.com/products/oracle_solutions.html, 2007.

230. T.B. Pedersen. *Aspects of Data Modeling and Query Processing for Complex Multidimensional Data*. PhD thesis, Aalborg University, Denmark, 2000.

231. T. Pedersen, C.S. Jensen, and C. Dyreson. A foundation for capturing and querying complex multidimensional data. *Information Systems*, 26(5):383–423, 2001.

232. T.B. Pedersen and N. Tryfona. Pre-aggregation in spatial data warehouses. In [134], pages 460–478.

233. N. Pelekis, B. Theodoulidis, I. Kopanakis, and Y. Theodoridis. Literature review of spatio-temporal database models. *Knowledge Engineering Review*, 19(3):235–274, 2004.

234. V. Peralta and R. Ruggia. Using design guidelines to improve data warehouse logical design. In [164].

235. W. Pereira and K. Becker. A methodology targeted at the insertion of data warehouse technology in corporations. In K. Becker, A. Augusto de Souza, and D. Yluska de Souza Fernandes, editors, *Proceedings of the 15th Simpósio Brasileiro de Banco de Dados, SBBD'00*, pages 316–330. CEFET-PB, 2000.

236. G. Pestana, M. Mira da Silva, and Y. Bédard. Spatial OLAP modeling: An overview base on spatial objects changing over time. In *Proceedings of the 3rd International Conference on Computational Cybernetics, ICCC'05*, 2005.

237. D. Pfoser, I. Cruz, and M. Ronthaler, editors. *Proceedings of the 12th ACM Symposium on Advances in Geographic Information Systems, ACM-GIS'04*. ACM Press, 2004.

238. C. Phipps and K.C. Davis. Automating data warehouse conceptual schema design and evaluation. In [157], pages 23–32.

239. V. Poe, P. Klauer, and S. Brobst. *Building a Data Warehouse for Decision Support*. Prentice Hall, second edition, 1997.

240. E. Pourabbas. Cooperation with geographic databases. In [249], pages 393–432.

241. E. Pourabbas and M. Rafanelli. A pictorial query language for querying geographic databases using positional and OLAP operators. *SIGMOD Record*, 31(2):22–27, 2002.

242. E. Pourabbas and M. Rafanelli. Hierarchies. In [249], pages 91–115.

243. N. Prakash and A. Gosain. Requirements driven data warehouse development. In J. Eder and T. Welzer, editors, *Short Paper Proceedings of the 15th International Conference on Advanced Information Systems Engineering, CAiSE'03*, pages 13–16. CEUR Workshop Proceedings, 2003.

244. N. Prat, J. Akoka, and I. Comyn-Wattiau. A UML-based data warehouse design method. *Decision Support Systems*, 42(3):1449–1473, 2006.

245. R.S. Pressman and R. Pressman. *Software Engineering: A Practitioner's Approach*. McGraw-Hill Education, sixth edition, 2004.

246. R. Price, N. Tryfona, and C.S. Jensen. Extended spatiotemporal UML: Motivations, requirements, and constructs. *Journal of Database Management*, 11(4):14–27, 2000.

247. R. Price, N. Tryfona, and C.S. Jensen. Modeling topological constraints in spatial part-whole relationships. In [155], pages 27–40.

248. M. Rafanelli. Basic notions. In [249], pages 1–45.

249. M. Rafanelli, editor. *Multidimensional Databases: Problems and Solutions*. Idea Group, 2003.

250. M. Rafanelli and M. Jarke, editors. *Proceedings of the 10th International Conference on Scientific and Statistical Database Management, SSDBM'98*. IEEE Computer Society Press, 1998.

251. M. Rafanelli and A. Shoshani. STORM: A statistical object representation model. In Z. Michalewicz, editor, *Proceedings of the 9th International Conference on Scientific and Statistical Database Management, SSDBM'90*, LNCS 420, pages 14–29. Springer, 1990.

252. F. Rao, L. Zhang, X. Lan, Y. Li, and Y. Chen. Spatial hierarchy and OLAP-favored search in spatial data warehouse. In *Proceedings of the 6th ACM International Workshop on Data Warehousing and OLAP, DOLAP'03*, pages 48–55. ACM Press, 2003.

253. F. Ravat and F. Teste. Supporting data changes in multidimensional data warehouses. *International Review on Computer and Software*, 1(3):251–259, 2006.

254. F. Ravat, O. Teste, and G. Zurfluh. A multiversion-based multidimensional model. In [198], pages 65–74.

255. M. Riedewals, D. Agrawal, and A. El Abbadi. Efficient integration and aggregation of historical information. In [75], pages 13–24.

256. P. Rigaux, M. Scholl, and A. Voisard. *Spatial Databases with Application to GIS*. Morgan Kaufmann, 2002.

257. S. Rivest, Y. Bédard, and P. Marchand. Toward better suppport for spatial decision making: Defining the characteristics of spatial on-line analytical processing (SOLAP). *Geomatica*, 55(4):539–555, 2001.

258. S. Rivest, Y. Bédard, M. Proulx, M. Nadeau, F. Hubert, and J. Pastor. SOLAP technology: Merging business intelligence with geospatial technology for interactive spatio-temporal exploration and analysis of data. *ISPRS Journal of Photogrammetry & Remote Sensing*, 60(1):17–33, 2005.

259. S. Rizzi. Open problems in data warehousing: 8 years later. In [164].

260. S. Rizzi and M. Golfarelli. What time is it in the data warehouse? In [198], pages 134–144.

261. S. Rizzi and E. Saltarelli. View materialization vs. indexing: Balancing space constraints in data warehouse design. In J. Eder and M. Missikoff, editors, *Proceedings of the 15th International Conference on Advanced Information Systems Engineering, CAiSE 2003*, LNCS 2681, pages 502–519. Springer, 2003.

262. M.A. Roth, H.F. Korth, and A. Silberschatz. Extended algebra and calculus for nested relational databases. *ACM Transactions on Database Systems*, 13(4):389–417, 1988.

263. C. Sapia, M. Blaschka, G. Höfling, and B. Dinter. Extending the E/R model for multidimensional paradigm. In T. Ling, S. Ram, and M. Lee, editors, *Proceedings of the 17th International Conference on Conceptual Modeling, ER'98*, LNCS 1507, pages 105–116. Springer, 1998.

264. N. Sarda. Temporal issues in data warehouse systems. In Y. Kambayashi and H. Takakura, editors, *Proceedings of the International Symposium on Database Applications in Non-Traditional Environments, DANTE'99*, pages 27–34. IEEE Computer Society Press, 1999.

265. H. Schek, F. Saltor, I. Ramos, and G. Alonso, editors. *Proceedings of the 6th International Conference on Extending Database Technology, EDBT'98*, LNCS 1377. Springer, 1998.

266. J. Schiefer, B. List, and R. Bruckner. A holistic approach for managing requirements of data warehouse systems. In *Proceedings of the 8th Americas' Conference on Information Systems, AMCIS'02*, pages 77–87, 2002.

267. M. Schneider and T. Behr. Topological relationships between complex spatial objects. *ACM Transactions on Database Systems*, 31(1):39–81, 2006.

268. S. Shekhar and S. Chawla. *Spatial Databases: A Tour*. Prentice Hall, 2003.

269. J. Shim, M. Warkentin, J. Courtney, D. Power, R. Sharda, and C. Carlsson. Past, present and future of decision support technology. *Decision Support Systems*, 32(2):111–126, 2002.

270. A. Shukla, P. Deshpande, J. Naughton, and K. Ramasamy. Storage estimation for multidimensional aggregates in the presence of hierarchies. In T.M. Vijayaraman, A.P. Buchmann, C. Mohan, and N.L. Sarda, editors, *Proceedings of the 22nd International Conference on Very Large Data Bases, VLDB'96*, pages 522–543. Morgan Kaufmann, 1996.

271. Y. Sismanis, A. Deligiannakis, N. Roussopoulos, and Y. Kotidis. Dwarf: Shrinking the PetaCube. In [75], pages 464–475.

272. J. Smith and D. Smith. Data abstractions: Aggregation and generalization. *ACM Transactions on Database Systems*, 2(2):105–133, 1977.

273. R.T. Snodgrass, editor. *The TSQL2 Temporal Query Language*. Kluwer Academic, 1995.

274. R.T. Snodgrass. *Developing Time-Oriented Database Applications in SQL*. Morgan Kaufmann, 2000.

275. R.T. Snodgrass, M. Böhlen, C.S. Jensen, and N. Kline. Adding valid time to SQL/Temporal. ANSI X3H2-96-501r2, ISO/IEC JTC1/SC21/WG3 DBL MAD-146r2, 1996.

276. R.T. Snodgrass, M. Böhlen, C.S. Jensen, and A. Steiner. Adding transaction time to SQL/Temporal: Temporal change proposal. ANSI X3H2-96-152r, ISO-ANSI SQL/ISO/IECJTC1/SC21/WG3 DBL MCI-143, 1996.

277. R.T. Snodgrass, M. Böhlen, C.S. Jensen, and A. Steiner. Transitioning temporal support in TSQL2 to SQL3. In [71], pages 150–194.
278. R.T. Snodgrass, S. Gomez, and L. McKenzie. Aggregates in the temporary query language TQuel. *IEEE Transactions on Knowledge and Data Engineering*, 5(5):826–842, 1993.
279. I. Sommerville. *Software Engineering*. Addison-Wesley, eighth edition, 2006.
280. I. Song, W. Rowen, C. Medsker, and E. Ewen. An analysis of many-to-many relationships between facts and dimension tables in dimensional modeling. In [290], page 6.
281. I.Y. Song and J. Trujillo, editors. *Proceedings of the 8h ACM International Workshop on Data Warehousing and OLAP, DOLAP'05*. ACM Press, 2005.
282. I.-Y. Song and T.J. Teorey, editors. *Proceedings of the 2nd ACM International Workshop on Data Warehousing and OLAP, DOLAP'99*. ACM Press, 1999.
283. S. Spaccapietra, C. Parent, M.L. Damiani, J. Macedo, F. Porto, and C. Vangenot. A conceptual view on trajectories. Technical Report LBD-REPORT-2007-001, Database Laboratory, Ecole Polytechnique Fédérale de Lausanne, 2008. Submited for publication.
284. N. Stefanovic, J. Han, and K. Koperski. Object-based selective materialization for efficient implementation of spatial data cubes. *IEEE Transactions on Knowledge and Data Engineering*, 12(6):938–958, 2000.
285. Y. Tao, D. Papadias, and Ch. Faloutsos. Approximate temporal aggregations. In *Proceedings of the 20th International Conference on Data Engineering, ICDE'04*, pages 190–201. IEEE Computer Society Press, 2004.
286. The Times Higher, Education Supplement. The world's top 200 universities, 2004.
287. The Times Higher, Education Supplement. The world's top 200 universities, 2005.
288. The Times Higher, Education Supplement. The world's top 200 universities, 2006.
289. D. Theodoratos, editor. *Proceedings of the 5th ACM International Workshop on Data Warehousing and OLAP, DOLAP'02*. ACM Press, 2002.
290. D. Theodoratos, J. Hammer, M. Jeusfeld, and M. Staudt, editors. *Proceedings of the 3rd International Workshop on Design and Management of Data Warehouses, DMDW'01*. CEUR Workshop Proceedings, 2001.
291. C.R. Thomas. Data definition for colleges and universities. Consortium for Higher Education Software Services, second edition, 2004. http://www.nchems.org/pubs/chess/DD2_Documentation.pdf.
292. E. Thomsen. *OLAP Solutions. Building Multidimensional Information Systems*. Wiley, 2002.
293. I. Timko and T. Pedersen. Capturing complex multidimensional data in location-based data warehouses. In [237], pages 147–156.
294. D. Tomlin. *Geographic Information Systems and Cartographic Modeling*. Prentice Hall, 1990.
295. R. Torlone. Conceptual multidimensional models. In [249], pages 69–90.
296. R. Torlone and I. Panella. Design and development of a tool for integrating heterogeneous data. In A. Min Tjoa and J. Trujillo, editors, *Proceedings of the 7th International Conference on Data Warehousing and Knowledge Discovery, DaWaK'05*, LNCS 3589, pages 105–114. Springer, 2005.
297. J. Trujillo, M. Palomar, J. Gomez, and I. Song. Designing data warehouses with OO conceptual models. *IEEE Computer*, 34(12):66–75, 2001.

408   References

298. N. Tryfona, F. Busborg, and J. Borch. StarER: A conceptual model for data warehouse design. In [282], pages 3–8.
299. N. Tryfona and M.J. Egenhofer. Consistency among parts and aggregates: A computational model. *Transactions in GIS*, 4(3):189–206, 1997.
300. N. Tryfona and C.S. Jensen. Conceptual data modelling for spatio-temporal applications. *GeoInformatica*, 3(3):245–268, 1999.
301. A. Tsois, N. Karayannidis, and T. Sellis. MAC: Conceptual data modelling for OLAP. In [290], page 5.
302. U.S. Census Bureau. Standard Hierarchy of Census Geographic Entities and Hierarchy of American Indian, Alaska Native, and Hawaiian Entities. http://www.census.gov/geo/www/geodiagram.pdf, 2004.
303. A. Vaisman. Data quality-based requirements elicitation for decision support systems. In [317], chapter 16, pages 58–86.
304. P. van Oosterom, W. Quak, and T. Tijssen. Testing current DBMS products with real spatial data. In *Proceedings of the 23rd Urban Data Management Symposium, UDMS'02*, pages VII.1–VII.18, 2002.
305. P. van Oosterom, J. Stoter, W. Quak, and S. Zlatanova. The balance between geometry and topology. In D. Richardson and P. van Oosterom, editors, *Proceedings of the 10th International Symposium on Spatial Data Handling, SDH'02*, pages 209–224. Springer, 2002.
306. P. Vassiliadis. Modeling multidimesional databases, cubes and cube operations. In [250], page 53.
307. P. Vassiliadis and T. Sellis. A survey on logical models for OLAP databases. In A. Delis, C. Faloutsos, S. Ghandeharizadeh, and E. Panagos, editors, *Proceedings of the ACM SIGMOD International Conference on Management of Data, SIGMOD'99*, pages 64–69. ACM Press, 1999.
308. I.F. Vega, R.T. Snodgrass, and B. Moon. Spatiotemporal aggregate computation: A survey. Technical Report TR-77, Time Center, 2004.
309. J.R.R. Viqueira, N.A. Lorentzos, and N.R. Brisaboa. Survey on spatial data modeling approaches. In Y. Manolopoulos, M. Papadopoulos, and M. Vassilakopoulos, editors, *Spatial Databases: Technologies, Techniques and Trends*, pages 1–22. Idea Group, 2005.
310. X. Wang, C. Bettini, A. Brodsky, and S. Jajodia. Logical design for temporal databases with multiple granularities. *ACM Transactions on Database Systems*, 22(2):115–170, 1997.
311. J. Wijsen. Design of temporal relational databases based on dynamic and temporal functional dependencies. In J. Clifford and A. Tuzhilin, editors, *Proceedings of the International Workshop on Temporal Databases*, pages 61–76. Springer, 1995.
312. J. Wijsen. Temporal FDs on complex objects. *ACM Transactions on Database Systems*, 24(1):127–176, 1999.
313. J. Wijsen. A string-based model for infinite granularities. In C. Bettini and A. Montanan, editors, *Proceedings of the AAAI Workshop on Spatial and Temporal Granularity*, pages 9–16. AAAI Press, 2000.
314. R. Winter and B. Strauch. A method for demand-driven information requirements analysis in data warehousing projects. In *Proceedings of the 36th Hawaii International Conference on System Sciences, HICSS-36*, pages 1359–1365, 2003.
315. M. Worboys and M. Duckham. *GIS A Computing Perspective*. CRC Press, 2004. Second edition.

316. R. Wrembel and B. Bebel. Metadata management in a multiversion data warehouse. In *Journal on Data Semantics VIII*, LNCS 4380, pages 118–157. Springer, 2007.

317. R. Wrembel and C. Koncilia, editors. *Data Warehouses and OLAP: Concepts, Architectures and Solutions*. IRM Press, 2007.

318. R. Wrembel and T. Morzy. Managing and querying version of multiversion data warehouse. In Y. Ioannidis, M. Scholl, J. Schmidt, et al., editors, *Proceedings of the 10th International Conference on Extending Database Technology, EDBT'06*, LNCS 3896, pages 1121–1124. Springer, 2006.

319. J. Yang and J. Widom. Maintaining temporal views over non-temporal information sources for data warehousing. In [265], pages 389–403.

320. E. Yu. *Modeling Strategic Relationships for Process Engineering*. PhD thesis, University of Toronto, 1995.

321. C. Zaniolo, S. Ceri, C. Faloutsos, R. Snodgross, V. Subrahmanian, and R. Zicari. *Advanced Database Systems*. Morgan Kaufmann, 1997.

322. D.C. Zilio, C. Zuzarte, S. Lightstone, W. Ma, G.M. Lohman, R.J. Cochrane, H. Pirahesh, L. Colby, J. Gryz, E. Alton, and G. Valentin. Recommending materialized views and indexes with the IBM DB2 Design Advisor. In *Proceedings of the 1st International Conference on Autonomic Computing, ICAC 2004*, pages 180–188. IEEE Computer Society, 2004.

323. E. Zimányi. Temporal aggregates and temporal universal quantifiers in standard SQL. *SIGMOD Record*, 32(2):16–21, 2006.

324. E. Zimányi and M. Minout. Preserving semantics when transforming conceptual spatio-temporal schemas. In [191], pages 1037–1046.

325. E. Zimányi, C. Parent, and S. Spaccapietra. TERC+: A temporal conceptual model. In M. Yoshikawa, editor, *Proceedings of the International Symposium on Digital Media Information*, 1997.

326. S. Zlatanova. On 3D topological relationships. In [112], pages 913–924.

Glossary

ad hoc query: A request for information that is created by a user as the need arises and about which the system has no prior knowledge. This is to be contrasted with a *predefined query*.

additive measure: A measure that can be meaningfully aggregated by addition along all of its dimensions. It is the most common type of measure. This is to be contrasted with *semiadditive* and *nonadditive measures*.

aggregation function: A function that computes an aggregated value from a set of values. In a data warehouse, aggregation functions are used for aggregating measures across dimensions and hierarchies.

analysis-driven design: An approach to designing a data warehouse based on the analysis requirements of the decision-making users or organizational processes. It is also called *requirements-* or *business-driven design*. This is to be contrasted with *source-driven design*.

analysis/source-driven design: An approach to designing a data warehouse that is a combination of the analysis-driven and source-driven approaches.

analytical application: An application that produces information for management decisions, usually involving issues such as demographic analysis, trend analysis, pattern recognition, and profiling.

attribute: A structural property of a type. In conceptual models, attributes are attached to entity or relationship types. In the relational model, attributes define the columns of a relation. An attribute has a name, a cardinality, and an associated data type.

back-end process: A process that populates a data warehouse with data from operational and external data sources. This is to be contrasted with the *front-end process*.

binary relationship type: A relationship type between two entity types or between two levels.

bitemporal time: A temporal specification that associates both valid time and transaction time with a particular schema element.

Boyce-Codd normal form (BCNF): A normal form that ensures that every functional dependency in a relation is implied by a key.

bridge table: A table with a composite key that represents a many-to-many relationship either between a fact table and a dimension table or between two dimension tables representing different hierarchy levels.

business intelligence: The process of collecting and analyzing information to derive strategic knowledge from business data. Sometimes used synonymously with *decision support*.

cardinality: The number of elements in a collection. In conceptual models, it is either a specification that constrains the collection of values that an attribute may take, or the collection of the instances of a given relationship type in which an entity type or a level may participate.

cell: A single point in a cube defined by a set of coordinates, one for each of the cube's dimensions.

child level: Given two related levels in a hierarchy, the lower level, containing more detailed data. This is to be contrasted with the *parent level*.

Common Warehouse Metamodel (CWM): A metamodel proposed as a standard by the OMG to enable interchange of warehouse metadata between tools, platforms, and metadata repositories in distributed heterogeneous environments.

complex attribute: An attribute that is composed of several other attributes. An attribute is either simple or complex.

composite key: A key of a relation that is composed of two or more attributes.

computer-aided software engineering (CASE) tool: A tool that supports the activities of software development. CASE tools may be used in all phases of the software development life cycle, including analysis, design, and implementation.

conceptual model: A set of modeling concepts and rules for describing conceptual schemas.

conceptual schema: A schema that is designed to be as close as possible to the users' perception of the data, not taking into account implementation considerations.

conformed dimension: A dimension whose semantics, structure, and use are agreed upon across an enterprise. Conformed dimensions are typically used in several facts of a data warehouse and/or in several data marts.

conformed fact: A fact whose semantics, dimensions, and units are agreed upon across an enterprise. Conformed facts are typically used in several data marts.

constellation schema: A relational schema for representing multidimensional data composed of multiple fact tables that share dimension tables.

conventional attribute: An attribute that has a conventional data type as its domain. This is to be contrasted with *spatial* and *temporal attributes*.

conventional data type: A data type that allows conventional alphanumeric information to be represented. Typical conventional data types include the Boolean, string, integer, and float types. This is to be contrasted with *spatial* and *temporal data types*.

conventional data warehouse: A data warehouse that manipulates only conventional data. This is to be contrasted with *spatial* and *temporal data warehouses*.

conventional dimension: A dimension composed of only conventional hierarchies. This is to be contrasted with *spatial* and *temporal dimensions*.

conventional fact relationship: A fact relationship that requires a classical join between its dimensions. This is to be contrasted with *spatial* and *temporal fact relationships*.

conventional hierarchy: A hierarchy composed of only conventional levels. This is to be contrasted with *spatial* and *temporal hierarchies*.

conventional level: A level that includes only conventional attributes. This is to be contrasted with *spatial* and *temporal levels*.

conventional measure: A measure that has a conventional data type as its domain. This is to be contrasted with *spatial* and *temporal measures*.

cube: A multidimensional structure that contains a set of measures at each cell. Cubes are used to implement online analytical processing (OLAP). Also called a *hypercube* or *multidimensional cube*.

current data: Data from the current time period that it is used for the daily operations of an organization. This is to be contrasted with *historical data*.

data aggregation: The process by which a set of data values are combined into a single value. Aggregation is typically done during OLAP operations but is also done as part of the ETL process. It is also called *summarization*.

data cleaning: The process of transforming source data by removing errors and inconsistencies or by converting it into a standardized format, typically done as part of the ETL process.

data extraction: The process of obtaining data from operational and external data sources in order to prepare the source data for a data warehouse. Extraction is typically done as part of the ETL process.

data integration: The process of reconciling data coming from different data sources, at both the schema and the data level. Integration is typically done as part of the ETL process.

data loading: The process of populating a data warehouse, typically done as part of the ETL process.

data mart: A specialized data warehouse that is targeted at a particular functional area or user group in an organization. The data in a data mart can either be derived from an enterprise-wide data warehouse or be collected directly from data sources. A data mart can be seen as a small, local data warehouse.

data mining: The process of analyzing large amounts of data to identify unsuspected or unknown relationships, trends, patterns, and associations that might be of value to an organization.

data model: A set of modeling concepts and rules for describing database or data warehouse schemas.

data quality: The degree of excellence of data. Various factors contribute to data quality, such as whether the data is consistent, is nonredundant, is complete, follows business rules, is timely, and is well understood.

data refreshing: The process of propagating updates from the data sources to the data warehouse at a specified frequency in order to provide up-to-date data for the decision-making process. Refreshing is typically done as part of the ETL process.

data source: A system from which data is collected in order to be integrated into a data warehouse. Typically, such a system could be a database, an application, a repository, or a file.

data staging area: An area where the ETL process is executed and where the source data is prepared in order to be introduced into a data warehouse or data mart.

data transformation: The manipulation of data to bring it into conformance with business rules, domain rules, integrity rules, and other data within the warehouse environment. Data transformation typically includes cleaning, aggregating, and integrating data from several data sources.

data type: A domain of values with associated operators. The data types in this book include conventional data types, spatial data types, and temporal data types.

data warehouse: A specific type of database that is targeted at analytical applications. It contains historical data about an organization obtained from operational and external data sources.

database: A shared collection of logically related data, and a description of this data, designed to meet the information needs of an organization and to support its activities.

database management system (DBMS): A software system that allows users to define, create, manipulate, and manage a database.

decision support system: A system that is primarily used to assist an organization in making decisions. It supports analytical operations on historical data. This is to be contrasted with an *operational system*.

denormalization: The process of modifying the schema of a relation so that it does not conform to a particular normal form. Denormalized relations are often used to improve access for specific user needs, but they result in some degree of data redundancy. This is to be contrasted with *normalization*.

derived attribute or measure: An attribute or measure whose value for each instance of a type is derived, by means of an expression, from other values and instances in the database.

dimension: One or several related levels that constitute a specific viewpoint for analyzing the facts of a multidimensional database. Dimensions may be composed of hierarchies.

dimension table: A relational table that contains dimension data.

drill-across: An OLAP operation that queries related data, moving from one fact to another through common dimensions.

drill-down: An OLAP operation that queries detailed data, moving down one or several hierarchy levels. This is to be contrasted with *roll-up*.

drill-through: An OLAP operation that queries detailed data, moving from OLAP data to the source data in a relational database.

enterprise data warehouse: A centralized data warehouse that encompasses an entire enterprise.

entity: In the ER model, an instance of an entity type.

entity-relationship (ER) model: A popular conceptual model defined by Peter Chen in a seminal paper of 1976.

entity type: In the ER model, a description of a set of entities that share the same attributes, relationships, and semantics.

external data source: A system or data file from an organization external to the organization under consideration that is used for providing data to a data warehouse. This is to be contrasted with an *operational data source*.

extraction-transformation-loading (ETL): The process that allows a data warehouse to be populated from one or several data sources. As the name indicates, it is a three-step process of extracting data from the data sources, transforming the data, and finally loading the data into a data warehouse. The ETL process also includes refreshing the data warehouse at a specified frequency in order to make it up to date.

fact: A central component of a multidimensional model that contains the measures to be analyzed. Facts are related to dimensions.

fact dimension: A dimension that does not have an associated dimension table because all its attributes are in a fact table or in other dimensions. It is also called a *degenerate dimension*.

fact relationship: In the MultiDim model, a relationship that contains fact data.

fact table: A relation that contains fact data. A fact table typically has two types of columns: those that contain measures and those that are foreign keys to dimension tables. The primary key of a fact table is usually a composite key that is made up of all of its foreign keys.

first normal form (1NF): A normal form that ensures that all underlying domains of a relation contain only atomic values. This is achieved by disallowing complex and multivalued attributes.

foreign key: One or several attributes in a relation that are related through referential integrity to other attributes of a relation. Foreign keys support navigation between relations.

fourth normal form (4NF): A normal form that ensures that every functional and multivalued dependency in a relation is implied by a key.

front-end process: A process that exploits the contents of a data warehouse. This can be done in many ways, including OLAP analysis, reporting, and data mining. This is to be contrasted with a *back-end process*.

functional dependency: A constraint between two sets of attributes in a relation stating that the tuples that have the same values for the first set of attributes also have the same values for the second set of attributes. Functional dependencies are used to define various normal forms of relations.

generalization hierarchy: A set of generalization relationships between entity types. It is also called an *is-a hierarchy.*

generalization relationship: A directed relationship, defined between a supertype and a subtype, stating that phenomena described in the subtype are the same as those described in the supertype, but at a more specific (less generic) abstraction level. Instances of the subtype are substitutable for instances of the supertype. It is also called an *is-a relationship.*

geographic information system (GIS): A software system that allows users to define, create, maintain, and control access to a geographic database.

geometry attribute: In the MultiDim model, a predefined attribute that stores the spatial extent of instances of a type or a level.

granularity: A specification that partitions a domain into groups of elements, where each group is perceived as an indivisible unit (a *granule*) at a particular abstraction level. In a multidimensional model, it is the level of detail at which data is captured in dimensions and facts.

hierarchy: Several related levels of a dimension that define aggregation paths for roll-up and drill-down operations.

historical data: Data from previous time periods that is used for trend analysis and for comparison with previous periods. This is to be contrasted with *current data.*

hybrid OLAP (HOLAP): A storage method in which OLAP applications are built on top of both a relational and a multidimensional database. In such an architecture, the detailed data is typically stored in the relational database, while the aggregated data is stored in the multidimensional database. This is to be contrasted with *relational OLAP* and *multidimensional OLAP.*

index: A mechanism to locate and access data within a database. An index may involve one or more columns and be a means of enforcing uniqueness of their values.

inheritance: The mechanism by which a subtype in a generalization hierarchy incorporates the properties of its supertypes.

instance: An element of an entity or a relationship type. In multidimensional models, instances of levels are members.

instant cardinality: A cardinality specification that is valid at each instant of the temporal extent of a database. It is also called the *snapshot cardinality.*

integrity constraint: A condition that restricts the possible states of a database in order to enforce their consistency with the rules of the applications using the database.

Java Database Connectivity (JDBC): An application programming interface that enables Java programs to access data in a database management system. JDBC is similar to ODBC, but ODBC is language-independent.

joining level: In the MultiDim model, a level in which two alternative aggregation paths are merged.

key: In a conceptual model, a set of attributes whose values uniquely identify an instance of an entity type or a level. In the relational model, a set of one or more columns in a relation whose values uniquely identify a row in that relation. It is also called a *user-defined identifier*.

leaf level: A level in a hierarchy that is not related to a child level, i.e., it contains the most detailed data. This is to be contrasted with a *root level*.

legacy system: An existing system that has been in place for several years and uses languages, platforms, and techniques prior to current technology. Legacy systems are difficult to modify and maintain.

level: In a multidimensional model, a type belonging to a dimension. A level defines a set of attributes and is typically related to other levels to define hierarchies.

lifespan (LS): The time frame that is associated with the membership of an instance into its type.

lifespan cardinality: A cardinality specification whose enforcement takes account of data over the whole temporal extent of a database.

loading time (LT): A temporal specification that stores information about when a data element was stored in a data warehouse.

logical model: A set of modeling concepts and rules for describing a logical schema. Some typical logical data models are the relational model, the object-oriented model, and the object-relational model.

logical schema: A schema whose design takes into account the specific functionality of the type of database management system used, such as relational or object-relational.

materialized view: A view which is physically stored in a database. Materialized views allow query performance to be enhanced by precalculating costly operations.

mandatory attribute: An attribute that must be given at least one value in the instances of a type. An attribute is either optional or mandatory.

mandatory role: A role in which instances of the corresponding entity type must participate. A role is either optional or mandatory.

measure: A particular piece of information that has to be analyzed in an analytical application. Measures are associated to cells in a cube.

member: In a multidimensional model, an instance of a level.

metadata: Literally, data about data. It is information about the contents and uses of a database or data warehouse.

metamodel: A modeling framework that allows one to represent the properties of a modeling language.

method: A sequence of proven processes followed in planning, defining, analyzing, designing, building, testing, and implementing a system.

model: A representation of the essential characteristics of a system, process, or phenomenon intended to enhance our ability to understand, predict, or control its behavior.

modeling: The process of constructing or modifying a model.

monovalued attribute: An attribute that may have at most one value in the instances of a type. An attribute is either monovalued or multivalued.

MultiDim model: A particular conceptual multidimensional model for data warehouse and OLAP applications.

multidimensional database: A database that represents data according to the multidimensional model.

multidimensional model: A model in which information is conceptually represented as facts, measures, dimensions, and hierarchies. The multidimensional model is used to represent the information requirements of analytical applications.

multidimensional OLAP (MOLAP): A storage method in which OLAP applications are built on top of a multidimensional database. This is to be contrasted with *relational OLAP* and *hybrid OLAP*.

multiple inheritance: The possibility for an entity type to have more than one direct supertype.

multivalued attribute: An attribute that may have several values in the instances of a type. An attribute is either multivalued or monovalued.

multivalued dependency: A constraint between two sets of attributes in a relation stating that the tuples that have the same values for the first set of attributes have a set of possible values for the second set of attributes independently of any other attributes of the relation. Multivalued dependencies are used to define various normal forms of relations.

***n*-ary relationship type:** A relationship type among three or more entity types.

non-first-normal-form (NF2) model: A logical model that removes from the relational model the restrictions of the first normal form. The object-relational model allows non-first-normal-form relations.

nonadditive measure: A measure that cannot be meaningfully aggregated by addition across any dimension. This is to be contrasted with *additive* and *semiadditive measures*.

nontemporal attribute: An attribute that does not keep track of the evolution of its values. This is to be contrasted with a *temporal attribute*.

nontemporal data warehouse: A data warehouse that allows only temporal support for measures. This is to be contrasted with a *temporal data warehouse*.

nontemporal dimension: A dimension that includes only nontemporal hierarchies. This is to be contrasted with a *temporal dimension*.

nontemporal hierarchy: A hierarchy that includes only nontemporal levels or relationships. This is to be contrasted with a *temporal hierarchy*.

nontemporal level: A level that does not keep the lifespan for its members. This is to be contrasted with a *temporal level*.

nontemporal relationship: A relationship that does not keep track of the evolution of the links between its instances. This is to be contrasted with a *temporal relationship*.

normal form: A set of conditions that a relation must satisfy to guarantee some desirable properties, typically to eliminate data redundancy. The most usual normal forms are the first, second, third, Boyce-Codd, and fourth normal forms.

normalization: The process of modifying the structure of a relational database so that the relations satisfy some normal forms. This is to be contrasted with *denormalization*.

null value: A particular marker that indicates that the value of an attribute is missing, unknown, or inapplicable.

object: In the real world, a phenomenon that is perceived as having some existence independently of other phenomena.

object identifier (oid): A system-generated value that is associated with each instance of a type in an object-oriented or an object-relational system. Each instance is given an identifier that is unique within a database. It is also called a *surrogate*.

Object Management Group (OMG): An international computer industry consortium that is developing enterprise integration standards for a wide range of technologies and industries.

object-oriented model: A logical model in which an application is modeled as a set of cooperating objects that exchange messages between them. Object-oriented models include features such as inheritance, encapsulation, polymorphism, complex types, and methods.

object-relational database: A database that represents data according to the object-relational model.

object-relational database management system (ORDBMS): A database management system in which data is organized according to the object-relational model.

object-relational model: A logical model that extends the relational model with object-oriented features, such as inheritance, complex types, and methods.

object-relational schema: A schema targeted at an object-relational database management system.

online analytical processing (OLAP): Interactive analysis of data contained in a data warehouse.

online transaction processing (OLTP): Transaction-oriented work on an operational system that supports daily business operations.

Open Database Connectivity (ODBC): A standardized application programming interface for accessing data in a database management system in a language-independent manner. ODBC is similar to JDBC, but JDBC is designed specifically for Java programs.

operational data source: An operational system that provides data for a data warehouse. This is to be contrasted with an *external data source*.

operational system: A system that supports the daily operations of an organization. This is to be contrasted with a *decision support system*.

optional attribute: An attribute that may have no value in the instances of a type. An attribute is either optional or mandatory.

optional role: A role in which instances of the corresponding entity type may not participate. A role is either optional or mandatory.

parent-child relationship: In a multidimensional model, a binary relationship type that links a parent and a child level in a hierarchy.

parent level: Given two related levels in a hierarchy, the upper level, containing more general data. This is to be contrasted with the *child level*.

partitioning: A technique to improve the performance or security of an application by splitting data in a relation into multiple objects that are stored separately. It is also called *fragmentation*.

physical model: A set of modeling concepts and rules for describing the physical schema of a database.

physical schema: A schema customized to maximize efficiency and performance on a particular database platform.

population: The set of instances of an entity or of a relationship type.

predefined query: A request for specific information in a specific format that is performed on a regular basis. This is to be contrasted with an *ad hoc query*.

primary key: A privileged key, selected from all the keys in a relation, that is used to represent the links of the relation to other relations. In the relational model, every relation must have a primary key.

recursive relationship type: A relationship type in which the same entity type is linked by two (or more) roles.

referential integrity: An integrity constraint of the relational model specifying that the values of one or several attributes of a relation (the foreign key) must be included in the set of values of some other attributes (typically the primary key) of the same or another relation.

relation: A two-dimensional structure for storing information in a relational database. A relation has a specified number of columns but can have any number of rows. Relations are also called *tables*.

relational database: A database that represents data according to the relational model.

relational database management system (RDBMS): A database management system in which data is organized according to the relational model.

relational model: A logical model in which information is represented using two-dimensional tables or relations.

relational OLAP (ROLAP): A storage method in which OLAP applications are built on top of a relational database. This is to be contrasted with *multidimensional OLAP* and *hybrid OLAP*.

relational schema: A schema targeted at a relational database management system.

relationship: In the ER model, an instance of a relationship type. Its existence is subject to the existence of the linked entities.

relationship type: A description of a set of relationships that share the same attributes, roles, and semantics. In the MultiDim model, a relationship type can be either a fact relationship or a parent-child relationship.

reporting: The process of extracting data from a database or a data warehouse and presenting it to users in reports containing graphs, charts, etc.

role: The participation of an entity type in a relationship type. Role names are mandatory when the same entity type is related more than once to a relationship type.

roll-up: An operation that queries summarized data, moving up one or several hierarchy levels. This is to be contrasted with *drill-down*.

root level: A level in a hierarchy that does not have a parent level, i.e., it contains the most general data. This is to be contrasted with a *leaf level*.

schema: A specification, according to a given data model, that includes a definition of how the data in a database or data warehouse is structured, the type of content that each data element can contain, and the rules that govern what data values may be entered in the database or data warehouse and how these values can evolve.

schema diagram: A diagram that illustrates a schema according to a diagrammatic notation that corresponds to the data model on which the schema is based.

second normal form (2NF): A normal form that ensures that a relation contains no partial key dependencies. This is the case when a relation has a composite key and an attribute is dependent upon only part of the key.

semiadditive measure: A measure that can be meaningfully aggregated by addition along some, but not all, of its dimensions. This is to be contrasted with *additive* and *nonadditive measures*.

simple attribute: An attribute that is not composed of other attributes, i.e., an attribute whose specification explicitly includes the associated data type. It is also called an *atomic attribute*. An attribute is either simple or complex.

slice-and-dice: An operation that allows one to select a portion of the data of a fact on the basis of specified values in one or several dimensions.

slowly changing dimension: A dimension whose data changes over time. The term *slowly* emphasizes the fact that the data in a dimension changes less frequently than the data in related facts.

snowflake schema: A relational schema for representing multidimensional data, composed of a single, central fact table related by normalized dimension hierarchies. Each dimension level is represented in a table. This is to be contrasted with *star* and *starflake schemas*.

source-driven design: An approach to designing a data warehouse based on the data available in the underlying source systems. It is also called *data-driven design*. This is to be contrasted with *analysis-driven design*.

spatial attribute: An attribute that has a spatial data type as its domain. This is to be contrasted with *conventional* and *temporal attributes*.

spatial constraint: An integrity constraint that imposes a restriction on a spatial extent.

spatial data type: A data type that allows one to represent geometric features of phenomena. The MultiDim model provides the following spatial data types: Geo, SimpleGeo, Point, Line, OrientedLine, Surface, SimpleSurface, ComplexGeo, PointSet, LineSet, OrientedLineSet, SurfaceSet, and SimpleSurfaceSet. This is to be contrasted with *conventional* and *temporal data types*.

spatial data warehouse: A data warehouse that manipulates spatial data, thus allowing spatial analysis. This is to be contrasted with *conventional* and *temporal data warehouses*.

spatial dimension: A dimension that includes at least one spatial hierarchy. This is to be contrasted with *conventional* and *temporal dimensions*.

spatial fact relationship: A fact relationship that requires a spatial join between two or more spatial dimensions. This is to be contrasted with *conventional* and *temporal fact relationships*.

spatial hierarchy: A hierarchy that includes at least one spatial level. This is to be contrasted with *conventional* and *temporal hierarchies*.

spatial level: A level whose members have a spatial extent that keeps track of their location. This is to be contrasted with *conventional* and *temporal levels*.

spatial measure: A measure that has a spatial data type as its domain. This is to be contrasted with *conventional* and *temporal measures*.

splitting level: In the MultiDim model, a level in which two alternative aggregation paths start.

star schema: A relational schema for representing multidimensional data, composed of a single, central fact table related to denormalized dimensions. Each dimension is represented in a single table. This is to be contrasted with *snowflake* and *starflake schemas*.

starflake schema: A relational schema for representing multidimensional data that is a combination of the star and snowflake schemas, i.e., it is composed of both normalized and denormalized dimensions.

subtype: In a generalization relationship, the most specific type. This is to be contrasted with the *supertype* of the relationship.

summarizability: A characteristic referring to the possibility of correctly aggregating measures in a higher hierarchy level by taking into account existing aggregations in a lower hierarchy level.

supertype: In a generalization relationship, the most generic type. This is to be contrasted with the *subtype* of the relationship.

surrogate key: A system-generated artificial primary key that is not derived from any data in the database. It is similar to the object identifier in an object-oriented system.

synchronization relationship: A relationship type that has an associated temporal constraint on the lifespans of the linked entities. In the Multi-Dim model, a synchronization relationship may be one of the following: meets, overlaps, intersects, contains, inside, covers, coveredBy, equals, disjoint, starts, finishes, precedes, or succeeds.

temporal attribute: An attribute that keeps track of the evolution of its values. This is to be contrasted with a *nontemporal attribute*.

temporal constraint: An integrity constraint that imposes a restriction on a temporal extent.

temporal data type: A data type that allows one to represent temporal features of phenomena. The MultiDim model provides the following temporal data types: Time, SimpleTime, Instant, Interval, ComplexTime, InstantSet, and IntervalSet. This is to be contrasted with *conventional* and *spatial data types*.

temporal data warehouse: A data warehouse that manages temporal (or time-varying) data, thus allowing temporal analysis. While a conventional data warehouse supports temporal measures, a temporal data warehouse also supports temporal levels, hierarchies, and dimensions. This is to be contrasted with *conventional* and *spatial data warehouses*.

temporal dimension: A dimension that includes at least one temporal hierarchy. This is to be contrasted with *conventional* and *spatial dimensions*.

temporal fact relationship: A fact relationship that requires a temporal join between two or more temporal dimensions. This is to be contrasted with *conventional* and *spatial fact relationships*.

temporal hierarchy: A hierarchy that includes at least one temporal level or one temporal relationship. This is to be contrasted with a *nontemporal hierarchy*.

temporal level: A level that stores the lifespans of its members. This is to be contrasted with *conventional* and *spatial levels*.

temporal measure: A measure that keeps track of the evolution of its values. Temporal measures are similar to temporal attributes for levels. This is to be contrasted with *conventional* and *spatial measures*.

temporal relationship: A relationship that keeps track of the evolution of the links between its instances. This is to be contrasted with a *nontemporal relationship*.

third normal form (3NF): A normal form that ensures that every attribute in a relation that it is not part of a key is dependent only on the primary key, and not on any other field in the relation.

topological relationship: A relationship that has an associated spatial constraint on the geometries of the linked entities. In the MultiDim model, a topological relationship may be one of the following: meets, overlaps, intersects, contains, inside, covers, coveredBy, equals, disjoint, or crosses.

transaction time (TT): A temporal specification that keeps information about when a data element is stored in and deleted from a database. This is to be contrasted with *valid time*.

trigger: A procedural code stored in a database management system that is automatically executed in response to certain events in a relation or a database.

valid time (VT): A temporal specification that keeps information about when a data element stored in a database is considered valid in the reality perceived from the point of view of the application. This is to be contrasted with *transaction time*.

view: In the relational model, a virtual relation that is derived from one or several relations or other views.

Index